BIOCHEMISTRY RESEARCH TRENDS

ANTHOCYANINS

STRUCTURE, BIOSYNTHESIS AND HEALTH BENEFITS

BIOCHEMISTRY RESEARCH TRENDS

Additional books in this series can be found on Nova's website
under the Series tab.

Additional e-books in this series can be found on Nova's website
under the e-book tab.

NUTRITION AND DIET RESEARCH PROGRESS

Additional books in this series can be found on Nova's website
under the Series tab.

Additional e-books in this series can be found on Nova's website
under the e-book tab.

BIOCHEMISTRY RESEARCH TRENDS

ANTHOCYANINS

STRUCTURE, BIOSYNTHESIS AND HEALTH BENEFITS

NOBORU MOTOHASHI
EDITOR

nova publishers
New York

Library of Congress Cataloging-in-Publication Data

ISBN 978-1-62257-329-5

Library of Congress Control Number: 2012941599

Published by Nova Science Publishers, Inc. † New York

Contents

PREFACE

Anthocyanins and flavonoids with their diverse colors have been distributed abundantly in especially higher plants of plant kingdom as natural pigments. Daily diets such as vegetables and fruits have also these pigments, which have the different functions for human health. The purpose of this book is to describe their distributions, extractions, separations, characterization, stability, functions, biosynthesis pathway, the regulation of the biosynthetic enzymes by anthocyanins and flavonoids, food colors, and molecular biology such as a relationship of biomolecule with anthocyanins or flavonoids in vegetables and fruits.

Chapter 1 - Pulps of kiwifruits have attractive green red or yellow colors. These natural pigments contain their many diverse anthocyanins and their aglycones such as delphinidin and cyanidin. Kiwifruit anthocyanins and flavonoids could represent to be beneficent as human daily safe and functional foods with their improvement of oxidative stressed brain, complex formation of flavonoid-metal-ascorbic acid, preventive resistance for endogenous oxidative damage, antioxidative activity, anti-hepatofibrotic activity and others. Based on these facts, this chapter describes mainly the kiwifruit flavonoids and anthocyanins, and their antioxidative health effects of the evidence-based kiwifruit flavonoids.

Chapter 2 - Muscadine berries (*Vitis rotundifolia* Michx.) are hardy grapes indigenous to the South Eastern United States of America with high antioxidant and anti-inflammatory properties. Noble fresh red muscadine berries were crushed, and the juice was separated. The crushed grapes contain different polyphenols such as proanthocyanidins, gallic acid, ellagic acid, resveratrol, and anthocyanins. The anthocyanins which give the color to the muscadine berries are antioxidants useful for cardio-protection, cholesterol-lowering and vision-improvement. Additionally, anthocyanins could suppress the DNA-damaging effects of topoisomerase poisons in human colon cancer cells. This chapter suggests muscadine berries might have important benefits in age-related diseases such as cardio-protection, cholesterol-lowering, vision-improvement, and possibly certain types of cancer.

Chapter 3 - Polyphenols are naturally occurring aromatic hydroxylated compounds which could be divided into different subclasses *viz*., phenylethanol derivatives, phenolic acids, flavones and flavonols, catechins, procyanidins and anthocyanins. Anthocyanins are glycosylated polyhydroxy and polymethoxy derivatives of flavylium salts responsible for the brilliant red, orange and blue colors of most fruits and flowers and have shown antioxidative, antimutagenic, antidiabetic and anticarcinogenic acitivities *in vitro*.

Electrochemical techniques *viz*., cyclic voltammetry, square wave voltammetry, differential pulse voltammetry and amperometry have been employed for the characterization of the antioxidant and electrochemical properties of anthocyanins from their diverse sources and for their assay in fruit juices and plant extracts. Further, the assay of anthocyanins and related compounds from various natural sources by analytical methods *viz.*, spectrophotometry, spectroflourimetry, voltammetry, HPLC, LC-MS, MS/MS, MALDI-TOF-MS and others has been incorporated. The chapter also reports the interactions of anthocyanins and related compounds with macromolecules by spectroscopic methods *viz.*, UV-absorption, fluorescence, FTIR, circular dichroism and 3-D fluorescence, and voltammetric techniques.

Chapter 4 – Rice anthocyanins are a kind of reddish-purple water-soluble flavonoids that are present in the red and black pigmented rice (*Oryza sativa* L.) grains or plants. Rice anthocyanins could show potent antioxidative effects against reactive oxygen radicals generated in the human body. By now, many studies indicate that consumption of black rice has a number of beneficial properties such as antioxidant, α-glucosidase inhibitory, hepatoprotective, antidiabetic, anti-inflammatory and anti-carcinogenic effects. Their biosynthetic pathway of anthocyanins in rices could provide a possible strategy to improve anthocyanins and other flavonoid contents in rice grains by the genetic engineering techniques. Black rice accumulating anthocyanins in the endosperm are strongly expected. This chapter summarizes the knowledge on the genetics, biosynthesis and health benefits of anthocyanins in rice grains. This chapter gives a suggestion for the production of further novel rice products with more anthocyanin.

Chapter 5 - Sickle cell disease is a genetic, hereditary, autosomale and recessive disease, where there is a replacement of the glutamic acid located in sixth position on the beta chain of normal haemoglobin by valin. This replacement of a polar amino acid by a less polar amino acid causes a polymerization of the haemoglobin S in the red globules. In Sub-Saharan Africa, the mainly affected area by this genetic disease, medicinal plants are intensely used to manage sickle anaemia. Some Congolese plants have shown *in vitro* and *in vivo* antisickling activities. These biological activities seem mainly to be due to anthocyanin extracts. Anthocyanin extracts of some species of *Justicia* such as *Justicia matammensis* and *Justicia laxa* displayed interesting antisickling activity *in vitro*.

Chapter 6 – Anthocyanins are susceptible to the heat and light which cause color deterioration in manufacturing process, transportation, and preservation of foods such as strawberry jam. In this study, the effects of temperature and the relative humidity during the preservation on the stability of anthocyanin in strawberry jam were evaluated. Their degradation was suppressed by the formation of the inclusion complex of anthocyanins with cyclodextrin due to its cyclic molecular structure or lowering the water activity in the jam. The impregnation of strawberry with sucrose in advance of the jam preparation stabilized anthocyanins in the jam. Moreover, the relationship between the two rate constants was almost linear. This chapter showed to prevent the oxidation or degradation of food anthocyanins in strawberry by using cyclodextrin and sucrose with the effective improvement of the stabilities of both anthocyanins and the colors in the jam.

Chapter 7 – The consumers are interested in healthier colorful products such as anthocyanins. Although these natural pigments are widely spread in nature, their extraction procedure demands special attention to avoid degradation. Due to this difficulty, the extraction method plays a fundamental role in the recovery of anthocyanins from natural sources. This chapter presents the recent work of our research group (LASEFI/DEA/FEA/UNICAMP) for obtaining anthocyanins from skins of a Brazilian fruit named jabuticaba (*Myrciaria cauliflora*). Several results from conventional and innovative extraction technologies are summarized and discussed with the aim of comparing extraction performances in terms of anthocyanins recovery. This chapter gaves a more complete understanding on the influence of different extraction methods on the recovery of anthocyanins.

Chapter 8 – Over 600 anthocyanins have been identified by now. These structures have been determined by high-performance liquid chromatography (HPLC) coupled to UV-Vis (UV) detector and diode array detector (DAD) and different mass spectrometry (MS) analyzer, such as ion trap (IT), quadrupole (Q), time-of-flight (TOF) and hybrid analyzers Q-TOF. MS has been generally coupled to different ion sources such as electrospray ionization (ESI) and matrix-assisted laser desorption/ionization (MALDI). Moreover, capillary electrophoresis (CE) coupled to DAD and MS detectors has also been used to characterize anthocyanins.Among these analytical methods, the UV-Vis spectra represents a valuable analytical tool for the identification of anthocyanins because they have an specific and intense absorbance band in the range of 520 to 560 nm; meanwhile, MS is a powerful tool which can provide information on molecular mass and also structural features of the different constituents. This chapter is to overview the uses of HPLC and CE coupled to different detection systems discussing the fundamentals of sample extraction and separation.

Chapter 9 – Since long ago, jujubes including berry fruits, seeds and leaves have been traditionally used in preventions and treatments for stress, antifungal, antibacterial, antiulcer, anti-inflammatory, sedative, antispastic, antifertility or contraception, hypotensive and antinephritic, cardiotonic, antioxidant, immunostimulant, wound healing properties and others in China, India, Korea and their surrounding countires. The jujube contains their phytochemicals with their highly beneficent functionality for humans such as sugars, dietary fiber, alkaloids, terpenoids, and flavonoids and anthocyanins. This chapter describes mainly the diverese anthocyanins and flavonoids, and their functions based on their evidences by the health effects of jujubes.

Chapter 10 – Anthocyanins are found in many fruits such as berries, grapes, cherries, pomegranates, plums, apples and some citrus. They have high antioxidant potency, and could act as antimicrobials agents and UV radiation absorbers. Anthocyanins structurally belong to the flavonoids, and are glycosydes containing a sugar moiety and an aglycone unit (the anthocyanidin). More than 635 different anthocyanins and 23 anthocyanidins have been reported. The distribution frequency of the six most common anthocyanidins in fruits ranges from 5 % (petunidin) up to 30 % (cyanidin). The conformation of anthocyanins has great impact in their color, stability, co-pigmentation capacity and antioxidant properties. This chapter describes the distribution of anthocyanins in fruits, and analyzes the effect of pre- and postharvest conditions and/or treatments on their accumulation and fate.

Chapter 11 – Resveratrol mainly in grape skin has been recognized for its cardiovascular protective effects. Based on "French paradox"the observation that the French suffer a relatively low incidence of coronary heart disease, despite having a diet relatively rich in saturated fats, might be directly related to the antioxidative activity of wine. Moreover, its physiological effects are widening based on many studies related to anti-cancer, anti-obesity and neuroprotective properties. Therefore, this chapter suggests that resveratrol might play a leading role as a substance for new drug development and in increasing economic value due to the increase in wine consumption worldwide.

Chapter 12 – Anthocyanins are a member of the bioflavonoid phytochemicals. Anthocyanins have their health-promoting functions for recent years. Among many berry fruits, bilberries have been particularly known as one of the richest sources of anthocyanins since long ago. The physiological effects of anthocyanins in humans might depend on the absorption after ingestion. The clinical studies on the bioavailability of anthocyanins were very low and highly variable because of their instability in physiological absorption conditions. Then, the targets for the development of improved anthocyanin products might elevate more absorption and reduced the absorption variability. A copigmentation of anthocyanins might be a natural occurring mechanism to stabilize anthocyanins. Based on these ideas, the anthocyanin extracts of bilberry fruits have been investigated for their copigmentation effects in solution. Bilberry Anthocyanin Extracts (BAEs) with the strong copigmentation effects were tested by both, an *in vitro* CaCo-2 cell culture system and an *in vivo* human clinical trial. This chapter discribes the copigmentation effects of BAEs might increase an increased absorption by the reduced absorption variability.

Chapter 13 – Some selected red wines from Morocco were studied on their antioxidant activities. These wines had high anthocyanin contents with antioxidant capacity for reactive oxygen species. There is a good relationship of their antioxidant capacity with two DPPH and FRAR assays on the wines. From these results, this chapter suggests that the intake of these wines in daily foods might be beneficial for human health.

Chapter 14 – Taros have been primarily used in their diverse culinary regions as major cereal crops in tropical and subtropical areas also including temperate region such as Japan and Korea. In these areas, taros also have been used as traditional medicines in preventions and treatments for hypertension, constipation, diuretic and other age-related diseases. By now, human has ingested their dietary anthocyanins with different colors in fruit, corm and vegetable to unknowingly. These having functionality could play a beneficial role in retarding or reversing the course of chronic degenerative diseases such as hypertension, diabetes and cardiac failure. This chapter represents their diverse functions of mainly taro anthocyanins to human health by taro as a vegetable or corm.

January 17, 2012

Noboru Motohashi, PhD
Supreme Adviser of Shandong Science
and Technology Association, Shandong, China,
and Former Professor and Director
of Meiji Pharmaceutical University, Tokyo, Japan

In: Anthocyanins: Structure, Biosynthesis and Health Benefits
Editor: Noboru Motohashi
ISBN: 978-1-62257-329-5

Chapter 1

Kiwifruit Flavonoids and Their Antioxidant Activity in Colorful *Actinidia spp.* Based on Their Evidences

Noboru Motohashi*
Meiji Pharmaceutical University, Noshio, Kiyose-shi, Tokyo, Japan

Abstract

Edible colorful pigmentary anthocyanins that originated from plant kingdom are essential **for** plant pigments such as chlorophyll, carotenes and betalains for human health. Additionally, kiwifruits also have abundant antioxidative phytochemicals such as carotenes, vitamin B, ascorbic acid, flavonoids, chlorophyll, and biomineral selenium. Recently, kiwifruit flavonoids, such as delphinidin and cyanidin glycosides, have been identified from the diverse kiwifruit species.

The purpose of this paper is to review the kiwifruit flavonoids and anthocyanins, and their antioxidative health effects of the evidence-based kiwifruit flavonoids.

Keywords: *Actinidia deliciosa*, kiwifruit genotypes, kiwifruit flavonoids, kiwifruit anthocyanins, antioxidative activity

Abbreviations

cyanidin 3-*O*-[2-*O*-(β-xylosyl)-β-galactoside] (1)
cyanidin 3-*O*-β-galactoside (2)
cyanidin 3-*O*-β-glucoside (3)
cyanidin 3-*O*-β-arabinoside (4)
delphinidin 3-*O*-[2-*O*-(β-xylosyl)-β-galactoside] (5)
delphinidin 3-*O*-β-galactoside (6)

* Corresponding author: noborumotohashi@jcom.home.ne.jp.

ascorbic acid (vitamin C. 7)
malondialdehyde (8)
quercetin (9)
(+)-catechin (10)
(l)-catechin (11)
tannic acid (12)
2,5-dihydroxybenzoic acid (13)
p-hydroxybenzoic acid (14)
chlorogenic acid (15)
benzoic acid derivatives (16)
p-coumaric acid (17)
caffeic acid (18)
2,2-diphenyl-1-picrylhydrazyl (DPPH. 19)
α-tocopherol (20)
β-carotene (21)

1. Introduction

Anthocyanin is one of the most important phytopigments alongside of carotenoids, chlorophylls and betalains. By now, over approximately 500 natural anthocyanins have been found [1]. Anthocyanins have been taken from daily human foods such as vegetable and fruits.

Flavonoids and anthocyanins are known to be effective for antioxidative, antimyopic, antinyctalopic, anti-goitrogenic [2], detoxication [3], type 2 diabetes mellitus, atherosclerosis (a stage of arteriosclerosis involving fatty deposits (atheromas) inside the arterial walls, thus narrowing the arteries), osteoporosis (abnormal loss of bony tissue resulting in fragile porous bones attributable to a lack of calcium), hypertension (high blood pressure), prostate cancer (cancer of the prostate gland), and chronic fatigue (not chronic fatigue syndrome) [4, 5].

Anthocyanins function diversely for human health, such as antitumor activity, antihepatotoxicity, skin care and beauty, and antioxidation [6].

In 2002 the National Institute of Education, Nanyang Technological University (NIE/NTU), Singapore, reviewed the health-promoting benefits of anthocyanins that could protect against a variety of oxidants of flavonoids by the most recent analytical techniques concerning anthocyanin isolation and identification [7].

Kiwifruits belong to the genus *Actinidia* and are known to have over 50 species (Photo 1). Their flesh or skin colors of the ripe fruits are green from *Actinidia arguta*, grayish green from *A. callosa*, green to brown from *A. chinensis*, *A. chrysantha*, *A. deliciosa* and *A. Latifolis*, gingery brown from *A. eriantha*, reddish or reddish brown from some clones of *A. arguta* and *A. melanandra*, orange from *A. polygama* and purplish red from *A. arguta* var. *purpurea*. Interestingly, it is believed that there are considerable variations within one taxa. For example, the skin of *A. chinensis* fruit is usually a light brown, however the skin could be reddish brown, yellowish brown, yellow green or ever dark green [8, 9, 10].

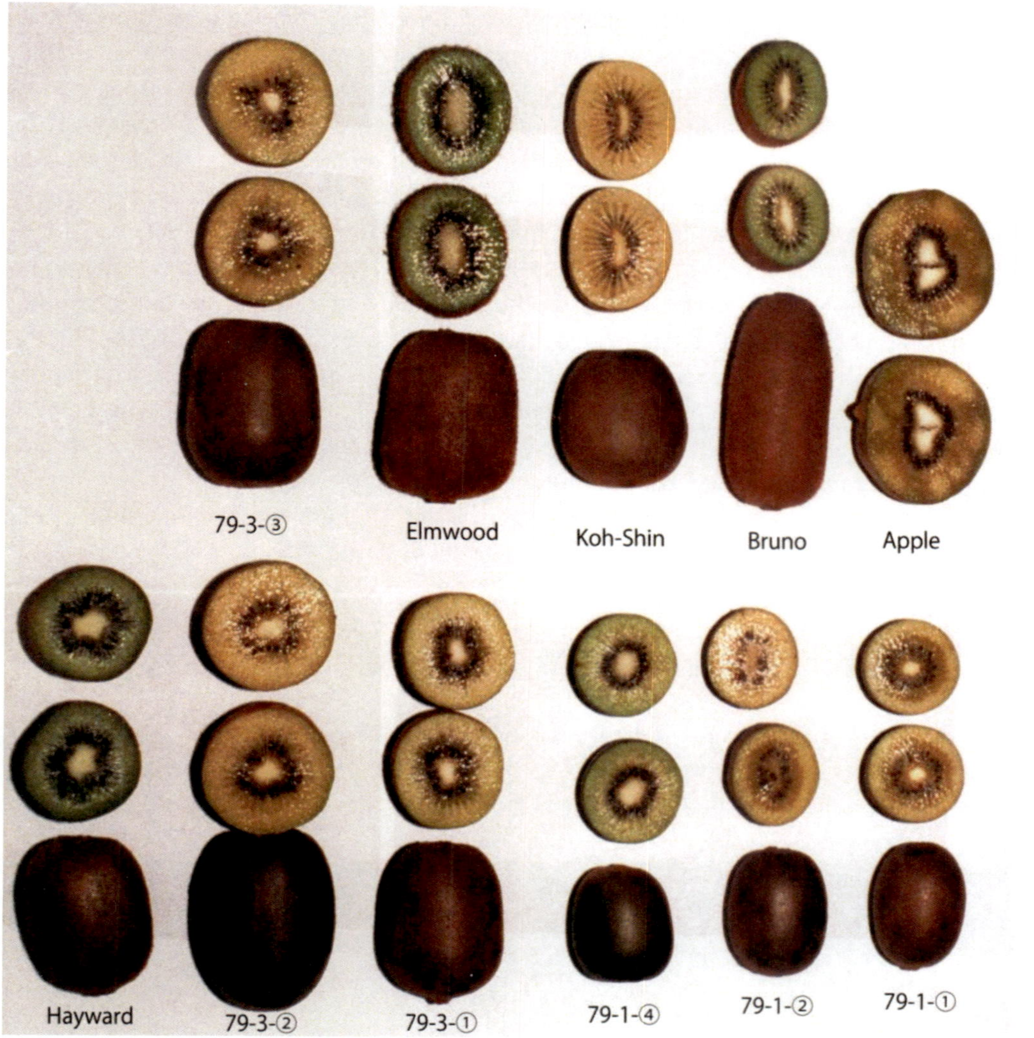

Photo 1. Different kiwifruit varieties (whole kiwifruits and their cross sections). Photographed by Noboru Motohashi. 1/10/1992 Fri.
fe24^070(different kiwifruit species and their cross sections, 1/10/1992 Fri).jpg.

The skin and flesh colors of most *Actinidia* species of Kiwifruit change little in their appearance when the kiwifruits are soft and ripe; however, some kiwifruits change their appearances exceptionally. The details are as follows: the fruits of *A. polygama* change from pale green to a deep greenish yellow. This color change is not particularly attractive and additionally, the color could change to orange when the kiwifruit has fully ripened. Similarly, the flesh colors of many kiwifruit species such as *A. kolomikta*, *A. macrosperma*, *A. tetramera* and *A. valvata* might change to yellow or orange when the kiwifruit fleshes had ripened. The fruits of some clones of *A. arguta* could develop a reddish hue when the kiwifruits have ripened. Therefore, such changes in the external skin colors could be used as a useful indicator of kiwifruits maturity and ripeness [8, 9].

Moreover, kiwifruits with high amounts of vitamin C, vitamin E and polyphenols might be beneficial to cardiovascular disease. In 2004 The Department of Nutrition at University of

Oslo in Norway evaluated whether consuming kiwifruits modulated the platelet activity and plasma lipids in human volunteers in a randomized cross-over study. Their consumption of two or three kiwifruits/day for 28 days could reduce the platelet aggregation response to two aggregating agents of collagen and adenosine diphosphate (ADP. a platelet agonist) that causes the platelet shape change and aggregation as well as the generation of thromboxane A_2, another platelet agonist, through its effects on P2Y1, P2Y12, and P2X1 receptors [11] by 18% when compared to the controls. In addition, their consumption of kiwifruits could lower the blood triglycerides levels by 15% when compared to control, whereas no such effects were observed in the case of cholesterol levels. This data suggests that the consumption of kiwifruits might be beneficial in age-related diseases, such as cardiovascular disease [12].

The flesh color might depend on the different kiwifruit species observed with rainbow colors, such as emerald green, red, purple, yellow and orange. Although the kiwifruit species with different colors have been known since long ago, their color phytochemicals, especially kiwifruit anthocyanins, have not been confirmed until recently.

The purpose of this review paper is to describe the structures of kiwifruit anthocyanins colors, and show the correlation of their functions with kiwifruit anthocyanins.

2. Flavonoids and Their Function of Colorful Kiwifruits

2.1. Red Color Anthocyanins in Red Kiwifruits

Many red kiwifruits are known and are commercially supplied (Photos 2, 3, 4, 5).

Photo 2. Ripe Ken Red hardy kiwifruit variety (red kiwifruit). At home orchard of an author in Tokyo. Photographed by Noboru Motohashi. 8/29/2009 Sat. 389. IMGP6710 (ripe Ken Red hardy kiwifruit. 8/29/2009 Sat).

Photo 3. Ripe Ken Red hardy kiwifruit variety (red kiwifruit. its cross section). At a home orchard of an author in Tokyo. Photographed by Noboru Motohashi. 9/2/2009 Wed.
390. IMGP6764 (cross section of ken red kiwifruit, 9/2/2009 Wed).

Photo 4. Kobayashi 39 kiwifruit (whole red kiwifruit and its cross section). A product from Kobayashi kiwifruit orchard of Fuji City, Shizuoka Prefecture in Japan. Kobayashi 39 was found as a variety registration by Toshio Kobayashi of kiwifruit reseacher. Photographed by Noboru Motohashi. 11/22/2000 Wed.
nv17-033(Kobayashi 39 and its cross section, 11/22/2000 Wed).jpg.

Photo 5. Commercial ripe Rainbow Red kiwifruit variety, *Actinidia chinensis* var. *rufopulpa* 'Hongyang' (red kiwifruit) and its cross section. A product of Fukuoka Prefecture in Japan. Photographed by Noboru Motohashi.11/9/2011 Wed.
487. MGP7881(Rainbow Red, 11/9/2011 Wed).JPG.

2.1.1. Red Color Anthocyanins in Genotypes Two Red-Fleshed Kiwifruits

In 2005 the Dipartimento di Colture Arboree at Università di Bologna in Bologna, Italy examined anthocyanins in two genotypes red-fleshed kiwifruits such as *Actinidia chinensis* and *Actinidia deliciosa.* The anthocyanins in two kiwifruit species were analyzed by high-pressure liquid chromatography (HPLC) and identified by the comparison with anthocyanin standards.

First, on *Actinidia chinensis* genotypes, the major anthocyanin was cyanidin 3-*O*-[2-*O*-(β-xylosyl)-β-galactoside] (1), followed by smaller amounts of cyanidin 3-*O*-β-galactoside (2).

Anthocyanin content (μg/g fresh edible part) of 15 red kiwifruits on *Actinidia chinensis* genotypes used in analysis were as follows: total anthocyanin 136.5, cyanidin 3-*O*-[2-*O*-(β-xylosyl)-β-galactoside] (1) 115.6, cyanidin 3-*O*-β-galactoside (2) 13.9, cyanidin 3-*O*-β-glucoside (3) 0.9 and cyanidin 3-*O*-β-arabinoside (4) 1.5 for CH-01 genotype; total anthocyanin 139.9, cyanidin 3-*O*-[2-*O*-(β-xylosyl)-β-galactoside] (1) 115.4, cyanidin 3-*O*-β-galactoside (2) 17.9, cyanidin 3-*O*-β-glucoside (3) 0.6 and cyanidin 3-*O*-β-arabinoside (4) 1.5 for CH-02 genotype; total anthocyanin 8.7, cyanidin 3-*O*-[2-*O*-(β-xylosyl)-β-galactoside] (1) 8.7, cyanidin 3-*O*-β-galactoside (2) nd (below the detection limit), cyanidin 3-*O*-β-glucoside (3) nd and cyanidin 3-*O*-β-arabinoside (4) nd for CH-03 genotype; total anthocyanin 22.5, cyanidin 3-*O*-[2-*O*-(β-xylosyl)-β-galactoside] (1) 17.4, cyanidin 3-*O*-β-galactoside (2) 4.4, cyanidin 3-*O*-β-glucoside (3) nd and cyanidin 3-*O*-β-arabinoside (4) nd for CH-26 genotype; total anthocyanin 27.6, cyanidin 3-*O*-[2-*O*-(β-xylosyl)-β-galactoside] (1) nd, cyanidin 3-*O*-β-galactoside (2) 26.7, cyanidin 3-*O*-β-glucoside (3) 0.9 and cyanidin 3-*O*-β-arabinoside (4) nd for CH-42 genotype; total anthocyanin 13.4, cyanidin 3-*O*-[2-*O*-(β-xylosyl)-β-galactoside] (1) 11.7, cyanidin 3-*O*-β-galactoside (2) 1.7, cyanidin 3-*O*-β-glucoside (3) nd and cyanidin 3-*O*-β-arabinoside (4) nd for CH-51 genotype; total anthocyanin 15.3, cyanidin 3-*O*-[2-*O*-(β-xylosyl)-β-galactoside] (1) 9.5, cyanidin 3-*O*-β-galactoside (2) 5.8, cyanidin 3-*O*-β-glucoside (3) nd and cyanidin 3-*O*-β-arabinoside (4) nd for CH-54 genotype; total anthocyanin 61.6, cyanidin 3-*O*-[2-*O*-(β-xylosyl)-β-galactoside] (1) 49.2, cyanidin 3-*O*-β-galactoside (2) 11.1, cyanidin 3-*O*-β-glucoside (3) 0.3 and cyanidin 3-*O*-β-arabinoside (4) nd for CH-59 genotype; total anthocyanin 2.9, cyanidin 3-*O*-[2-*O*-(β-xylosyl)-β-galactoside] (1) 2.9, cyanidin 3-*O*-β-galactoside (2) nd, cyanidin 3-*O*-β-glucoside (3) nd and cyanidin 3-*O*-β-arabinoside (4) nd for CH-69 genotype; total anthocyanin 67.5, cyanidin 3-*O*-[2-*O*-(β-xylosyl)-β-galactoside] (1) 44.7, cyanidin 3-*O*-β-galactoside (2) 20.1, cyanidin 3-*O*-β-glucoside (3) 0.8 and cyanidin 3-*O*-β-arabinoside (4) nd for CH-77 genotype; total anthocyanin 6.2, cyanidin 3-*O*-[2-*O*-(β-xylosyl)-β-galactoside] (1) 5.0, cyanidin 3-*O*-β-galactoside (2) 1.2, cyanidin 3-*O*-β-glucoside (3) nd and cyanidin 3-*O*-β-arabinoside (4) nd for CH-79 genotype; total anthocyanin 32.1, cyanidin 3-*O*-[2-*O*-(β-xylosyl)-β-galactoside] (1) 29.4, cyanidin 3-*O*-β-galactoside (2) 2.1, cyanidin 3-*O*-β-glucoside (3) nd and cyanidin 3-*O*-β-arabinoside (4) nd for CH-88 genotype; total anthocyanin 13.6, cyanidin 3-*O*-[2-*O*-(β-xylosyl)-β-galactoside] (1) 10.9, cyanidin 3-*O*-β-galactoside (2) 2.7, cyanidin 3-*O*-β-glucoside (3) nd and cyanidin 3-*O*-β-arabinoside (4) nd for CH-89 genotype; total anthocyanin 0.9, cyanidin 3-*O*-[2-*O*-(β-xylosyl)-β-galactoside] (1) 0.9, cyanidin 3-*O*-β-galactoside (2) nd, cyanidin 3-*O*-β-glucoside (3) nd and cyanidin 3-*O*-β-arabinoside (4) nd for CH-90 genotype; and total anthocyanin 29.9, cyanidin 3-*O*-[2-*O*-(β-xylosyl)-β-galactoside] (1) 29.4, cyanidin 3-*O*-β-galactoside (2) nd, cyanidin 3-*O*-β-glucoside (3) 0.5 and cyanidin 3-*O*-β-arabinoside (4) nd for Hongyang genotype, respectively.

Second, on 10 red kiwifruits of *Actinidia deliciosa* genotypes, two major anthocyanins were found as cyanidin 3-*O*-β-galactoside (2) and cyanidin 3-*O*-β-glucoside (3). However, cyanidin 3-*O*-[2-*O*-(β-xylosyl)-β-galactoside] (1) was not found.

Anthocyanin content (μg/g fresh edible part) of 10 red kiwifruits of *Actinidia deliciosa* genotypes used in analysis were as follows: total anthocyanin 6.7, cyanidin 3-*O*-[2-*O*-(β-xylosyl)-β-galactoside] (1) nd (below the detection limit), cyanidin 3-*O*-β-galactoside (2) 3.1, cyanidin 3-*O*-β-glucoside (3) 3.6 and cyanidin 3-*O*-β-arabinoside (4) nd for DE-08 genotype; total anthocyanin 4.0, cyanidin 3-*O*-[2-*O*-(β-xylosyl)-β-galactoside] (1) nd, cyanidin 3-*O*-β-galactoside (2) 2.0, cyanidin 3-*O*-β-glucoside (3) 2.0 and cyanidin 3-*O*-β-arabinoside (4) nd for DE-09 genotype; total anthocyanin 5.8, cyanidin 3-*O*-[2-*O*-(β-xylosyl)-β-galactoside] (1) nd, cyanidin 3-*O*-β-galactoside (2) 3.3, cyanidin 3-*O*-β-glucoside (3) 2.5 and cyanidin 3-*O*-β-arabinoside (4) nd for DE-10 genotype; total anthocyanin 6.2, cyanidin 3-*O*-[2-*O*-(β-xylosyl)-β-galactoside] (1) nd, cyanidin 3-*O*-β-galactoside (2) 1.5, cyanidin 3-*O*-β-glucoside (3) 4.7 and cyanidin 3-*O*-β-arabinoside (4) nd for DE-11 genotype; total anthocyanin 3.2, cyanidin 3-*O*-[2-*O*-(β-xylosyl)-β-galactoside] (1) nd, cyanidin 3-*O*-β-galactoside (2) 1.5, cyanidin 3-*O*-β-glucoside (3) 1.7 and cyanidin 3-*O*-β-arabinoside (4) nd for DE-12 genotype; total anthocyanin 2.6, cyanidin 3-*O*-[2-*O*-(β-xylosyl)-β-galactoside] (1) nd, cyanidin 3-*O*-β-galactoside (2) 1.1, cyanidin 3-*O*-β-glucoside (3) 1.5 and cyanidin 3-*O*-β-arabinoside (4) nd for DE-13 genotype; total anthocyanin 4.3, cyanidin 3-*O*-[2-*O*-(β-xylosyl)-β-galactoside] (1) nd, cyanidin 3-*O*-β-galactoside (2) 1.8, cyanidin 3-*O*-β-glucoside (3) 2.5 and cyanidin 3-*O*-β-arabinoside (4) nd for DE-14 genotype; total anthocyanin 7.2, cyanidin 3-*O*-[2-*O*-(β-xylosyl)-β-galactoside] (1) nd, cyanidin 3-*O*-β-galactoside (2) 3.5, cyanidin 3-*O*-β-glucoside (3) 3.7 and cyanidin 3-*O*-β-arabinoside (4) nd for DE-15 genotype; total anthocyanin 23.9, cyanidin 3-*O*-[2-*O*-(β-xylosyl)-β-galactoside] (1) nd, cyanidin 3-*O*-β-galactoside (2) 11.6, cyanidin 3-*O*-β-glucoside (3) 12.3 and cyanidin 3-*O*-β-arabinoside (4) nd for DE-18 genotype; and total anthocyanin nd, cyanidin 3-*O*-[2-*O*-(β-xylosyl)-β-galactoside] (1) nd, cyanidin 3-*O*-β-galactoside (2) nd, cyanidin 3-*O*-β-glucoside (3) nd and cyanidin 3-*O*-β-arabinoside (4) nd for DE-01 genotype, respectively (Figure 1).

Their differences of anthocyanin compositions basically could not be found among these two kiwifruit genotypes [9].

2.1.2. Red Color Anthocyanins in Four Red Kiwifruits

In 2009, a group of The New Zealand Institute for Plant and Food Research Ltd has examined their red anthocyanions in four red kiwifruit species with red color in the flesh or skin of *Actinidia arguta* var. *purpurea* (Rehd.), *A. melanandra* French, *A. deliciosa* (A. Chev.) C. F. Liang at A.R. Ferguson and *A. chinensis* Planch by the extraction of acidified ethanol and isolated by solid phase extraction, followed using preparative high-pressure liquid chromatography (HPLC). Among these four red kiwifruit species, red kiwifruit only contained major anthocyanins such as two delphinidin glycosides of delphinidin 3-*O*-[2-*O*-(β-xylosyl)-β-galactoside] (5) and delphinidin 3-*O*-β-galactoside (6) (Figure 2), and three cyanidin glycosides of cyanidin 3-*O*-[2-*O*-(β-xylosyl)-β-galactoside] (1), cyanidin 3-*O*-β-galactoside (2) and cyanidin 3-*O*-β-glucoside (3) responsible for the red color of red kiwifruits.

Among four red kiwifruit species, *A. purpurea* (PU 33) contained rich cyanidin 3-*O*-[2-*O*-(β-xylosyl)-β-galactoside] (1), followed by delphinidin 3-*O*-[2-*O*-(β-xylosyl)-β-galactoside] (5). *A. melanandra* (ME 12) contained cyanidin 3-*O*-[2-*O*-(β-xylosyl)-β-galactoside] (1), followed by cyanidin 3-*O*-β-galactoside (2), delphinidin 3-*O*-[2-*O*-(β-xylosyl)-β-galactoside] (5) and delphinidin 3-*O*-β-galactoside (6). *A. deliciosa* (DE 02) was cyanidin 3-*O*-β-glucoside (3), followed by cyanidin 3-*O*-β-galactoside (2). *A. chinensis* (CH 77) contained cyanidin 3-*O*-[2-*O*-(β-xylosyl)-β-galactoside] (1), followed by cyanidin 3-*O*-β-galactoside (2). Two above anthocyanins of delphinidin 3-*O*-[2-*O*-(β-xylosyl)-β-galactoside] (5) and delphinidin 3-*O*-β-galactoside (6) were for the first time found in the *Actinidia* species.

Interestingly, from these results, no single kiwifruit species contained all the five anthocyanins as major components [13].

2.1.3. Improvement of Oxidative Stressed Brain by Red Kiwifruit Anthocyanins

In 2006, a group of Nutrition and Food Science Department-CeRTA, Pharmacy Faculty, University of Barcelona, Barcelona, Spain studied their effective in reversing their age-related deficits in neuronal signaling and behavioral parameters by blueberry anthocyanins. Antioxidative anthocyanins such as cyanidin 3-*O*-β-galactoside (2), cyanidin 3-*O*-β-glucoside (3), cyanidin 3-*O*-β-arabinoside (4) and delphinidin 3-*O*-β-galactoside (6) were found in the cerebellum, cortex (the layer of unmyelinated neurons (the grey matter) forming the cortex of the cerebrum), hippocampus (a complex neural structure (shaped like a sea horse) consisting of grey matter and located on the floor of each lateral ventricle) or striatum of the BBS rats, but not the controls.

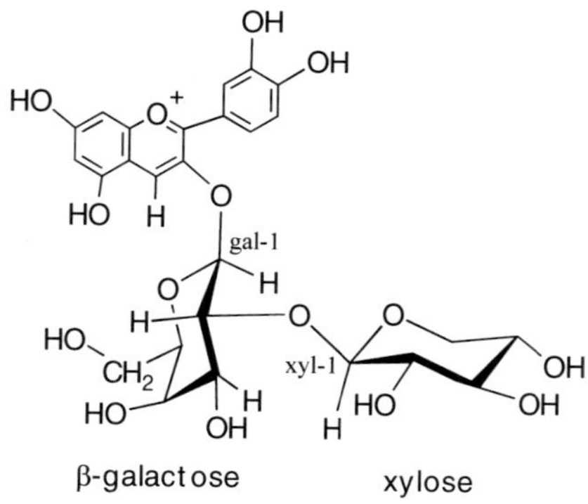

cyanidin 3-*O*-[2-*O*-(β-xylosyl)-β-galactoside] (**1**)

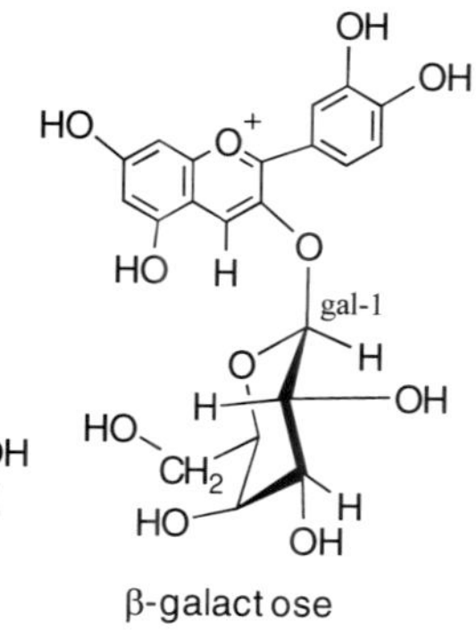

cyanidin 3-*O*-β-galactoside (**2**)

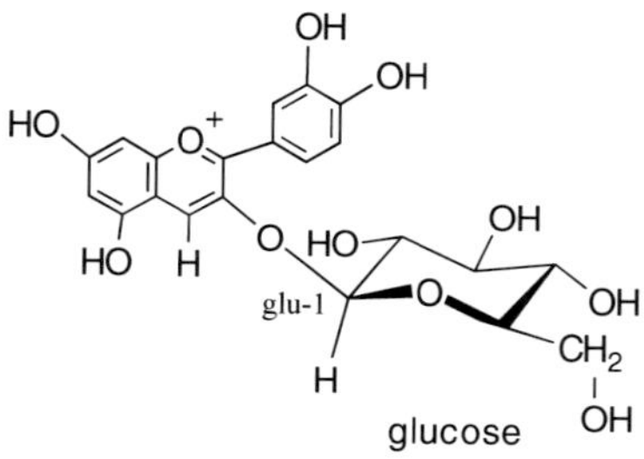

cyanidin 3-*O*-β-glucoside (**3**)

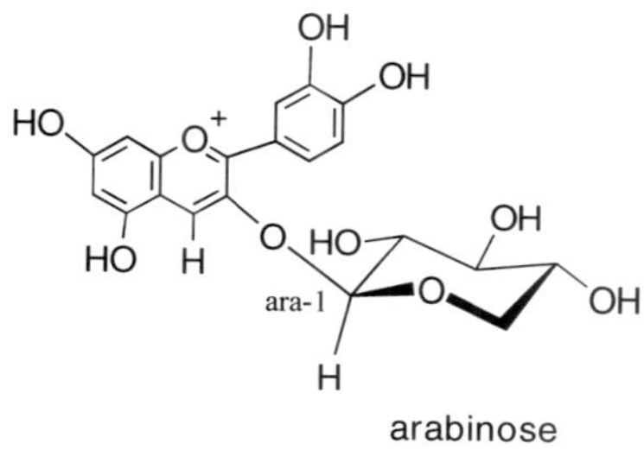

cyanidin 3-*O*-β-arabinoside (**4**)

Figure 1. Four cyanidin glycosides of cyanidin 3-O-[2-O-(®-xylosyl)-®-galactoside] (**1**), cyanidin 3-O-®-galactoside (**2**), cyanidin 3-O-®-glucoside (**3**), and cyanidin 3-O-®-arabinoside (**4**) from four red kiwifruit species.

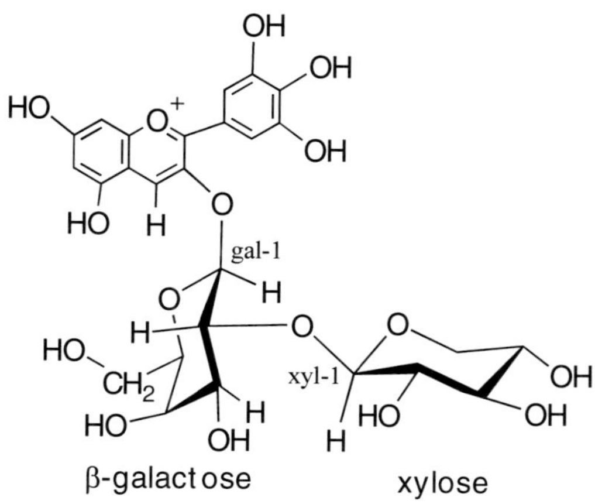

delphinidin 3-*O*-[2-*O*-(-xylosyl)-β-galactoside] (**5**)

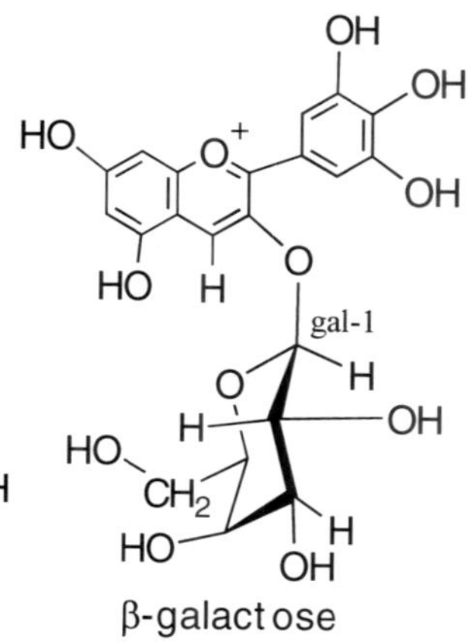

delphinidin 3-*O*-β-galactoside (**6**)

Figure 2. Two delphinidin glycosides of delphinidin 3-O-[2-O-(-xylosyl)-®-galactoside (**5**) and delphinidin 3-O-®-galactoside (**6**) from four red kiwifruit species.

As the results, it might be suggested that antioxidative flanonoids could cross the blood brain barrier (BBB) and localize in the various brain regions important for learning and memory. There is a relationship between Morris water mazer performance (MWM test) in BBS rats and the total number of anthocyanins found in the cortex. It could suggest that these anthocyanins with antioxidative activity might be centrally delivered in cortex and represent their signaling modifying capabilities [14].

2.1.4. Triplet Complex Formation of Flavonoid-Metal-Ascorbic Acid

It is known that the metal-complex formation ability of flavonoids could protect the ascorbate oxidation. In 1996, a group of Department of Plant Sciences, School of Life Sciences, University of Hyderabad, Hyderabad-500046 India studied the mechanism on the antioxidant ability of anthocyanins against ascorbic acid oxidation. The oxidation of ascorbic acid (vitamin C. **7**) by metal ions like copper (Cu) could be prevented by addition of anthocyanin (cyanidin derivative). It was confirmed by an evidence that anthocyanin not only chelated the metal ions, but also formed an ascorbic acid (copigment)-metal-anthocyanin complex, and this complex could be the possible protection mechanism (Figure 3) [3, 15].

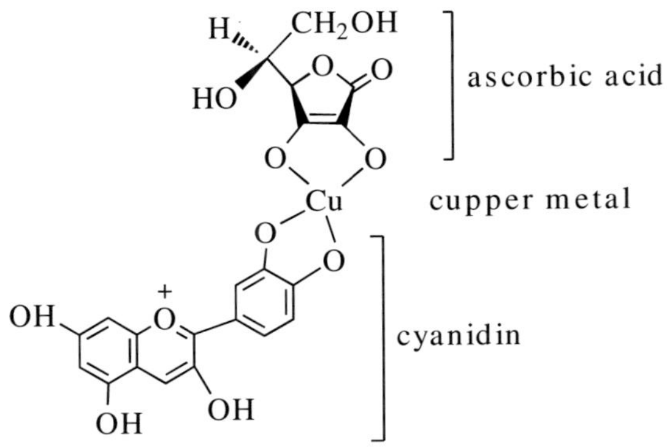

Figure 3. Triplet complex formation of cyanidin-cupper-ascorbic acid.

2.2. Yellow Color Anthocyanins in Yellow Kiwifruits

Generally, it is thought that their origin of most yellow kiwifruits have mainly been in China. Now, some yellow kiwifruits also were available as seasonable kiwifruits in a market of Japan (Photo 6).

2.2.1. Preventive Resistance for Endogenous Oxidative Damage

Actinidia chinensis var. Hort16A (Hort16A) (Photos 7, 8) of a yellow flesh kiwifruit contains significantly around 20% higher ascorbic acid (vitamin C. 7) content in phytochemical when compared to that of their commercial kiwifruits [16], and also had the higher ferric reducing activity of plasma (FRAP) [12] when compared to that of the green kiwifruit. Based on these properties, it could suggest that Hort16A might represent the stronger protection against the effects of oxidative damage *in vivo*.

In 2011, for confirmation of this hypothesis, a human dietary intervention trial with Hort16A was studied by examining their potential effects on the platelet function, plasma antioxidant status, DNA oxidation, base excision repair (BER) and nucleotide excision repair (NER) activity. The plasma malondialdehyde (8) as a marker for oxidative stress was measured by high-pressure liquid chromatography (HPLC). A product of lipid peroxidation might act as a marker for overall oxidative stress. From the results of their potential effects, it was suggested that Hort16A might enhance a resistance for the endogenous oxidative damage by reactive oxygen species (ROS) in human. However, green kiwifruit variety also might have similar resistance because the typical commercial green kiwifruits contains around 85 mg ascorbic acid (vitamin C. 7)/100 g fresh edible weight [16]. In fact, a kiwifruit extract was highly antioxidative against the oxidative DNA damage in human cells and *in vitro* [17]. The regular consumption of kiwifruits in human inhibited epidemiologically the platelet aggregation [18], decreased endogenous oxidation of lymphocyte DNA, protected lymphocyte DNA from oxidation *in vitro*, and enhanced the capacity of lymphocytes to repair the DNA oxidation damage [17, 19, 20, 21, 22, 23, 24]. From these results, the consumption of Hort16A enhanced the resistance against the endogenous oxidative damage by reactive oxygen species (ROS). This inhibition might be apparently by the radical quenching effects of the antioxidative ascorbic acid (vitamin C. 7) (Figure 4).

Photo 6. Ripe Apple kiwifruit variety (yellow kiwifruit) and its cross section. at home orchard of author in Tokyo. Photographed by Noboru Motohashi. 11/27/2007 Tue.
220. IMGP2416 (cross section of Apple kiwifruit, 11/27/2007 Tue).JPG.

Photo 7. *Actinidia chinensis* var. Hort16A (yellow kiwifruit). at kiwifruit orchard of Tauranga in New Zealand. Photographed by Noboru Motohashi. 4/24/1996 Wed.
ja08^075(golden kiwifruit, 4/24/1996).jpg.

Photo 8. Commercial ripe *Actinidia chinensis* var. Hort16A (yellow kiwifruit) and its cross section. An imported product from NZ. Photographed by Noboru Motohashi. 11/9/2011 Wed.
487. IMGP7879 (Hort16A, 11/9/2011 Wed).JPG.

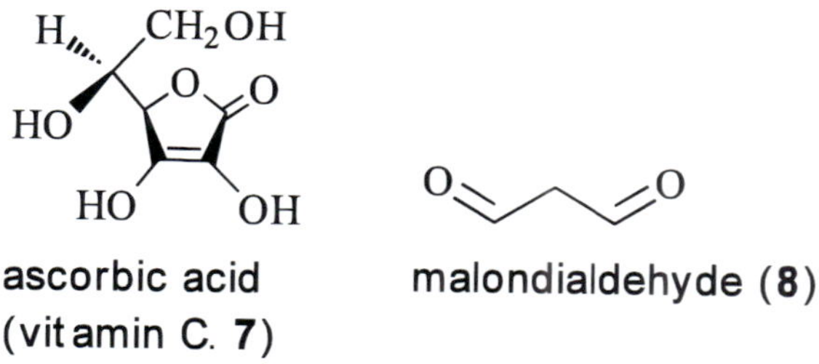

Figure 4. Ascorbic acid (7) and malondialdehyde (**8**).

Photo 9. Ripe Elmwood kiwifruit variety (green kiwifruit) and its cross section. at home orchard of author in Tokyo. Photographed by Noboru Motohashi. 11/27/2007 Tue. 220. IMGP2400 (Elmwood, 11/27/2007 Tue).JPG.

Moreover, kiwifruit could prevent the DNA damage in lymphocytes by the comet assay single-cell gel electrophoresis [17].

This suggests that Hort16 also might contain some anthocyanins, however any anthocyanins have not been found in Hort16 yet.

2.3. Anthocyanins in Miscellaneous Kiwifruits

Green kiwifruits (Photos 9, 10, 11, 12) are very popular in daily food life worldwide. Some green kiwifruits. Especially green kiwifruits are very high in ascorbic acid (vitamin C. ビタミン C. 7).

2.3.1. Anthocyanins in Six Kiwifruits

In 2010, a group of Department of Environmental Protection, Faculty of Horticulture and Landscape Architecture, Warsaw University of Life Sciences-SGGW, Poland studied the content and their antioxidative activity on anthocyanins in six *Actinidia* species fruits such as *Actinidia arguta* "Ananasnaya" genotype (fruit origin: Weihenstephan, Freising of Germany)

)(Photo 13), *A. arguta* “Weiki” genotype (fruit origin: Weihenstephan, Freising of Germany), *A. deliciosa* “Hayward” genotype (fruit origin: local market of Chile)(Photo 14), *A. kolomikta* “Dr Szymanowski” genotype (fruit origin: Clematis nursery of Poland) (Photo 15), *A. macrosperma* genotype (fruit origin: NCGR Corvallis, acc. 132.001, USA), *A. melanandra* “1064-79” genotype (fruit origin: NCGR Corvallis, acc. 102.001, USA) and *A. purpurea* genotype (fruit origin: Fruits du Paradise, Switzerland).

Photo 10. Commercial ripe Hisui of *Actinidia spp.* (hardy green kiwifruit) and its cross section. Photographed by Noboru Motohashi. 9/17/2000 Sun. nv20-012(Hisui. 9/17/2000 Sun).jpg.

Photo 11. Kokuwa, one of *Actinidia arguta* (hardy green kiwifruit). at home orchard of author in Tokyo. Photographed by Noboru Motohashi. 5/26/2002 Sun. nv18-008(Kokuwa, 5/26/2002 Sun).jpg

Photo 12. Kokuwa, one of *Actinidia arguta* (hardy green kiwifruit) and its cross section. at home orchard of author in Tokyo. Photographed by Noboru Motohashi. 8/6/2002 Tue. nv18-387(Kokuwa, 8/6/2002 Tue).jpg.

Their contents (mg/100 g fresh edible part) of quercetin (9), (+)-catechin (10), (l)-catechin (11)(Figure 5) and total phenolics of 7 phenolics (tannic acid (12), 2,5-dihydroxybenzoic acid (13), *p*-hydroxybenzoic acid (14), chlorogenic acid (15), benzoic acid derivatives (16), *p*-coumaric acid (17) and caffeic acid (18)) (Figure 6) on three anthocyanins in seven *Actinidia* species fruits were follows: 0.87, 0.18, 0.17 and 22.4 for *Actinidia arguta* "Ananasnaya" genotype; 1.00, 0.16, 0.05 and 23.9 for *A. arguta* "Weiki" genotype; 0.29, ND (not detected), 0.11 and 8.70 for *A. deliciosa* "Hayward" genotype; 0.80, 0.01, 1.04 and 443.13 for *A. kolomikta* "Dr Szymanowski" genotype; 10.64, 2.73, 0.24 and 131.10 for *A. macrosperma* genotype; ND, ND, 0.11 and 22.25 for *A. melanandra* "1064-79" genotype, and 1.99, ND, 0.04 and 34.62 for *A. purpurea* genotype, respectively.

Photo 13. *Actinidia arguta* (hardy green kiwifruit) and its cross section. a reference unripe Hayward (right). at home orchard of author in Tokyo. Photographed by Noboru Motohashi. 8/29/1988 Mon. nv25-104(cross sections of *A. arguta* and Hayward, 8/29/1988 Mon).jpg.

Photo 14. Commercial ripe Hayward variety, *Actinidia deliciosa* cv. Hayward (green kiwifruit) and its cross section. A product of Fukuoka Prefecture in Japan. Photographed by Noboru Motohashi. 11/9/2011 Wed.
487. IMGP7889 (kiwi.Hayward, 11/9/2011 Wed).JPG.

Photo 15. *Actinidia kolomikta* (hardy green kiwifruit) and its cross section. a reference unripe Hayward (right). at home orchard of author in Tokyo. Photographed by Noboru Motohashi. 8/29/1988 Mon. nv25-101(*A. kolomikta* & Hayward(right), 8/29/1988 Mon).jpg.

Their contents (mg/100 g fresh edible part) of flavonoids depended on their species. First, on the content of quercetin (9) among seven kiwifruit species, six kiwifruits ranged from 10.64 of *A. macrosperma* genotype to 0.29 of *A. deliciosa* "Hayward" genotype except ND of *A. melanandra* "1064-79" genotype. Second, on the content (mg/100 g fresh edible part) of (+)-catechin (10) among six kiwifruit species, 4 kiwifruits ranged from 2.73 of *A. macrosperma* genotype to 0.01 of *A. kolomikta* "Dr Szymanowski" genotype except ND of other three genotypes. Third, on the content (mg/100 g fresh edible part) of (l)-catechin (11) among six kiwifruit species, all kiwifruits ranged from 1.04 of *A. kolomikta* "Dr Szymanowski" genotype to 0.04 of *A. purpurea* genotype. Fourth, on the content (mg/100 g fresh edible part) of total phenolics, among seven kiwifruit species, all kiwifruits ranged from

443.13 of *A. kolomikta* "Dr Szymanowski" genotype to 8.70 of *A. deliciosa* "Hayward" genotype. From the above results on these seven kiwifruit species, it might be suggested that *A. macrosperma* genotype and *A. kolomikta* "Dr Szymanowski" genotype are especially a good resource of anthocyanins [25].

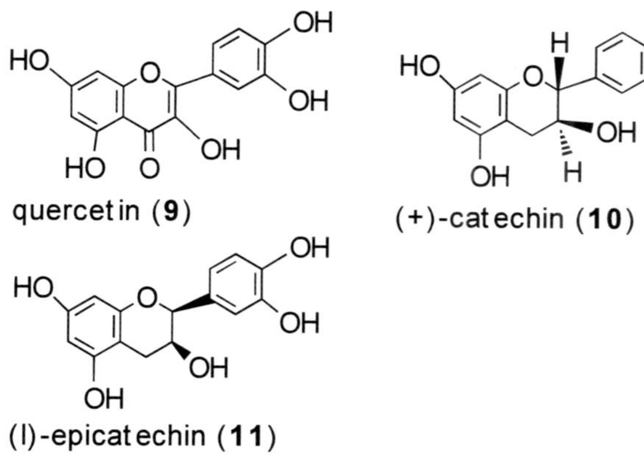

Figure 5. Quercetin (9), (+)-catechin (10) and (l)-epicatechin (11) from some kiwifruits.

2.3.2. Antioxidative Activity of Some Kiwifruits

On their antioxidative activity (AAC) (mg ascorbic acid equivalent (AAE)/g, fresh weight) of seven kiwifruit species by authentic 2,2-diphenyl-1-picrylhydrazyl (DPPH. 19) radicals, 6.87 for *A. kolomikta* "Dr Szymanowski" genotype was the highest activity, followed by 1.44 for *A. purpurea* genotype, 1.35 for *A. melanandra* "1064-79" genotype, 1.20 for *A. macrosperma* genotype, 1.10 for *A. arguta* "Weiki" genotype, 1.01 for *A. arguta* "Ananasnaya" genotype, and 0.98 for *A. deliciosa* "Hayward" genotype (Figure 7) [25].

In 2008, a group of Dipartimento di Chimica e Biotecnologie Agrarie (DCBA), University of Pisa, Pisa, Italy studied their antioxidant capacity, ascorbic acid (vitamin C. 7), total phenol content, carotenoids, soluble solids content and flesh firmness at harvest time and storage of kiwifruit such as *Actinidia deliciosa*. From an oxidative analysis, their antioxidative activity of kiwifruits in storage was similar or higher when compared to those of kiwifruits at the harvest time. Then, their antioxidative activity depended on the total content of antioxidative components such as ascorbic acid (vitamin C. 7) and total phenol [26].

In 2009, a group of College of Horticulture, Northwest A&F University, Yangling, Shaanxi 712100, China evaluated their antioxidant potential on the fresh kiwifruit fleshes of eight *Actinidia* genotypes by total 6 assays including 2,2-diphenyl-1-picrylhydrazyl (DPPH. 19) method and tested for their polyphenol composition (phenolics and flavonoids) and ascorbic acid (vitamin C. 7) contents. The antioxidant capacity of *Actinidia eriantha* and *Actinidia latifolia* were significantly higher when compared to that of other genotypes in 4 assays including 2,2-diphenyl-1-picrylhydrazyl (DPPH. 19) method. Especially, their antioxidant capacity of *Actinidia eriantha* and *Actinidia latifolia* was significantly about 3.3–8.7-fold higher when compared to that of *Actinidia deliciosa* cv. Hayward in 4 assays. The total polyphenol and vitamin C contents showed a great variety amongst *Actinidia* genotypes and highly correlation with the total antioxidant capacity. From these results, it might be sudgested that these significant differences of genotypes depend on their total antioxidant capacity of *Actinidia* species. Their antioxidant capacities of two wild *A. eriantha* and *A.*

latifolia species were significantly higher when compared to that of the both cultivars of *A. chinensis* and *A. deliciosa*. Two basic antioxidative phytochemicals of polyphenols (phenolics and flavonoids) and ascorbic acid (vitamin C. 7) might be greatly contribute to their total antioxidant capacity of *Actinidia* species [27].

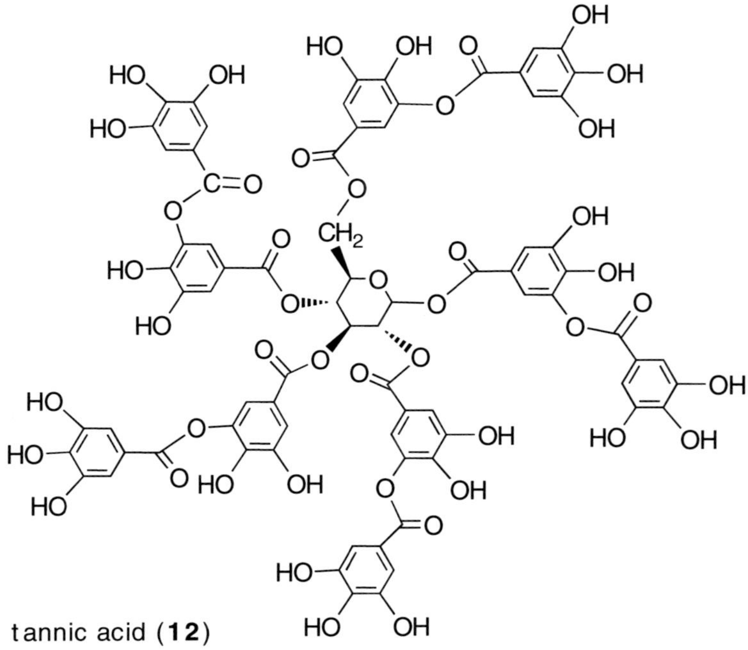

tannic acid (**12**)

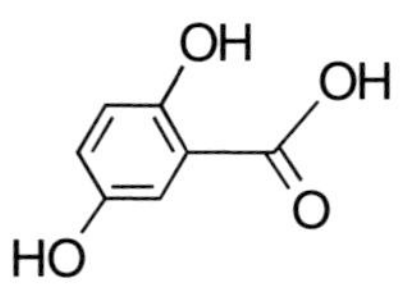

2,5-dihydroxybenzoic acid (**13**)

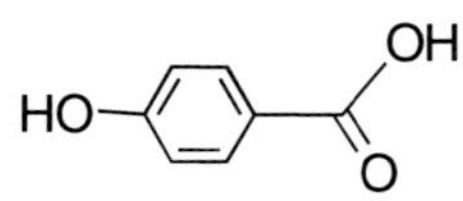

p-hydroxybenzoic acid (**14**)

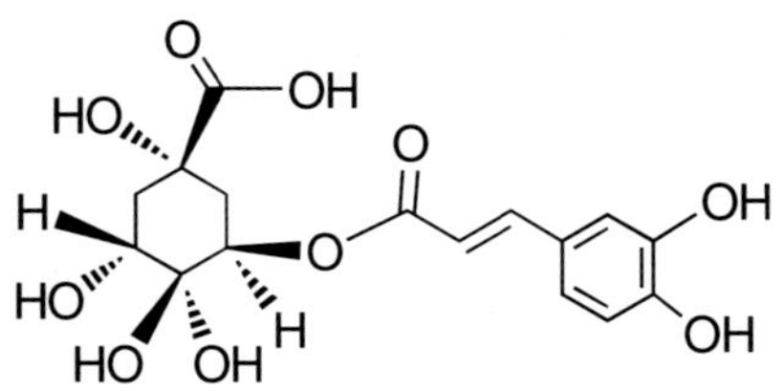

chlorogenic acid (**15**)

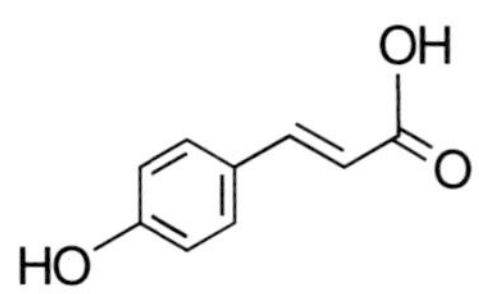

p-coumaric acid (**17**)

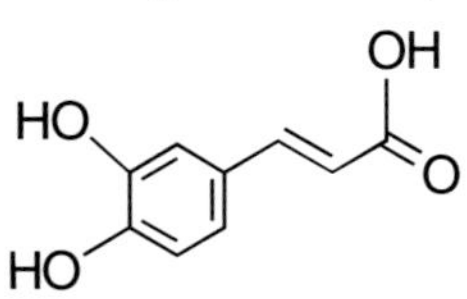

caffeic acid (**18**)

Figure 6. Tannic acid (12), 2,5-dihydroxybenzoic acid (13), *p*-hydroxybenzoic acid (14), chlorogenic acid (15), *p*-coumaric acid (17) and caffeic acid (18) from some kiwifruits.

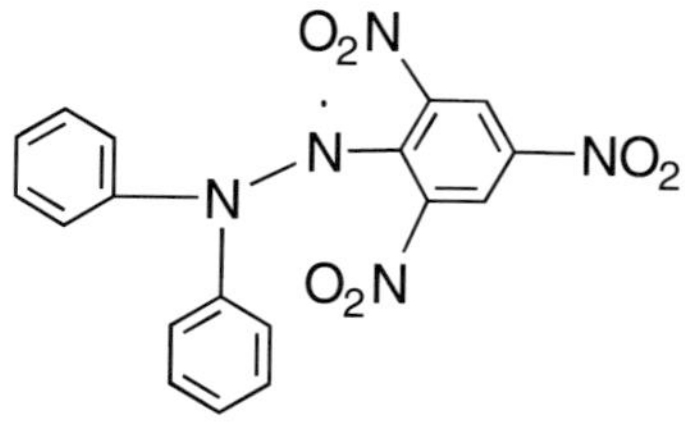

2,2-diphenyl-1-picrylhydrazyl (DPPH. **19**)

Figure 7. 2,2-Diphenyl-1-picrylhydrazyl (DPPH. 19).

2.3.3. Anti-hepatofibrotic Activity of Quercetin

In 2011, a group of Department of Applied Chemistry, National Chi Nan University, Taiwan studied their anti-hepatofibrotic mechanisms of quercetin (9) in the activated hepatic stellate cells were isolated from Sprague Dawley rats. The activated hepatic stellate cells were spontaneously activated *in vitro*. However, quercetin (9) restrained the proliferation of the activated hepatic stellate cells. Additionally, quercetin (9) also suppressed the activation of hepatic stellate cells by the reduction on the levels of interferon-inducible protein 10 (CXCL10. chemokine CXCL10) of inflammatory cytokines. Thus, quercetin (9) might work the diverse mechanisms for their inhibition against the growth of the activated hepatic stellate cells. From these results, it suggests that the intake of quercetin (9) in daily dietary life might be essential to control the progression of heptofibrosis [28].

In 1998, a group of Fisicoquímica, Facultad de Farmacia y Bioquímica, Universidad de Buenos Aires, Argentina studied the role of (+)-catechin (10) on the protection of human plasma (ヒト血漿) against the oxidation. The addition of (+)-catechin (10) could dose-dependently prevent or delay their formation of 2-thiobarbituric acid reactive substances, and the depletion of α-tocopherol (20) and β-carotene (21). This antioxidant effect was dependent on the concentration of (+)-catechin (10) and on the physical characteristics of the radical initiator. From these results, it was suggested that (+)-catechin (10) might be an effective antioxidant in human blood plasma which could delay the consumption of the endogenous lipid soluble antioxidants such as α-tocopherol (20) and β-carotene (21), and their lipid oxidation in human blood plasma could be inhibited (Figure 8).

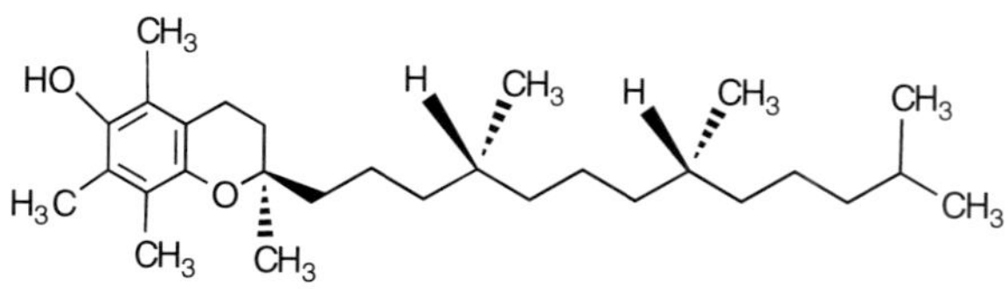

α-tocopherol (**20**)

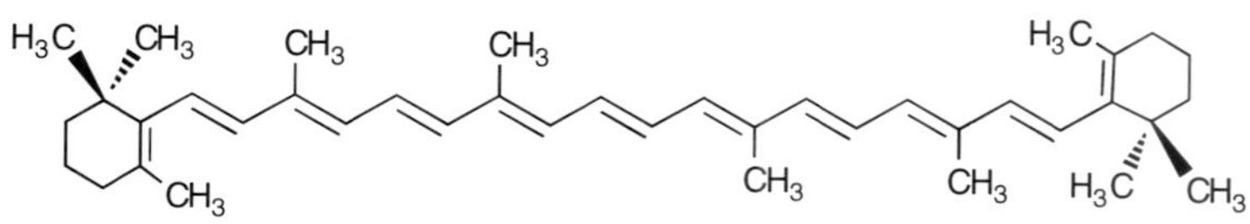

β-carotene (**21**)

Figure 8. ⟨α-Tocopherol (20) and β-carotene (21)

CONCLUSION

Kiwifruits are a good source of antioxidative anthocyanins and flavonoids. These colorful kiwifruit pigments with anthocyanins and flavonoids might be beneficial to prevent and improve age-related diseases such as hypertension (high blood pressure), hyperlipemia, oxidative stressed brain, diabetes, hepatic insufficiency and heart failure.

REFERENCES

[1] Andersen OM, Jordheim M. *The Anthocyanins. In Flavonoids Chemistry, Biochemistry and Applications*; (eds) Anderson OM, Markham KR, CRC Press, Taylor and Francis, Boca Raton, FL, pp. 471-551, 2006.

[2] Gary Null, *The Clinician's Handbook of Natural Healing.* Kensington Books, NY, USA, p.658, 1997.

[3] Delgado-Vargas F, Jiménez AR, Paredes-López O. Natural pigments: carotenoids, anthocyanins, and betalains--characteristics, biosynthesis, processing, and stability. *Crit Rev Food Sci Nutr* 40(3), 173-289, 2000.

[4] Motohashi N. (ed) *Bioactive Heterocycles VI, Flavonoids and Anthocyanins in Plants, and Latest Bioactive Heterocycles I.* Springer-Verlag Berlin Heidelberg, Germany. 2008.

[5] Motohashi N. (ed) *Bioactive Heterocycles VII, Flavonoids and Anthocyanins in Plants, and Latest Bioactive Heterocycles II.* Springer-Verlag Berlin Heidelberg, Germany. 2009.

[6] Motohashi N, Sakagami H. Functionality of anthocyanins as alternative medicine. *In (ed) Motohashi N. Bioactive Heterocycles VI*, Springer-Verlag Berlin Heidelberg, Germany. pp.1-48, 2008.

[7] Kong JM, Chia LS, Goh NK, Chia TF, Brouillard R. Analysis and biological activities of anthocyanins. *Phytochemistry* 64(5), 923-933, 2003.

[8] Ferguson AR. Kiwifruit (*Actinidia*). *Acta Hortic* 290, 603-653, 1990.

[9] Montefiori M, McGhie TK, Costa G, Ferguson AR. Pigments in the fruit of red-fleshed kiwifruit (*Actinidia chinensis* and *Actinidia deliciosa*). *J Agric Food Chem* 53(24), 9526-9530, 2005.

[10] Cui Zhi Xue (ed). *Actinidia in China* (Chinese kiwifruit). (in Chinese) Shandong Technology Publishers, China, 1993.

[11] Jin J, Quinton TM, Zhang J, Rittenhouse SE, Kunapuli SP. Adenosine diphosphate (ADP)-induced thromboxane A(2) generation in human platelets requires coordinated signaling through integrin alpha(IIb)beta(3) and ADP receptors. Blood. 99(1), 193-198, 2002.

[12] Duttaroy AK, Jørgensen A. Effects of kiwi fruit consumption on platelet aggregation and plasma lipids in healthy human volunteers. *Platelets* 15(5), 287-292, 2004.

[13] Comeskey DJ, Montefiori M, Edwards PJ, McGhie TK. Isolation and structural identification of the anthocyanin components of red kiwifruit. *J Agric Food Chem* 57, 2035-2039, 2009.

[14] Andres-Lacueva C, Shukitt-Hale B, Galli RL, Jauregui O, Lamuela-Raventos RM, Joseph JA. Anthocyanins in aged blueberry-fed rats are found centrally and may enhance memory. *Nutr Neurosci* 8(2), 111-120, 2005.

[15] Annamraju D. Sarma, Yellamraju Sreelakshmi, Rameshwar Sharma. Antioxidant ability of anthocyanins against ascorbic acid oxidation. *Phytochemistry* 45, 671-674, 1997.

[16] Ferguson AR, Fillion L. Are kiwifruit really good for you? *Acta Horticulturae* 610, 131-135, 2003.

[17] Collins BH, Horská A, Hotten PM, Riddoch C, Collins AR. Kiwifruit protects against oxidative DNA damage in human cells and *in vitro*. *Nutr Cancer* 39(1), 148-153, 2001.

[18] Dutta-Roy AK. Dietary components and human platelet activity. *Platelets* 13(2), 67-75, 2002.

[19] Fillion L, Collins A, Southon S. Beta-carotene enhances the recovery of lymphocytes from oxidative DNA damage. *Acta Biochim Pol* 45(1), 183-190, 1998.

[20] Guarnieri S, Loft S, Riso P, Porrini M, Risom L, Poulsen HE, Dragsted LO, Møller P. DNA repair phenotype and dietary antioxidant supplementation. *Br J Nutr* 99(5), 1018-1024, 2008.

[21] Spormann TM, Albert FW, Rath T, Dietrich H, Will F, Stockis JP, Eisenbrand G, Janzowski C. Anthocyanin/polyphenolic-rich fruit juice reduces oxidative cell damage in an intervention study with patients on hemodialysis. *Cancer Epidemiol Biomarkers Prev* 17(12), 3372-3380, 2008.

[22] Stoner GD, Wang LS, Casto BC. Laboratory and clinical studies of cancer chemoprevention by antioxidants in berries. *Carcinogenesis* 29(9), 1665-1674, 2008.

[23] Salganik RI. The benefits and hazards of antioxidants: controlling apoptosis and other protective mechanisms in cancer patients and the human population. *J Am Coll Nutr* 20(5 Suppl), 464S-472S; discussion 473S-475S, 2001.

[24] Brevik A, Gaivão I, Medin T, Jørgenesen A, Piasek A, Elilasson J, Karlsen A, Blomhoff R, Veggan T, Duttaroy AK, Collins AR. Supplementation of a western diet with golden kiwifruits (*Actinidia chinensis* var.'Hort 16A':) effects on biomarkers of oxidation damage and antioxidant protection. *Nutr J.* 2011 May 18;10:54.

[25] Latocha P, Krupa T, Wołosiak R, Worobiej E, Wilczak J. Antioxidant activity and chemical difference in fruit of different *Actinidia* sp. *Int J Food Sci Nutr* 61(4), 381-394, 2010.

[26] Silvia Tavarini, Elena Degl'Innocenti, Damiano Remorini, Rossano Massai, Lucia Guidi. Antioxidant capacity, ascorbic acid, total phenols and carotenoids changes during harvest and after storage of Hayward kiwifruit. *Food Chem* 107(1), 282-288, 2008.

[27] Du G, Li M, Ma F, Liang D. Antioxidant capacity and the relationship with polyphenol and vitamin C in *Actinidia* fruits. *Food Chem.*, 113, 557-562, 2009.

[28] Wu LC, Lu IW, Chung CF, Wu HY, Liu YT. Antiproliferative mechanisms of quercetin in rat activated hepatic stellate cells. *Food Funct* 2(3-4), 204-212, 2011.

[29] Lotito SB, Fraga CG. (+)-Catechin prevents human plasma oxidation. *Free Radic Biol Med* 24(3), 435-441, 1998.

In: Anthocyanins: Structure, Biosynthesis and Health Benefits ISBN: 978-1-62257-329-5
Editor: Noboru Motohashi

Chapter 2

ANTHOCYANINS OF RED NOBLE MUSCADINE BERRIES

Youssef W. Mirhom*[*], *Yuping V. Williamson*, *Paula L. Schulbaum*, *Erin B. Krueger* and *Frank S. D'Amelio, Sr.
Bio-Botanica, Inc., Hauppauge, NY, US

ABSTRACT

Muscadine berries (*Vitis rotundifolia* Michx.) are hardy grapes indigenous to the South Eastern United States of America with high antioxidant and anti-inflammatory properties. Noble fresh red muscadine berries were crushed, and the juice was separated. The crushed grapes (skin, pulp, and seed) were extracted following novel extraction protocols. The collected liquid extract was concentrated to 70% solid content under reduced pressure below 45°C. The actives in the extract were studied using HPLC-PDA and LC-ESI-MS. The total polyphenols, proanthocyanidins (OPC), gallic acid, ellagic acid, and resveratrol were previously determined [1]. The anthocyanins that are responsible for the color of the muscadine berries, are important natural antioxidants, and proved to be beneficial for cardio-protection, cholesterol-lowering and vision-improvement [2]. Recently, it was found that anthocyanins suppress the DNA-damaging effects of topoisomerase poisons in human colon cancer cells [3]. Therefore, it was deemed necessary to determine the composition and relative amount of the major anthocyanins present. The anthocyanins from the muscadine extract were further analyzed by HPLC-QTOF MS. The composition and structure of the major anthocyanins were tentatively elucidated to be delphinidin, cyanidin, petunidin, peonidin, and malvidin-3,5-diglucosides. Their definite structures will be confirmed by total synthesis.

[*] E-mail: ymirhom@bio-botanica.com.

INTRODUCTION

Muscadine berries (*Vitis rotundifolia* Michx.) are hardy grapes indigenous to the South Eastern United States (SE U.S.A.). Whereas common grapes have 19 pairs of chromosomes, the muscadines have 20 pairs of chromosomes [4]. This surely accounts for the ability to thrive in the high heat and humidity of SE U.S.A.

Muscadines make an excellent contribution to disease prevention diet strategy due to their content of wide array of groups of constituents as explained thereafter [5]:

a. *Trans-resveratrol* is a phytoalexin, antibiotic compound (trans-3,5,4'-trihydroxystilbene) produced by the grapes in response to external invasion, infection or disease. It is found in the skin [6]. Resveratrol has shown potent anti-aging effects and therapeutic application in oxidative stress related diseases including cancer, arteriosderosis, heart diseases, etc. [7].
b. *Flavonols and flavanols* help cells to regulate gene expression in very beneficial ways. They are therefore useful in diseases that involve chronically adverse patterns of gene expression like atherosclerosis, cancer, diabetes or inflammatory and degenerative diseases [8,9].
c. *Oligomeric Proanthocyanidins (OPC)* or simply procyanidins are abundant in the seed and are one of nature's most powerful and readily available antioxidants and cytoprotective [10] used for venous insufficiency and ophthalmologic complaints.
d. *Ellagic and gallotannins (Hydrolysable Tannins)* are a group of tannins which contributes to cancer chemoprevention from occurring, growing or metastasis. It also induces gene expression of the enzyme NADP, H/quinone reductase involved in the detoxification of carcinogens and thereby an induction of gene expression for this enzyme will decrease the mutagenesis and tumorigenesis of a battery of carcinogens. This group of tannins has been shown to be anti-proliferative in colon cancer cells, breast cancer and prostatic cancer cells [11].
e. *Anthocyanins* are a group of compounds responsible for the color of the muscadine berries with very important physiological activities being natural antioxidants which proved to be beneficial for cardio-protection, cholesterol-lowering and vision improvement [2]. Recently, it was found that anthocyanins suppress the DNA-damaging effects of Topoisomerase poisons in human colon cancer cells [3].

In a previous communication we reported on gallic and ellagic acids, proanthocyanidins, resveratrol and polyphenols of red muscadine grapes [1]. We would like to report now in detail on the anthocyanins of muscadine berries using mass/spectrometric approach.

EXPERIMENTAL

Plant Material

Fresh red noble muscadine berries were supplied and identified by Duplin Winery (Rose Hill, NC, USA).

Extraction

Muscadines were brought fresh in refrigerated trucks to the lab and expressed to give the juice and pomace (skin, seed and pulp). The pomace was extracted two times with 100% methanol and two times with 50% aqueous methanol. The pooled extracts were concentrated under reduced pressure below 45°C to give a dark brown liquid with about 70% solid content.

Instrumentation

The LC-MS/MS system consists of Agilent 6520 QTOF Mass Spectrometer (G6510A) (Agilent Technologies, Santa Clara, CA, USA), Agilent 1200 HPLC with a binary pump (G1312B), autosampler (G1367D), Diode Array Detector (G1315C), and column heated compartment (G1316B). The mobile phases used for analysis are: mobile phase A: 0.1% w/w Formic acid in water; Mobile phase B: MeOH - ACN - 0.1% w/w Formic acid - H_2O (25:25:10:40 by vol.). HPLC column 4.6x150mm Agilent Extend C18 (3.5um, Waters, Milford, MA) was used. For MS, the ESI positive (ESI capillary voltage at 3.5KV) source was used with the gas flow of 12 ml/min. The gas temperature was adjusted at 35°C. The nebulizer gas was adjusted at 55psi. MS scan range was set at 50-1700 m/z, with the scan rate of 2.5. Ref. Mass 121.050873 and 922.009798 m/z were enabled throughout the analysis.

RESULTS

Figures 1 and 2 are the HPLC chromatograms of the pomace extract at two different wavelengths, namely, 517nm for the anthocyanins and 267nm, to show its other most important components. There are five major peaks in the UV chromatogram at 517nm, and they are the major anthocyanin peaks as they show absorbance at wavelength around 520nm (Figure 1). The retention times are 6.347, 9.237, 12.207, 16.140, and 19.103 minutes respectively. While at 267nm, major peaks were gallic acid, proanthocyanidins, ellagic acid, and resveratrol, in the order shown in Figure 2.

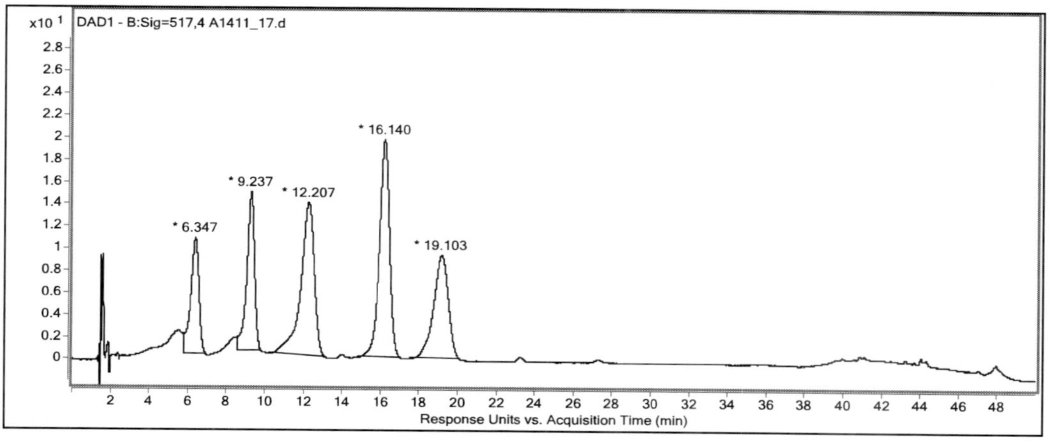

Figure 1. Chromatogram of the extract at 517nm.

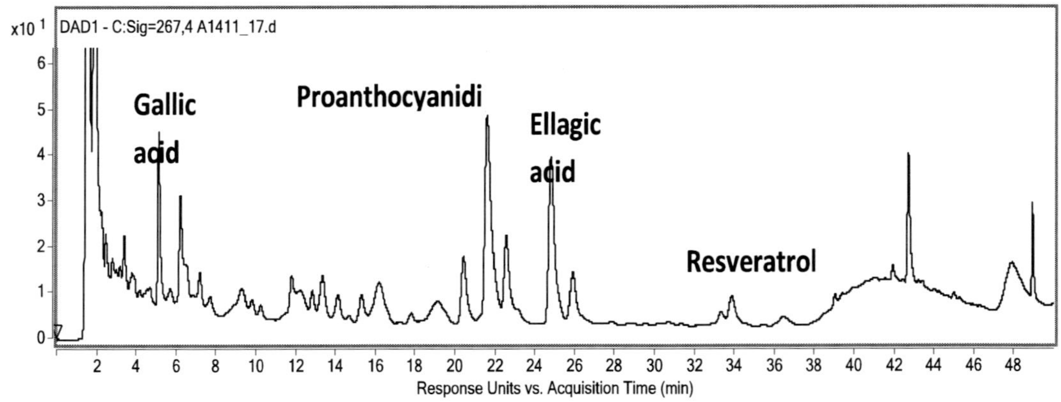

Figure 2. UV chromatogram of the extract at 267nm.

The structures of small molecules such as gallic acid, ellagic acid, and resveratrol were characterized by single quadrupole, while the structures of anthocyanins Tandem MS was needed for structure elucidation due to the relative complexity of their structure. The first MS was used to select the mother ion, then by collision-induced dissociation (CID), the daughter ions

could be detected in the second MS.

Peak 1

For the first peak with the retention time of 6.347min, the first MS and the MS after CID were illustrated in figures 3 and 4, respectively. The m/z of the mother ion was 627.1557 (Figure 3), which corresponds to ion formula $C_{27}H_{31}O_{17}^{+}$. The m/z of the two daughter ions after CID were found to be 465.1039 and 303.0492 (Figure 4), corresponding to ion formulas of $C_{21}H_{21}O_{12}^{+}$ and $C_{15}H_{11}O_{7}^{+}$ respectively.

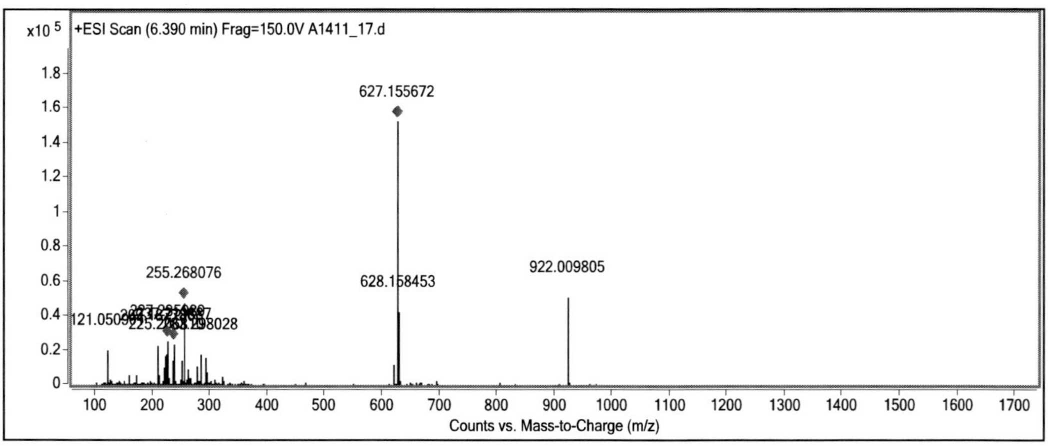

Figure 3. Positive ESI mass spectrum of peak at 6.347 minutes.

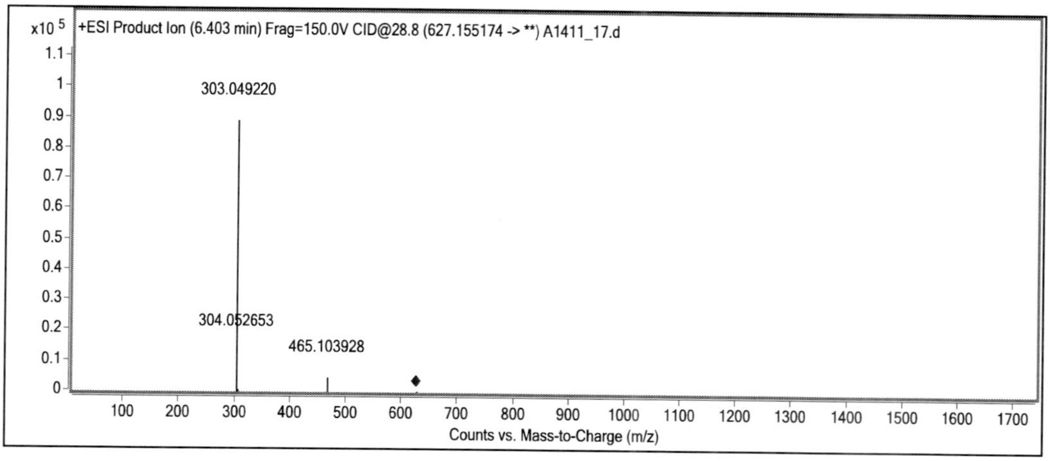

Figure 4. CID mass spectrum of peak at 6.347 minutes.

After calculation and identification, the mother ion was found to be delphinidin 3,5-diglucosides, and the two daughter ions were found to be delphinidin 3-glucoside and delphindin (Figure 5).

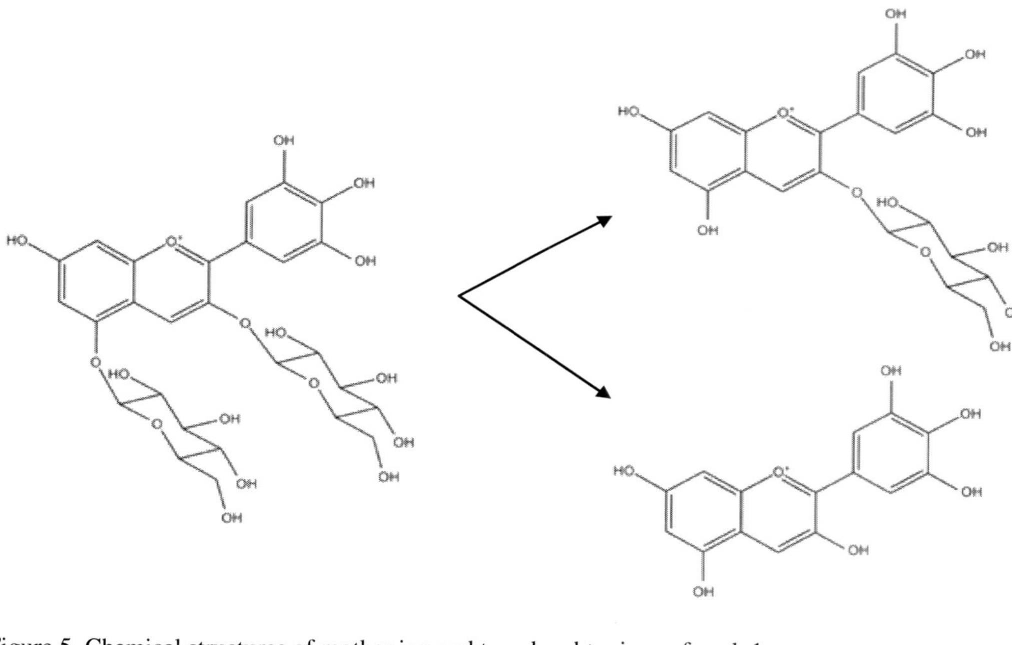

Figure 5. Chemical structures of mother ion and two daughter ions of peak 1.

Peak 2

For the second peak with the retention time of 9.237min, the first MS and the MS after CID were illustrated in figures 6 and 7, respectively. The m/z of the mother ion was 611.1609 (Figure 6), which corresponds to ion formula $C_{27}H_{31}O_{16}^{+}$. The m/z of the two daughter ions after CID were found to be 449.1082 and 287.0548 (Figure 7), corresponding to ion formulas of $C_{21}H_{21}O_{11}^{+}$ and $C_{15}H_{11}O_{6}^{+}$ respectively.

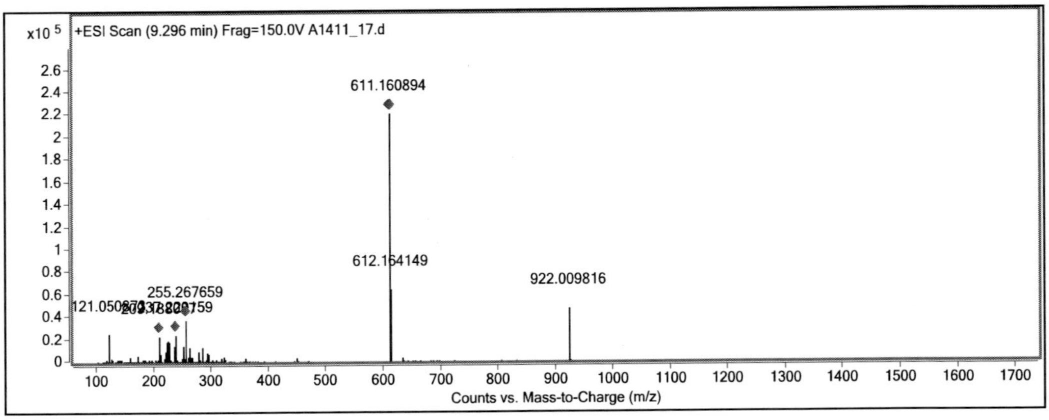

Figure 6. Positive ESI mass spectrum of peak at 9.237 minutes.

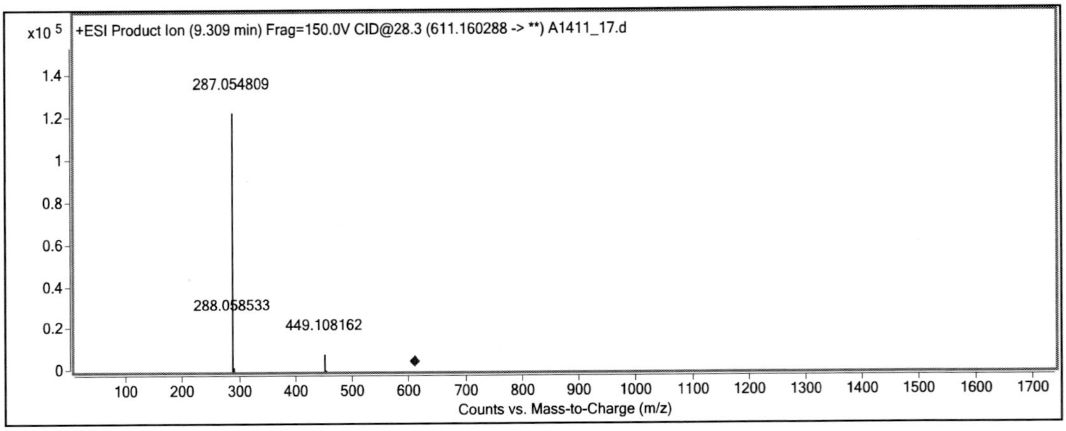

Figure 7. CID mass spectrum of peak at 9.237 minutes.

After calculation and identification, the mother ion was found to be cyanidin 3,5-diglucosides, and the two daughter ions were found to be cyanidin 3-glucoside and cyanidin (Figure 8).

Peak 3

For the third peak with the retention time of 12.207min, the first MS and the MS after CID were illustrated in figure 9 and 10 respectively. The m/z of the mother ion was 641.1708 (Figure 9), which corresponds to ion formula $C_{28}H_{33}O_{17}^+$. The m/z of the two daughter ions after CID were found to be 479.1172 and 317.0645 (Figure 10), corresponding to ion formulas of $C_{22}H_{23}O_{12}^+$ and $C_{16}H_{13}O_7^+$ respectively.

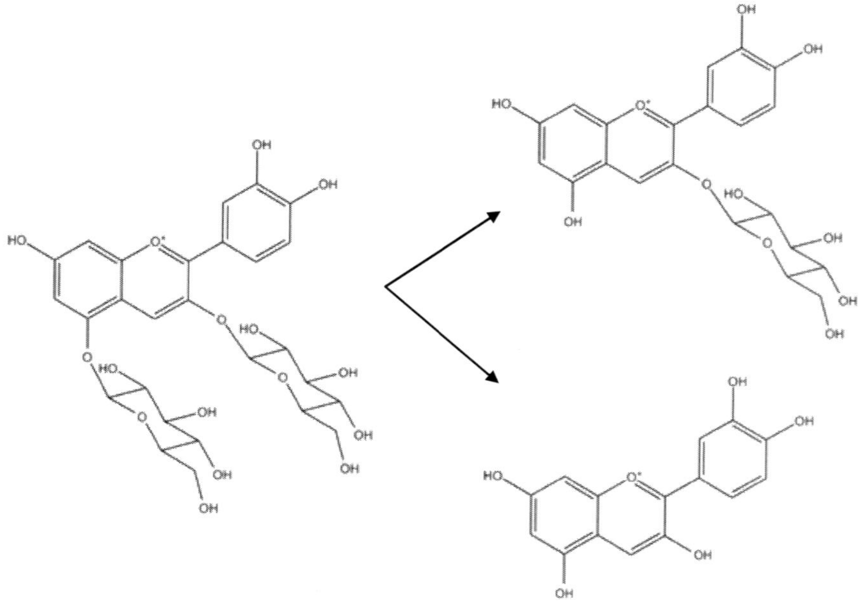

Figure 8. Chemical structures of mother ion and two daughter ions of peak 2.

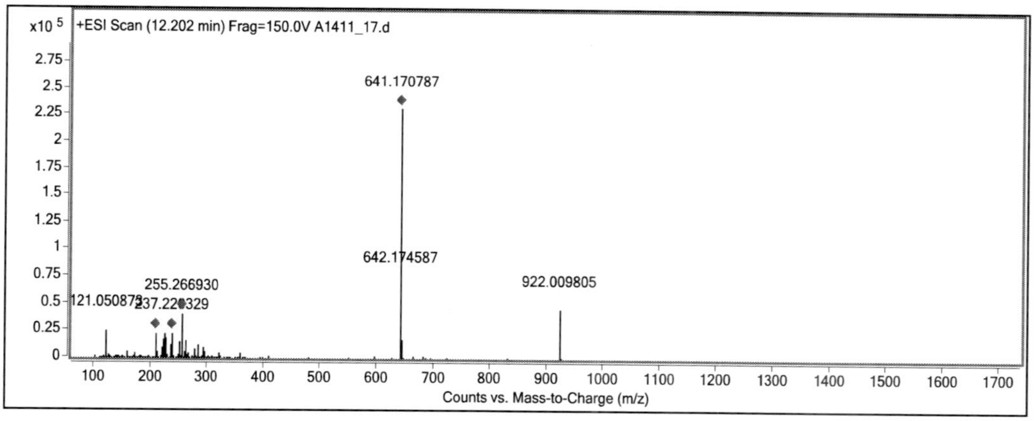

Figure 9. Positive ESI mass spectrum of peak at 12.207 minutes.

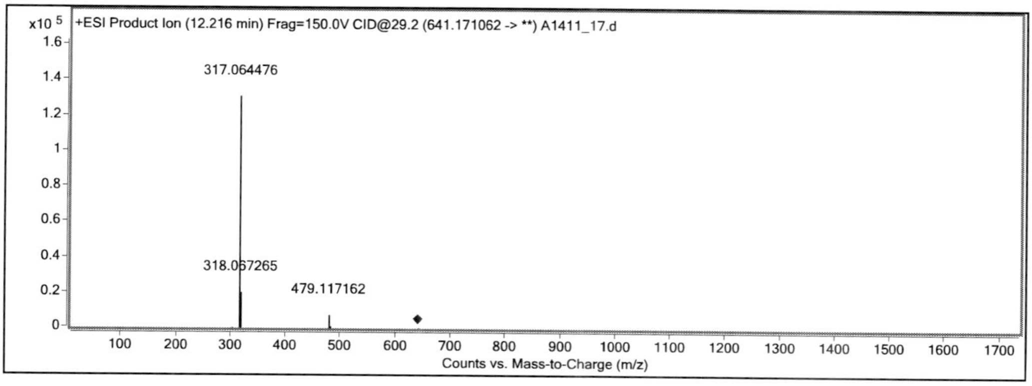

Figure 10. CID mass spectrum of peak at 12.207 minutes.

After calculation and identification, the mother ion was found to be petunidin 3,5-diglucosides, and the two daughter ions were found to be petunidin 3-glucoside and petunidin (Figure 11).

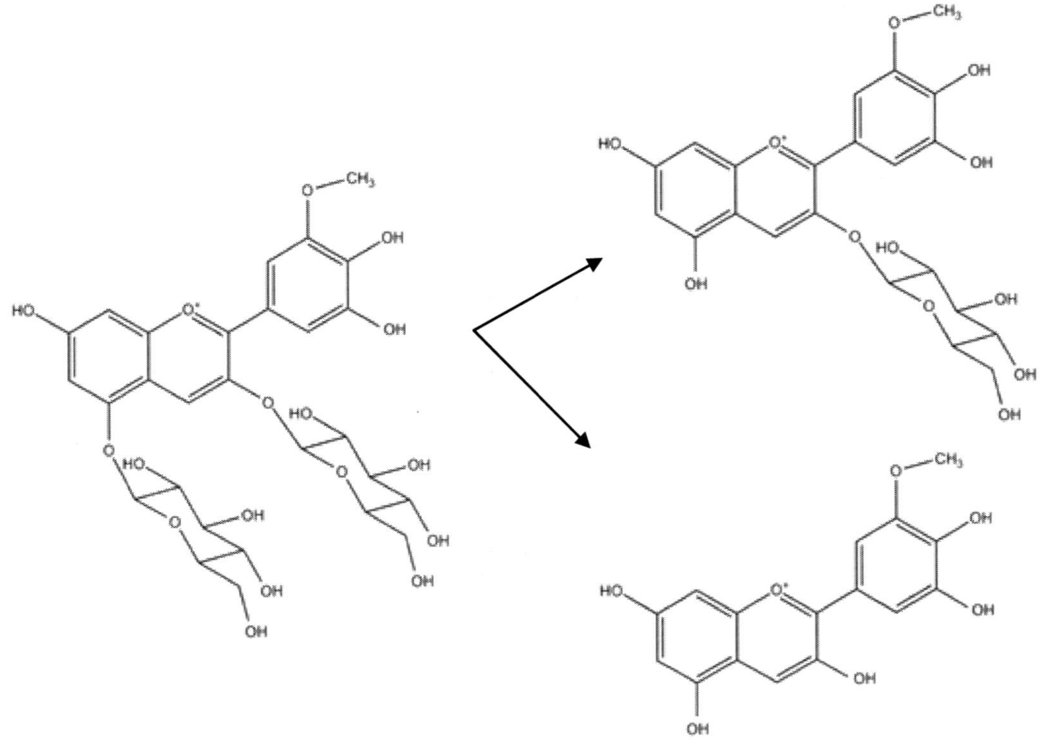

Figure 11. Chemical structures of mother ion and two daughter ions of peak 3.

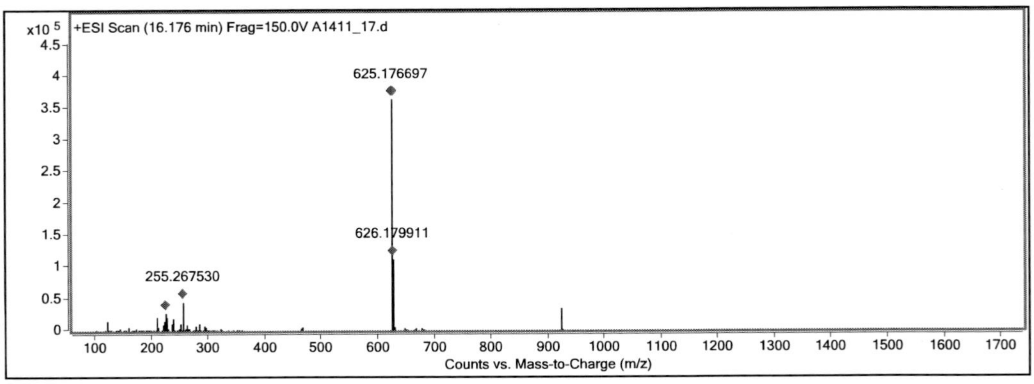

Figure 12. Positive ESI mass spectrum of peak at 16.140 minutes.

Peak 4

For the fourth peak with the retention time of 16.140min, the first MS and the MS after CID were illustrated in figures 12 and 13, respectively. The m/z of the mother ion was 625.1767 (Figure 12), which corresponds to ion formula $C_{28}H_{33}O_{16}^{+}$. The m/z of the two daughter ions after CID were found to be 463.1237 and 301.0702 (Figure 13), corresponding to ion formulas of $C_{22}H_{23}O_{11}^+$ and $C_{16}H_{13}O_6^+$ respectively.

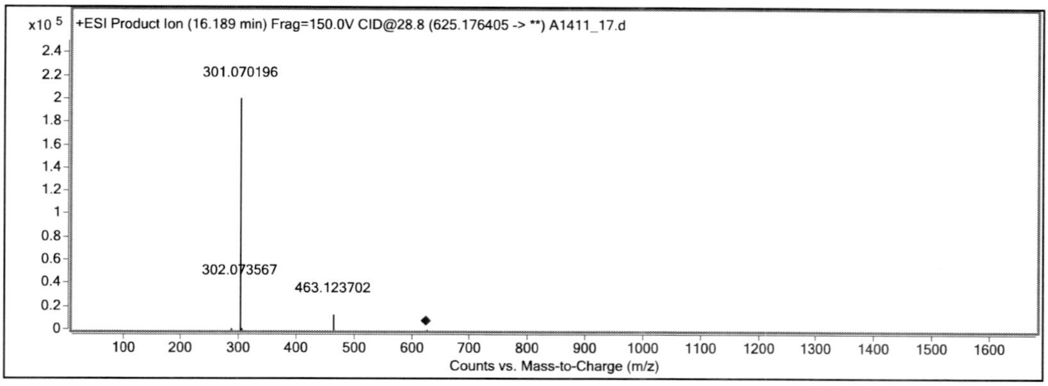

Figure 13. CID mass spectrum of peak at 16.140 minutes.

After calculation and identification, the mother ion was found to be peonidin 3,5-diglucosides, and the two daughter ions were found to be peonidin 3-glucoside and peonidin, respectively (Figure 14).

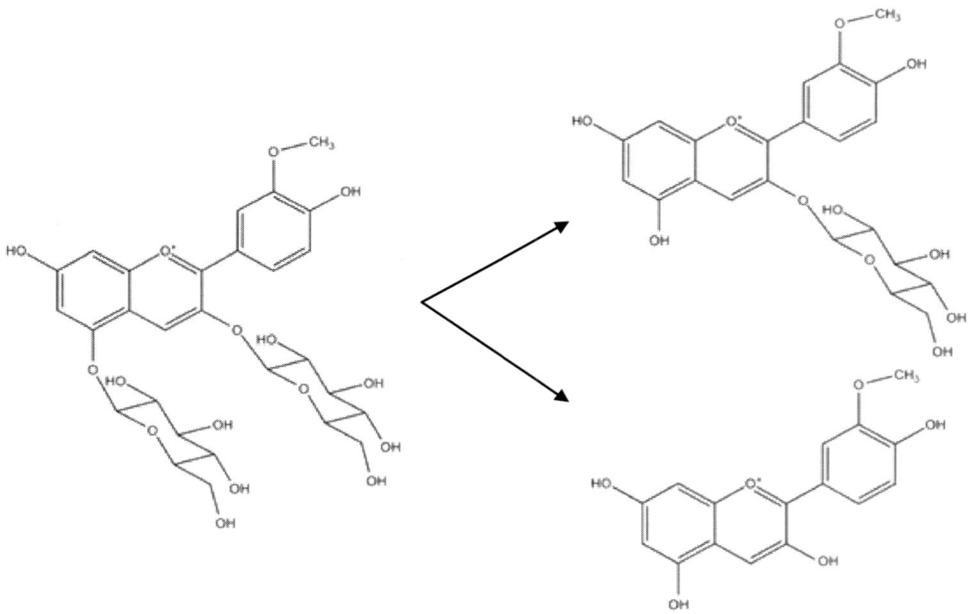

Figure 14. Chemical structures of mother ion and two daughter ions of peak 4.

Peak 5

For the last peak with the retention time of 19.103min, the first MS and the MS after CID were illustrated in figures 15 and 16, respectively. The m/z of the mother ion was 655.1871 (Figure 15), which corresponds to ion formula $C_{29}H_{35}O_{17}^+$. The m/z of the two daughter ions after CID were found to be 493.1334 and 331.0806, respectively (Figure 16), corresponding to ion formulas of $C_{23}H_{25}O_{12}^+$ and $C_{17}H_{15}O_7^+$ respectively.

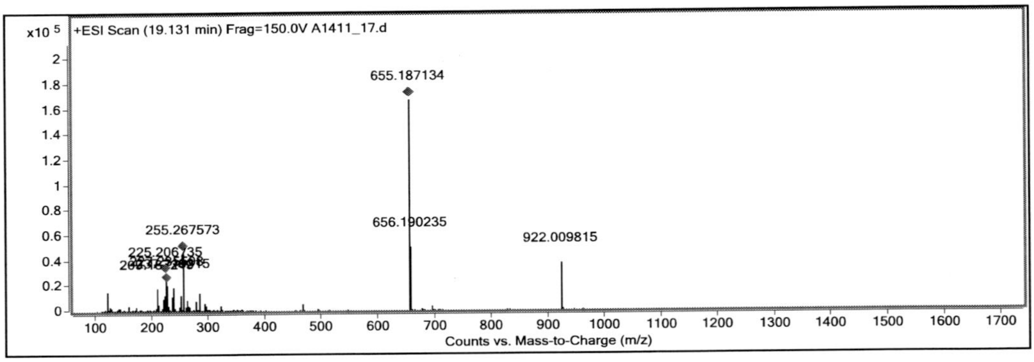

Figure 15. Positive ESI mass spectrum of peak at 19.103 minutes.

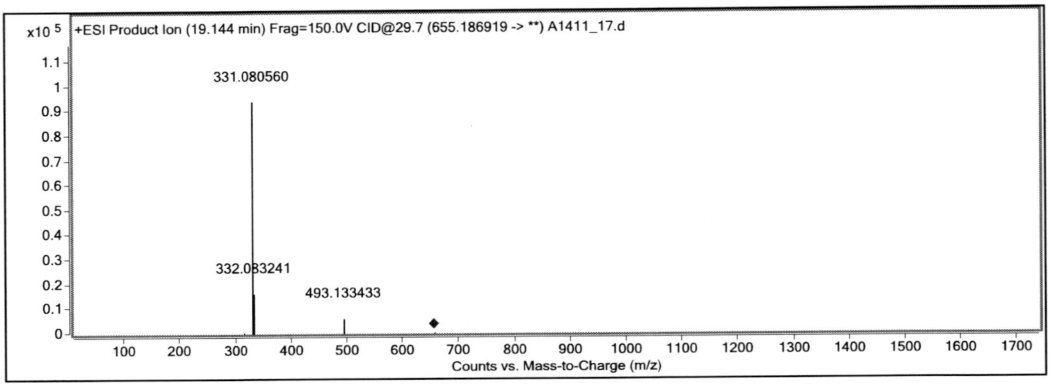

Figure 16. CID mass spectrum of peak at 19.103 minutes.

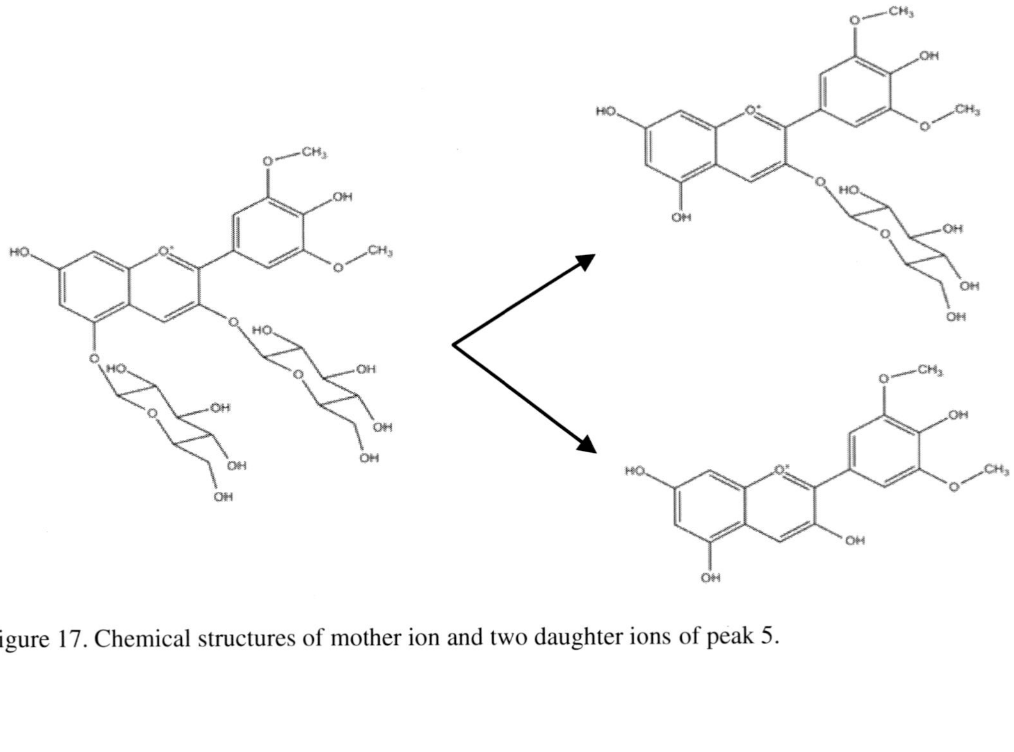

Figure 17. Chemical structures of mother ion and two daughter ions of peak 5.

After calculation and identification, the mother ion was found to be malvidin 3,5-diglucosides, and the two daughter ions were found to be malvidin 3-glucoside and malvidin, respectively (Figure 17).

Table 1 shows the comparative structures of certain closed related anthocyanin- -aglycones and Table 2 shows the structures of the five elucidated 3,5-diglucosides.

Table 1. Structures of certain closely related anthocyanin-aglycones

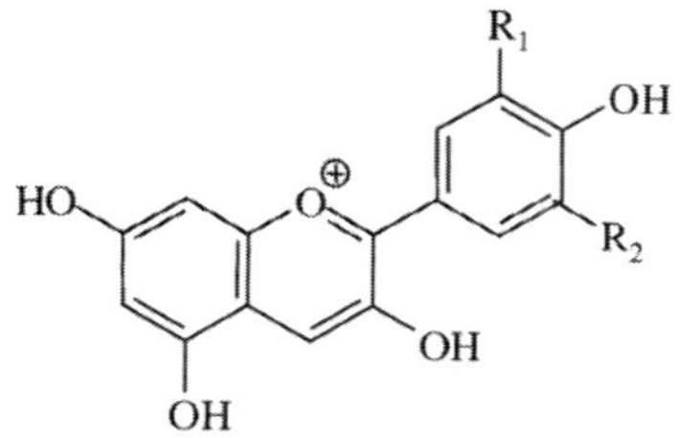

Aglycone	R_1	R_2
Pelargonidin	H	H
Cyanidin	OH	H
Peonidin	OCH_3	H
Delphinidin	OH	OH
Petunidin	OH	OCH_3
Malvidin	OCH_3	OCH_3

Table 2. Structures of the five elucidated 3,5-diglucosides

tr (min)	Ion	m/z	CID ion formula	m/z	CID ion formula	m/z	Name
6.347	$C_{27}H_{31}O_{17}+$	627.155	$C_{21}H_{21}O_{12}+$	465.103	$C_{15}H_{11}O_7+$	303.049	Delphinidin 3,5-diglucoside
9.237	$C_{27}H_{31}O_{16}+$	611.160	$C_{21}H_{21}O_{11}+$	449.108	$C_{15}H_{11}O_6+$	287.054	Cyanidin 3,5-diglucoside
12.207	$C_{28}H_{33}O_{17}+$	641.170	$C_{22}H_{23}O_{12}+$	479.117	$C_{16}H_{13}O_7+$	317.064	Petunidin 3,5-diglucoside
16.140	$C_{28}H_{33}O_{16}+$	625.176	$C_{22}H_{23}O_{11}+$	463.123	$C_{16}H_{13}O_6+$	301.070	Peonidin 3,5-diglucoside
19.103	$C_{29}H_{35}O_{17}+$	655.187	$C_{23}H_{25}O_{12}+$	493.133	$C_{17}H_{15}O_7+$	331.080	Malvidin 3,5-diglucoside

CONCLUSION

Q-TOF is an excellent tool for determining the molecular formula and structure elucidation of natural products, and it was successfully used in this study for the analysis of major anthocyanins in the red noble muscadine berry extracts. The advantage of Q-TOF mass spectrometer is that it gives exact mass of a component and its molecular formula which helps out significantly during the structure interpretation. However, for the final confirmation of the structure we may need other techniques and total synthesis. Methods established in this research can also be applied to the analysis of anthocyanins in other available berries.

ACKNOWLEDGMENT

The authors wish to thank Steve Baugh and Christopher Hardwick from ChromaDex for running the QTOF and the technical personnel of Duplin Winery for collecting and identifying the red noble muscadine berries.

REFERENCES

[1] F.S. D'Amelio, Sr., Y.W. Mirhom, and Y.V. Williamson. Red muscadine species as the grape of choice for improving the quality of human life, 51st ASP meeting, P-60, St. *Petersburg Beach,* FL, July, 2010.

[2] M.T. Meunier. E. Duroux, P. Bastide. Free radical scavenger activity of procyanidolic oligomers and anthocyanosides with respect to superoxide anion and lipid peroxidation. *Planta Med. Phytother. 23(4)*, 267-74 (1989).

[3] M. Esselen, J. Fritz, M. Hutter, N. Teller, S. Baechler, U. Boettler, T.H. Marczylo, A.J. Gescher, D. Marko. Anthocyanin-rich extracts suppress the DNA-damaging effects of topoisomerase poisons in human colon cancer cells. *Mol Nutr Food Res. 55 (suppl 1)*, 143-53 (2011).

[4] G.I. Patel, H.P. Olmo, Cytogenetics of *Vitis.*: I. the hybrid *V. vinifera* times *V. rotundifolia*. *Amer J. Botany 42*, 141-55 (1955).

[5] D.K. Hartle, P. Greenspan, J. Hargrove. in *Muscadine Medicine, Phytochemical power of muscadine grapes, skins, seeds and extracts,* pp, 27-46, publisher, Blue Heron Nutraceuticals, LLC, FL, USA (2005).

[6] M. Rubilar, M. Pinelo, C. Shene, J. Sineiro and M. Nunez, Separation and HPLC-MS identification of phenolic antioxidants from agricultural residues: almond hulls and grape pomace. *J. Agric. Food Chem. 55*, 10101-9 (2007).

[7] A. Ali, J. Strommer. A simple extraction and chromatographic system for the simultaneous analysis of anthocyanins and stilbenes of *Vitis* species. *J. Agric. Food Chem. 51*, 7246-51 (2003).

[8] L. Rivera, R. Morón, M. Sánchez, A. Zarzuelo, M. Galisteo. *Quercetin ameliorates metabolic syndrome and improves the inflammatory status in obese Zucker rats.* Obesity (Silver Spring) *16(9)*, 2081-7 (2008).

[9] J.L. Barger, T.D. Pugh, R. Weindruch. Short-term consumption of a resveratrol-containing nutraceutical mixture mimics gene expression of long-term caloric restriction in mouse heart. *Exp. Gerontol. 43(9)*, 859-66 (2008).

[10] J. Masquelier, J.Michaud, J. Laparra, M.C. Dumon. Flavonoids and pycnogenols. *J. Vit. Nutr. Res 49(3)*, 307-11 (1979).

[11] J.N. Losso, R.R. Bansode, A. Trappey, H.A. Bawadi, R.Truax, *In vitro* anti-proliferative activities of ellagic acid, *J. Nutr. Biochem. 15(11)*, 672-8 (2004).

In: Anthocyanins: Structure, Biosynthesis and Health Benefits ISBN: 978-1-62257-329-5
Editor: Noboru Motohashi

Chapter 3

ELECTROCHEMICAL, BIOACTIVE MECHANISM OF INTERACTION AND ANALYTICAL STUDIES OF ANTHOCYANINS AND RELATED COMPOUNDS

Seetharamappa Jaldappagari[1], Ashwini H. Hegde[1], Prashanth Shankar Narayan[1], Nagappa L. Teradal[1] and Noboru Motohashi[2]

[1]Department of Chemistry, Karnatak University, Dharwad, India
[2]Meiji Pharmaceutical University, Noshio, Kiyose-shi, Tokyo, Japan

ABSTRACT

Polyphenols are natural molecules containing one or more hydroxylated rings. They can be divided into different subclasses *viz.*, phenylethanol derivatives, phenolic acids, flavones and flavonols, catechins, procyanidins and anthocyanins. Anthocyanins are glycosylated polyhydroxy and polymethoxy derivatives of flavylium salts, which are a group of naturally occurring phenolic compounds responsible for the brilliant red, orange and blue colors of most fruits and flowers. They have shown antioxidative, antimutagenic, antidiabetic, and anticarcinogenic acitivities *in vitro*.

Electrochemical techniques *viz.*, cyclic voltammetry, square wave voltammetry, differential pulse voltammetry and amperometry have been employed for the characterization of the antioxidative properties of anthocyanins from various sources and for their assay in fruit juices and plant extracts.

The binding ability of anthocyanins to DNA and protein has attracted attention of researchers, recently as anthocyanins are proven to be bioactive. Such investigations are being studied by employing different analytical/spectroscopic techniques.

The present review also reports the assay of anthocyanins in various natural sources by analytical methods *viz.,* spectrophotometry, voltammetry, HPLC, LC-MS, MS/MS, MALDI-TOF-MS and others.

Keywords: Anthocyanins, electrochemical studies, interactions, macromolecules, spectroscopic techniques, analytical methods

ABBREVIATIONS

epicatechin equivalents	ece
Gallic acid equivalents	Gae
Total anthocyanins	TA
Total phenols	TP
Total flavonoids	TF
Acetic Acid	AC
Formic acid	FC
Acetonitrile	ACN
Trifluoroacetic acid	TFA
Quercetin	Quer
Quercetin-3-*O*-glucoside	quer-3-*O*-gluc
Kaempferol	Kaem
Kaempferol-3-*O*-glucoside	Kaem-3-*O*-gluc
Icaritin	Icar
desmethylicaritin-3-*O*-di- rhamnosyl -7-*O*-glucoside	desmethylicar-3-*O*-di-rha-7-*O*-gluc
xylosyl	Xyl
Delphinidin	Delp
Delphinidin-3-galactoside	delp-3-gala
Delphinidin-3-*O*-glucoside	delp-3-*O*-gluc
Delphinidin-3-hexose	delp-3-hex
Delphinidin-oxalyl-hexose	delp-oxa-hex
Delphinidin -3-arabinoside	delp-3-ara
Delphinidin-3-rutinoside	delp -3-rut
Delphinidin 3-*O*-acetyl-glucoside	delp-3-*O*-ace-gluc
Delphinidin 3-O-coumaryl-glucoside	delp-3-*O*-coum-gluc
Delphinidin 3-*O*-(6-*O*-acetyl)- glucoside	delp-3-*O*-(6-*O*-ace)-gluc
Delphinidin 3-*O*-(6-O-coumaryl)-glucoside	delp-3-*O*-(6-*O*-coum)-gluc
Delphinidin-3-*O*-*p*-coumaryl-rutinoside-5O-glucoside	delp 3-*O*-*p*-coum-rut-5-*O*-gluc
Delphinidin-3-*O*-caffeylglucoside	delp 3-*O*-caf-gluc
Cyanidine	Cya
Cyanidine-3-galactoside	cya-3-gala
Cyanidine-3-propionyl-galactoside	cya-3-prop-gala
Cyanidine-3-glucoside	cya-3-gluc
Cyanidine-3- arabinoside	cya-3-ara
Cyanidine-3-malonyl-galactoside	cya-3-malo-gala
Cyanidin-3,5-di-*O*-glucoside	cya-3,5-di-*O*-gluc
Cyanidin-3-rutinoside	cya-3-rut
Cyanidin-3-*O*-coumaryl-glucoside	cya-3-*O*-coum-gluc
Cyanidin 3-*O*-(6-*O*-acetyl)-glucoside	cya-3-*O*-(6-*O*-ace)-gluc
Cyanidin 3-*O*-(6-*O*-coumaryl)-glucoside	cya-3-*O*-(6-*O*-coum)-gluc
Petunidin	Pet

Petunidin-3-galactoside	pet-3- gala
Petunidin-3-propionyl-glucoside	pet-3-prop-gala
Petunidin-3-propionyl-hexose	pet-3-prop-hex
Petunidin-3-acetyl-glucoside	pet-3-ace-gluc
Petunidin-3-glucoside	pet-3-gluc
Petunidin-3-arabinoside	pet-3-ara
Petunidin -3-rutinoside	pet-3-rut
Petudidin-3-*O*-coumarylglucoside	pet-3-*O*-coum-gluc
Petunidin -3-rutinoside-5-*O*-glucoside	pet 3-*O*-rut-5-*O*-gluc
Petunidin-3-*O*-(6-*O*-acetyl)-glucoside	pet-3-*O*-(6-*O*-ace)-gluc
Petunidin-3-*O*-(6-*O*-coumaryl)-glucoside	pet-3-*O*-(6-*O*-coum)-gluc
Petunidin-3-*O*-caffeoyl-5- rutinoside -glucoside	pet 3-*O*-caf-rut-5-*O*-gluc
Petunidin -3-*O*-*p*-coumaryl-rutinoside-5-*O*-glucoside	pet-3-*O*-*p*-coum-rut-5-*O*-gluc
Petunidin -3-*O*-feruloyl- rutinoside-5-*O*-glucoside	pet 3-*O*-ferul-rut-5-*O*-gluc
Peonidin	Peo
Peonidin-3-succinyl-arabinoside	peo-3-succ-ara
Peonidin-3-galactoside	peo-3-gala
Peonidin-3-oxalyl-galactoside	peo-3-oxa-gala
Peonidin-3-glucoside	peo-3-gluc
Peonidin-3-arabinoside	peo-3-ara
Peonidin -3-rutinoside	peo-3-rut
Peonidin-3-*O*-coumarylglucoside	peo-3-*O*-coum-gluc
Peonidin -3-rutinoside-5-O-glucoside	pet 3-*O*-rut-5-*O*-gluc
Peonidin-3-*O*-glucoside-4-vinylphenol	peo-3-*O*-gluc-4-vinylphen
Peonidin-3,5-di-*O*-glucoside	peo-3,5-di-*O*-gluc
Peonidin 3-*O*-(6-*O*-acetyl)-glucoside	peo-3-*O*-(6-*O*-ace)-gluc
Peonidin 3-*O*-(6-*O*-coumaryl)-glucoside	peo-3-*O*-(6-*O*-coum)-gluc
Peonidin-3-*O*-*p*-coumaryl-rutinoside-5-*O*-glucoside	peo 3-*O*-*p*-coum-rut-5-*O*-gluc
Peonidin-3-*O*-feruloyl-rutinoside-5-*O*-glucoside	peo 3-*O*-ferul-rut-5-O-gluc
Pelargonidin 3-*O*-glucoside	pel -3-*O*-gluc
Pelargonidin 3-*O*-rutinoside	pel -3-*O*-rut
Pelargonidin 3-*O*-*p*-coumaryl-rutinoside	pel 3-*O*-*p*-coum-rut
Pelargonidin-3-*O*-feruloyl-rutinoside	pel 3-*O*-ferul-rut
Pelargonidin-3-*O*-glucoside-malonate	pel-3-*O*-gluc-mal
Pelargonidin-3,5-di-*O*-glucoside	pel-3,5-di-*O*-gluc
Pelargonidin 3-*O*-*p*-coumaryl-glucoside	pel -3-*O*-*p*-coum-gluc
Pelargonidin-3-*O*-rutinoside-5-*O*-rutinoside-glucoside	pel 3-*O*-rut-5-*O*-gluc
Pelargonidin-3-*O*-caffeoyl-rutinoside-5-*O*-glucoside	pel 3-*O*-caf-rut-5-*O*-gluc
Pelargonidin-3-*O*-feruloyl-rutinoside-5-*O*-glucoside	pel 3-*O*-ferul-rut-5-*O*-gluc
Pelargonidin-3-*O*-*cis*-*p*-coumaryl-rutinoside-5-*O*-glucoside	pel-3-*O*-*cis*-*p*-coum-rut-5-*O*-gluc

Pelargonidin-3-*O*-*p*-coumaryl-rutinoside-5-*O*-glucoside	pel-3-*O*-*p*-coum-rut-5-*O*-gluc
Malvidin	Mal
Malvidin hexose	mal hex
Malvidin-3-acetyl-hexose	mal-3-ace-hex
Malvidin-3-galactoside	mal-3-gala
Malvidin-3-glucoside	mal-3-gluc
Malvidin-3-arabinoside	mal-3-ara
Malvidin 3-*O*-acety-glucoside	mal-3-*O*-ace-gluc
Malvidin-3-*O*-coumarylglucoside	mal-3-*O*-coum-gluc
Malvidin 3-*O*-(6-*O*-acetyl)-glucoside	mal-3-*O*-(6-*O*-ace)-gluc
Malvidin 3-*O*-(6-*O*-coumaryl)-glucoside	mal-3-*O*-(6-*O*-coum)-gluc
Malvidin 3-*O*-(6-*O*-caffeoyl)-glucoside	mal-3-*O*-(6-*O*-caf)-gluc
Malvidin-3 rutinoside	mal-3-rut
Malvidin-3-*O*-glucoside-catechin	mal-delp-3-gluc-cat
Malvidin-3-*O*-glucoside-petunidin-3-*O*-glucoside	mal-3-*O*-gluc-pet-3-*O*-gluc
Malvidin-3-*O*-rutinoside-5-glucoside	mal 3-*O*-rut-5-*O*-gluc
Malvidin-3-*O*-*p*-coumaryl-rutinoside-5-*O*-glucoside	mal 3-*O*-*p*-coum-rut-5-*O*-gluc
Malvidin-3-*O*-glucoside-delphinidin-6-*O*-coumaroyl 3-*O*-glucoside	mal-3-*O*-gluc-delp-6-*O*-coum-3-O-gluc
Malvidin 3-feruloyl-rutinoside-5-*O*-glucoside	mal 3-*O*-ferul-rut-5-*O*-gluc
Malvidin 3-caffeoyl-rutinoside-5-*O*-glucoside	mal-3-*O*-caf-rut-5-O-gluc
Malvidin-3-*O*-glucoside-4-vinyl-epicatechin	mal-3-*O*-gluc-4-vinyl-epic
Malvidin-6-*O*-acetyl-3-*O*-glucoside	mal-6-*O*-ace-3-*O*-gluc
Malvidin-6-*O*-coumaroyl-3-*O*-glucoside–catechin	mal-6-*O*-coum-3-*O*-gluc–cat
Malvidin-6-*O*-*p*-coumaroyl-3-*O*-glucoside-pyruvate	mal-6-*O*-*p*-coum-3-*O*-gluc-pyr
malvidin-3-*O*-glucoside-petunidin-6-*O*-coumaroyl-3-*O*-glucoside	mal-3-*O*-gluc-pet-6-*O*-coum-3-*O*-gluc
Malvidin-6-*O*-*p*-coumaroyl-3-*O*-glucoside	mal-6-*O*-*p*-coum-3-*O*-gluc
Malvidin-3-*O*-glucoside-4-vinylphenol	mal-3-*O*-gluc-4-vinyl-phen
Malvidin-6-*O*-acetyl-3-*O*-glucoside-epicatechin	mal-6-*O*-acel-3-*O*-gluc-epi
Malvidin-6-*O*-acetyl-3-*O*-glucoside–gallocatechin	mal-6-*O*-ace-3-*O*-gluc–gcat
Malvidin-3-*O*-diglucoside	mal-3-*O*-digluc
Malvidin-6-*O*-acetyl-3-*O*-glucoside-4-vinyl-epicatechin	mal-6-*O*-ace-3-*O*-gluc-4-vinyl-epi
Malvidin-6-*O*-acetyl-3-*O*-glucoside-4-vinyl-diepicatechin	mal-6-*O*-ace-3-*O*-gluc-4-vinyl-diepi
Epicatechin	Epi
Gallocatechin	Gcat
Afzelechin	Afz

1. INTRODUCTION

Flavonoids, a group of polyphenolic compounds with low molecular weights, are widely distributed in vascular plants and have been identified as potentially important therapeutic drugs with both high potency and low systemic toxicity [1, 2]. They are present in a wide range of fruits, vegetables, nuts, seeds, citrus fruits, olive oil and beverages, including wine and tea and recognized as the pigments responsible for the different colors of leaves, especially in autumn. Over 5000 different flavonoids have been identified to date and they are classified into at least 10 chemical groups and are categorized as flavones, flavonols, flavonones, anthocyanins and isoflavones commonly found in our diet [3, 4].

The basic structure of flavonoids is usually characterized by three rings. There are two aromatic rings (A and B), which are joined by a three-carbon linked γ-pyrone ring (C), where side group is usually hydroxyl, methoxyl or glycosyl. Their variation in the two benzene rings of the basic structure gives rise to flavonols, flavones, catechins, flavanones and isoflavones (Scheme 1). Some of the naturally occurring flavonoids and their glycosides are listed in Table 1. They are well recognized for their antioxidative activities like scavenging the free radicals, thus breaking the free radical chain reaction of lipid peroxidation [5,6]. They also have been found to have antiviral, antiproliferative, antiallergic, anticancer, antitumor and anti-inflammatory effects [7-9]. Because of differences in their chemical structure, bioavailability, distribution, and metabolism [10] different flavonoid compounds might have different effects on human health. As is true with many agents, excessive amounts of polyphenols can be toxic, even though they are from dietary sources. The limited absorption of most of the dietary polyphenols may be considered as a protective mechanism for age-related diseases.

Anthocyanins (of the Greek anthos = flower and kianos = blue) are the most important pigments of the vascular plants; they are harmless and of easy incorporation in aqueous media, which makes them interesting for its use as natural water-soluble colorants [11]. These pigments are responsible of the shiny orange, pink, red, violet and blue colours in the flowers and fruits of some plants. Another significant property of anthocyanins is their antioxidative activity, which plays a vital role in the prevention of neuronal and cardiovascular illnesses, cancer and diabetes, among others [12]. There are several reports focused on the effect of anthocyanins in cancer treatments [13], human nutrition [14], and its biological activity. Anthocyanins are the glycosides of flavonoids with polyhydroxy and polymethoxy derivatives of 2-phenylbenzopyrilium or flavylium cation (Schemes 2 and 3). Anthocyanins are the conjugated systems, which are often positively charged. An anthocyanidin, termed as the aglycone, does not have the sugar at 3 position. Some of naturally occurring anthocyanidins and their glycosides are tabulated in Tables 2 and 3. Anthocyanins can be found in different chemical forms which depend on the pH of the solution [15-17]. At pH 1, the flavylium cation (red colour) is the predominant species and contributes to purple and red colours. Between pH values 2 and 4, the quinoidal blue species are predominant while two colourless species (a carbinol pseudobase and a chalcone) can be observed between pH values 5 and 6. At pH values higher than 7, the anthocyanins are degraded depending on their substituent groups.

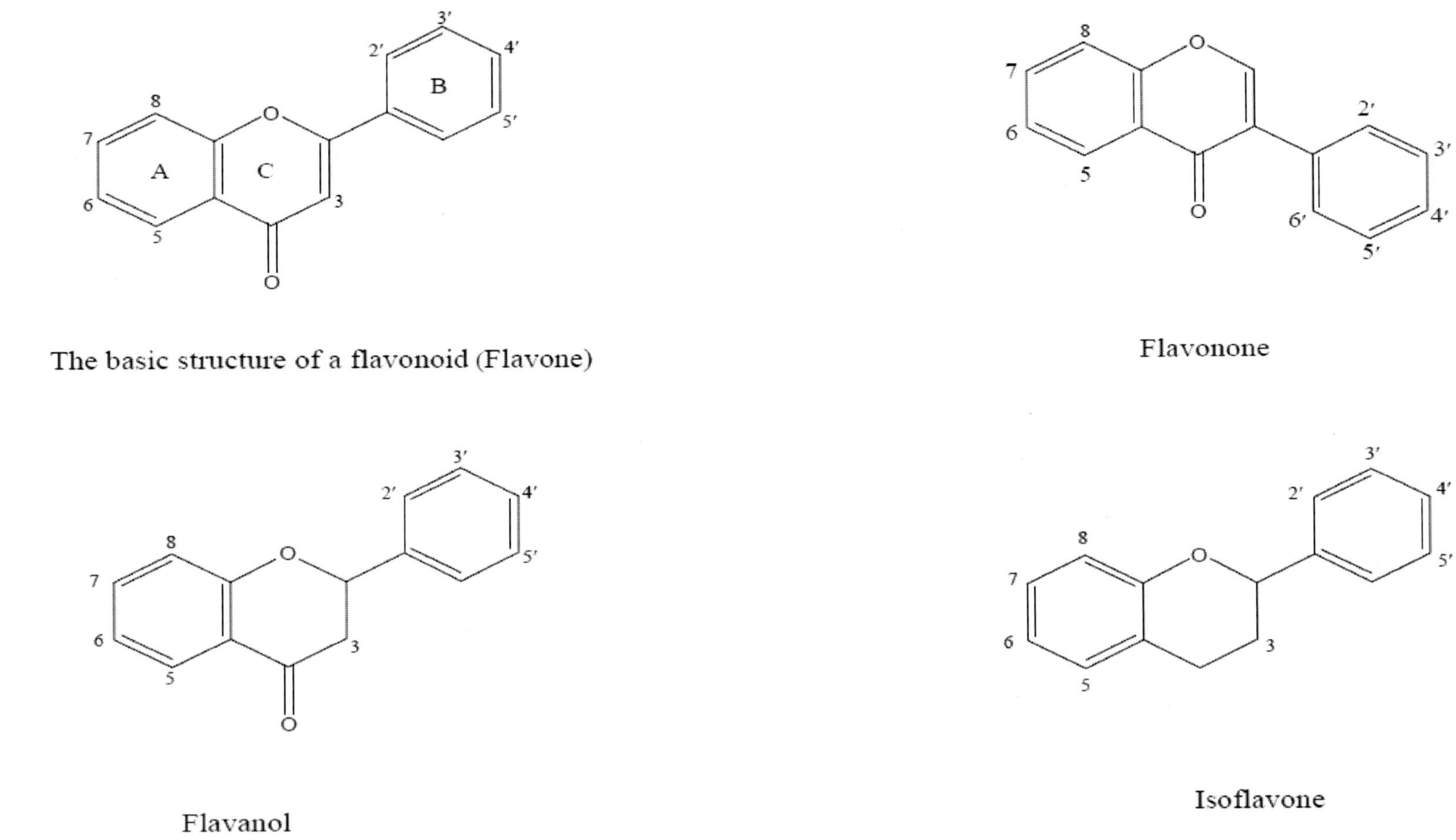

Scheme 1. The basic structure of a flavonoid (Flavone).

Table 1. Structural identification of flavonoids and their glycosides

	Name	Substitution pattern								
		3	5	6	7	8	2’	3’	4’	5’
Flavonones	Flavonone	H	H	H	H	H	H	H	H	H
	Naringenin	H	OH	H	OH	H	H	H	OH	H
	Eriodictyol	H	OH	H	OH	H	H	OH	OH	H
	Hesperetin	H	OH	H	OH	H	H	OH	OMe	H
	Taxifolin	OH	OH	H	OH	H	H	OH	OH	H
	Alpinetin	H	OMe	H	OH	H	H	H	H	H
	Naringenin	H	OH	H	OH	H	H	H	OH	H
	Nobiletin	H	OMe	OMe	OMe	OMe	H	OMe	OMe	H
	Tangeretin	H	OMe	OMe	OMe	OMe	H	OMe	H	H
Flavanols	Catechin	OH	OH	H	OH	H	H	OH	OH	H
	Epicatechin	OH	OH	H	OH	H	H	OH	OH	H
	Epigallocatechin	OH	OH	H	OH	H	H	OH	OH	OH
	Epicatechin gallate	G	OMe	H	OH	H	H	OH	OH	H
	Epigallocatechin gallate	G	OH	H	OH	H	H	OH	OH	OH
Flavone/ols	Flavone	H	H	H	H	H	H	H	H	H
	Flavonol	OH	H	H	H	H	H	H	H	H
	Baicalein	H	OH	OH	OH	H	H	H	H	H
	Apigenin	H	OH	H	OH	H	H	H	OH	H
	Chrysin	H	OH	H	OH	H	H	H	H	H
	Galangin	OH	OH	H	OH	H	H	H	H	H
	Kaempferol	OH	OH	H	OH	H	H	H	OH	H
	Morin	OH	OH	H	OH	H	OH	H	OH	H
	Quercetin	OH	OH	H	OH	H	H	OH	OH	H
	Myricetin	OH	OH	H	OH	H	H	OH	OH	OH
	Luteolin	H	OH	H	OH	H	H	OH	OH	H
	Wogonin	H	OH	H	OH	OMe	H	H	H	H
	Hispidulin	H	OH	OMe	OH	H	H	H	OH	H

Table 1. (Continued)

	Name	Substitution pattern								
		3	5	6	7	8	2’	3’	4’	5’
Isoflavones	Daidzein		H	H	OH	H	H	H	OH	H
	Genistein		OH	H	OH	H	H	H	OH	H
	Formononetin		H	H	OH	H	H	H	OMe	H
	7,3′,4′THIF		H	H	OH	H	H	OH	OH	H
Glycosides	Rutin	sugar	OH	H	OH	H	H	H	OH	OH
	Tetuin	H	OH	sugar	OH	H	H	H	H	H
	Hyperin	sugar	OH	H	OH	H	H	OH	OH	H
	Hesperidin	H	OH	H	sugar	H	H	OH	OMe	H
	Naringin	H	OH	H	sugar	H	H	H	OH	H
	Baicalin	H	OH	OH	sugar	H	H	H	H	H

Table 2. Naturally occurring anthocyanidins

Name	Abbrevi-ation	Substitution pattern							Color
		3	5	6	7	3’	4’	5’	
Apigeninidin	Ap	H	OH	H	OH	H	OH	H	Orange
Aurantinidin	Au	OH	OH	OH	OH	H	OH	H	Orange
Arrabidin	Ab	H	H	OH	OH	H	OH	OMe	N.R[a]
Capensinidin	Cp	OH	OMe	H	OH	OMe	OH	OMe	Bluish-red
Carajurin	Cj	H	H	OH	OH	OMe	OMe	OMe	N.R[a]
Cyanidin	Cy	OH	OH	H	OH	OH	OH	H	Orange-red
Delphinidin	Dp	OH	OH	H	OH	OH	OH	OH	Bluish-red
Europinidin	Eu	OH	OMe	H	OH	OMe	OH	OH	Bluish-red
Hirsutidin	Hs	OH	OH	H	OMe	OMe	OH	OMe	Bluish-red
3′-HydroxyAb	3’OHAb	H	H	OH	OH	OH	OH	OMe	N.R[a]
6-HydroxyCy	6OHCy	OH	OH	OH	OH	OH	OH	H	Red
6-HydroxyDp	6OHDp	OH	OH	OH	OH	OH	OH	OH	Blue-red
6-HydroxyPg	6OHPg	OH	OH	OH	OH	H	OH	H	N.R[a]
Luteolinidin	Lt	H	OH	H	OH	OH	OH	H	Orange
Malvidin	Mv	OH	OH	H	OH	OMe	OH	OMe	Bluish-red
5-Methylcyanidin	5-MCy	OH	OMe	H	OH	OH	OH	H	Orange-red
Pelargonidin	Pg	OH	OH	H	OH	H	OH	H	Orange
Peonidin	Pn	OH	OH	H	OH	OMe	OH	H	Orange-red
Petunidin	Pt	OH	OH	H	OH	OMe	OH	OH	Bluish-red
Pulchellidin	Pl	OH	OMe	H	OH	OH	OH	OH	Bluish-red
Riccionidin A	RiA	OH	H	OH	OH	H	OH	H	N.R[a]
Rosinidin	Rs	OH	OH	H	OMe	OMe	OH	H	Red
Tricetinidin	Tr	H	OH	H	OH	OH	OH	OH	Red

N.R[a] : Not Reported.

2. Extraction Methods

Literature survey revealed that the solvent extraction has been the most common method for extraction of diverse compounds generally found in plants. The methods of extraction of anthocyanins employed by various groups have been reported in the literature [18]. Since anthocyanins are polar molecules, the most common solvents used in the extractions include aqueous mixtures of ethanol, methanol or acetone [19]. These extractions necessarily involve the coextraction of non-phenolic compounds like sugars, organic acids and proteins requiring subsequent purification methods. Among the most common methods are those which use acidified methanol or ethanol as extractants [20-25]. From these methods, the extraction with methanol was reported to be the most efficient [26] in fact, it was observed that in anthocyanin extractions from grape pulp, the extraction with methanol was 20% more effective than with ethanol, and 73% more effective than water alone [27]. Generally, food industries prefer ethanol for extraction as methanol is toxic.

Table 3. Glycosides of anthocyanidins

	R	R1	R2	R3
Cyanidin 3-*O*-glycoside	OH	OH	H	Sugar
Delphinidin 3-*O*-glycoside	OH	OH	OH	Sugar
Malvidin 3-*O*-glycoside	OCH_3	OH	OCH_3	Sugar
Peonidin 3-*O*-glycoside	OCH_3	OH	H	Sugar
Petunidin 3-*O*-glycoside	OH	OH	OCH_3	Sugar
Pelargonidin 3-*O*-glucoside	H	OH	H	Sugar

In another study [28], anthocyanins were extracted from Cabernet Sauvignon red wine, skins of *Vitis vinifera* grapes and raspberry fruits (*Rubus idaeus*). 16 anthocyanins were identified from red wine, of which malvidin-3-*O*-glucoside was the main anthocyanin, which constituted 53.6% of the total anthocyanin content in grape extract. Delphinidin-3-*O*-glucoside and cyanidin-3-*O*-glucoside were found as the major anthocyanins in raspberry extract. The antioxidant capacity of the extracts and red wine was determined by differential pulse voltammetry (DPV) since best resolution of oxidation peaks were observed in this technique. 20 times diluted wine exhibited peak potentials of 443 mV and 676 mV for peak 1 and peak 2, respectively, similar to those corresponding to anthocyanin malvidin-3-*O*-glucoside and wine extract. The linear pulse voltammetry curve of a 100 ppm solution of malvidin-3-*O*-glucoside in acetate buffer of pH 3.6 exhibited two peaks at 490 and 674 mV. The first peak (P1) was attributed to the oxidation of OH of B-ring while the second oxidation peak (P2) was believed to be due to 5,7-dihydroxyl moiety of A-ring. Further, the antioxidant capacity of anthocyanins in these extracts was also determined by the Trolox equivalent antioxidant capacity (TEAC) method.

Korotkova *et al.* have studied the antioxidant activity of extracts and their individual fractions of the extracts obtained from most promising medicinal plants found in Siberia by cathode voltammetry (CVA) based on the reaction of oxygen reduction [29]. For this an electrochemical cell comprising of a glass cylinder with a background electrolyte solution, a working mercury-film electrode (Hg film on a silver substrate), a reference silver chloride electrode, and an auxiliary glassy carbon composite electrode (in the case of aqueous ethanol media) was employed [30-32]. A rapid method for the evaluation of the antioxidant activity of herb extracts from the Labiatae family was described [33]. Analysis of the methanolic extracts were analysed amperometrically at a potential of 0.5 V (vs. Ag/AgCl) by using a flow injection (FI) system with an electrochemical detector equipped with a glassy carbon working electrode. A very good correlation was observed between the results obtained by this method and those obtained by using 2,2-diphenyl-1-picrylhydrazyl (DPPH) test.

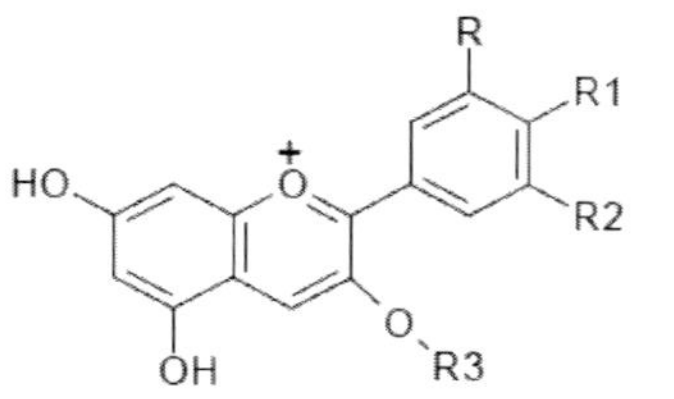

Chemical structure of anthocyanins

Scheme 2. Chemical structure of anthocyanins.

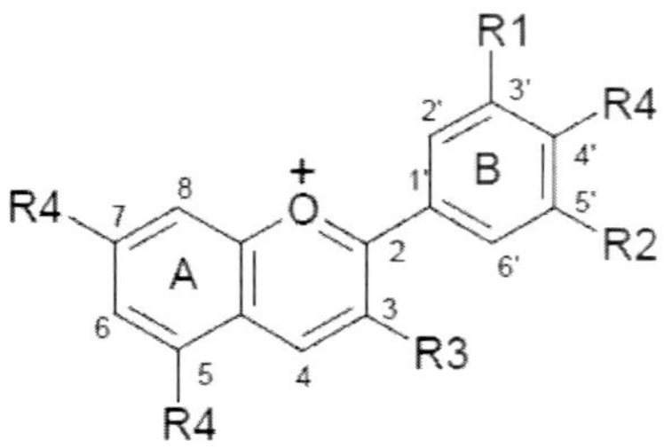

The flavylium cation: R1 and R2 are H, OH, or OCH_3; R3 is a glycosyl or H; and R4 is OH or a glycosyl.

Scheme 3. The flavylium cation: R1 and R2 are H, OH, or OCH3; R3 is a glycosyl or H; and R4 is OH or a glycosyl.

Red wines from three different regions in Croatia were analysed for total polyphenols (TP) content, antioxidant activity and concentration of individual polyphenols [34]. Measurement of TP was done by DPV, Folin–Ciocalteu (FC) and high-pressure liquid chromatography (HPLC) methods. Measurements of individual polyphenols were done using HPLC and antioxidant activity (AA) was determined chemically by 2,2'-azinobis(3-ethylbenzothiazoline-6-sulphonic acid (ABTS) and DPPH methods. From the results, it is evident that DPV is a very sensitive method for the determination of TP. A very good correlation was found between the TP content determined by the above three methods. The major polyphenols in the wine samples were gallic acid and (+)-catechin. Individual polyphenols, AA and TP varied among wines from different regions.

3. Electrochemical Studies

Flavonoids are benzo-γ-pyrone derivatives with several hydroxyl groups attached to the ring structures C_6-C_3-C_6; of which some contain *o*-dihydroxy catechols (the B-ring) such as quercetin, rutin, catechin and luteolin. Owing to their polyphenolic nature, flavonoids exhibit strong antioxidant properties. All flavonoids are electroactive, easily subject to either oxidation or reduction electron transfer reactions, hence they can be investigated by electrochemical methods [35, 36]. Electrochemical studies have shown general trends in the electron-donating abilities of flavonoids [37]. Determination of physicochemical parameters for antioxidants by electrochemical measurements (*e.g.*, redox potential, number of electrons transferred, electrode reaction rate constant, *etc*), are very significant not only for evaluating the antioxidative abilities of flavonoids, but also to understand their reaction mechanisms [38]. It was demonstrated that the catechol in the B-ring is more easily oxidizable than the resorcinol in the A-ring. The reduction potentials of flavonoids depend strongly on the electron donating properties of the substituents in the ring B [39], and on the ring B, the most oxidizable phenolic group is the more basic site [37]. The pKa values were assigned to ring A and ring B [40]. The mechanism of phenolic compound action as antioxidants seems to involve the ability of phenols to scavenge radicals by an H-atom or electron transfer process by which the phenol is converted into a phenoxyl radical. The ease of oxidation of the phenol is of importance for its effectiveness as an antioxidant [41, 42].

Oxidation potentials of flavonoids in aqueous buffers ranged from 0.0 to 1.2 V depending on the pH of the solution and hydroxylation pattern of the flavonoid. Flavonoids with 3',4',5'-trihydroxy (pyrogallol-type) and 3',4'-dihydroxy (catechol-type) substitutions were the more readily oxidized, while flavonoids with 1',3',5'-hydroxy (phloroglucinol- type) and phenol-type substitutions were much harder to oxidize. It was observed that the oxidation of some flavonoids is chemically reversible while that of others is irreversible [36]. This property was used to distinguish catechol-containing flavonoids that showed reversible electrochemistry from flavonoids with phloroglucinol or pyrogallol- type hydroxylation where the electrochemistry was chemically irreversible [43]. Detailed electrochemical studies of flavonoids, with identification of their oxidation products were performed by Hendrickson *et al.* [44] and a correlation revealed between their oxidation potential and their effect on the activity of various oxygenases. The applicability of nujol-based carbon paste electrodes to accumulate certain flavonoids has been established and rutin, a flavonoid glycoside known also as vitamin P, has been determined in a pharmaceutical preparation by adsorptive stripping voltammetry (AdSV) using batch measurements [45]. It has been stated that "depending on the experimental conditions, all hydroxyl groups can be electrochemically oxidized and the anthocyanins studied revealed a complex and pH dependent oxidation process" in a recent study which included detailed voltammetric behaviour of several anthocyanidins and anthocyanins, including kuromanin [46, 47]. The redox behaviour of anthocyanins was also studied by Janeiro and Oliveira Brett [47] by voltammetric techniques. They found that all the phenolic hydroxyl groups can be electrochemically oxidized.

Vestergaard *et al.* have studied the electrochemistry of eight flavonoids *viz,,* myricertin, catechin gallate, quercetin, delphinidin, baicalein, cyanidin, catechin and luteolin belonging to four flavonoid sub-classes: flavone, flavonol, flavanol and anthocyanidin using pencil graphite electrodes (PGEs) [48]. Electrochemical oxidation of a synthetic anthocyanin analog of 4-methyl-7,8-dihydroxyflavylium salt was studied using cyclic voltammetry (CV) and square wave voltammetry (SWV). These results were also compared with kuromanin that is found in raspberry fruits. The studies were carried out in acidic condition (pH=1) since the dominant species at this pH would be flavylium cation. The electrochemical signal observed both in kuromanin and 4-methyl-7,8-dihydroxyflavylium salt was due to the *o*-dihydoxy groups which have undergone two electron, two proton redox process coupled with an irreversible homogeneous chemical reaction. Since the oxidation potential of the synthetic molecule occurred at a potential lower than that of the natural compound, its antioxidant activity might be relevant as to a possible antioxidant role of the synthetic analogues [49]. Voltammetric and spectral characterization of two flavonols, kaempferol and morin for assay-dependent antioxidant activity was reported by Jian-Bo He *et al.* [50]. In this study, they have investigated the electro-oxidation of the flavonols using CV and in situ UV–vis spectroelectrochemical technique with a long-optical-path thin-layer electrolytic cell. The spectral changes recorded in different oxidation potentials were compared with those in the chemical oxidation either by H_2O_2 or ABTS•+ radical in the same thin-layer cell. In the oxidation of 4'-OH group, kaempferol was noticed to be somewhat more active than morin. The 4'-OH group was first oxidized at lower potentials or induced by H_2O_2. Morin was found to show antioxidant activity twice to that of kaempferol. Hence, it was concluded that the difference in the requirement of oxidants of different reactivities or difference in their oxidation potentials in electrochemical oxidation was due to the difference in the number of oxidizable –OH groups and which in turn was responsible for the difference in antioxidant capacities.

Electrooxidation studies of some flavonoids (quercetin, morin hydrate, apigenin, chrysin, xanthone, rutin hydrate, hesperidin and falvone) were investigated by CV using platinum electrode [51]. Electrooxidation of flavonoids was noticed to be an irreversible process. The first step of electrooxidation involved exchange of two electrons during the oxidation of hydroxyl groups in ring B of morin hydrate, rutin, dihydroxyflavone, trihydroxyflavone, hesperidin and quercetin. In subsequent steps, hydroxyl groups in the rings A and C were probably oxidized. The heterogeneous rate constants for the flavonoid oxidation was determined and observed to be 3.59×10^{-4} , 4.42×10^{-4} , 4.54×10^{-4} , 4.19×10^{-4} , 4.50×10^{-4} , 4.63×10^{-4} , 7.08×10^{-4} and 6.46×10^{-4} cms^{-1} for morin, rutin, chrysin, apigenin, hesperidin, quercetin, xanthone and flavone, respectively. Higher heterogeneous rate constants indicated that the xanthone and flavones were oxidized easily compared to other substrates on platinum electrodes.

Electrochemical oxidation mechanism of quercetin in non aqueous solution was studied so as to know the degradation pathways by Sokolovaa *et al.* [52]. The final oxidation product at the first oxidation wave was identified as 2-(3',4'-dihydroxybenzoyl)-2,4,6-trihydroxybenzofuran-3(2*H*)-one by HPLC-DAD and GC–MS techniques. The oxidation mechanism proceeding through a short-lived anion radical was evident from results of *in situ* spectroelectrochemistry measurements. Two oxidation mechanisms were proposed-the presence of a disproportionation chemical reaction following the first one electron transfer, or two one electron transfer reactions, which did not have identical, but similar redox potentials. The quinone formed in either case was short lived and yielded 2-(3,4-dihydroxybenzoyl)-2,4,6-trihydroxybenzofuran-3-one by undergoing fast hydroxylation reaction. Oxidation of this compound gave rise to the second oxidation wave of quercetin.

The electrochemical oxidation of three flavones, [3-hydroxyflavone, 6-hydroxyflavone and 2',3,4',5,7-pentahydroxyflavone (morin)] was studied on a GCE using CV and SWV techniques, at different pH [53]. The oxidation of 3-hydroxyflavone showed two peaks in acidic medium corresponding to the oxidation of OH group at C-3 and OH group at C-4. The 6-hydroxyflavone exhibited only one peak in all pH range, corresponding to the oxidation of OH group at C-6. However, the oxidation mechanism of morin was found to proceed in steps, which occured at the OH groups in the resorcinol moiety at ring B and OH group at C-3 of the flavonoid structure. This study revealed that the mechanism of electron transfer was a complex phenomenon and pH dependent process.

The mechanism of electrochemical oxidation of quercetin on GCE was studied using cyclic, differential pulse and square-wave voltammetric techniques at different pH [54]. The electrooxidation occured at two catechol hydroxyl groups and the other three hydroxyl groups which are all electroactive, and the oxidation was pH dependent. Quercetin displayed the strong adsorption on electrode surface and the final oxidation product also showed the strong adsorption on electrode surface, which is not electroactive and thus blocked the electrode surface. The oxidation of the catechol 3',4'-dihydroxyl electron donating groups occured first at very low positive potentials, and was a two electron two proton reversible reaction. The next hydroxyl group could form a intermolecular hydrogen bond with the neighbouring oxygen and was oxidized by an irreversible oxidation reaction. The other two hydroxyl groups also showed an electron donating effect and further, their oxidation was reversible. A dependence of peak current on pH was observed showing a maximum around neutral pH values.

Electrooxidation of quercetin was studied by Nematollahi and Malakzadeh in the absence and presence of benzenesulfinic acids as nucleophiles in an acetonitrile-water mixture by CV and controlled potential coulometry [55]. The results clearly indicated that the *o*-quinone derived from quercetin participated in a Michael addition reaction with the benzenesulfinic acids yielding the corresponding sulfonyl derivatives. The electrochemical synthesis of these derivatives was successfully performed at carbon rod electrodes in an undivided cell in good yield and purity. The electrochemical behavior of the quercetin was investigated at a carbon paste electrode (CPE) containing 15% paraffin oil [56]. Quercetin was determined using square wave anodic stripping voltammetry in the presence of 0.04M phosphate buffer (pH ~4) containing 0.1M KCl with 15 s accumulation time at 25 ± 1^{0}C. Under the optimal experimental conditions, irreversible oxidation of quercetin was observed. The linearity range was from 67.66 to 338.3 ppb, with a detection limit of 6.77 ppb at 15 s accumulation time and the relative standard deviation of 0.45 (n = 5) was observed.

Bulk electrolysis of quercetin and kaempferol was carried out in acetonitrile, both yielded a single oxidation product each in two-electron processes [57]. The oxidation products were identified as 2-(3,4-dihydroxybenzoyl)-2,4,6-trihydroxy-3(2*H*)-benzofuranone and 2-(4-hydroxybenzoyl)-2,4,6-trihydroxy-3(2*H*)-benzofuranone for quercetin and kaempferol, respectively. Half-peak potentials: $E_{P/2}$ = 0.97 V, 0.98 V and 1.17 V *vs.* NHE for quercetin, kaempferol and luteolin, respectively were measured by CV in acetonitrile. No well-defined oxidation product of luteolin (lacking the 3-OH group) was isolated despite rather similar half-peak potentials. The electrochemical investigations of quercetin on a CPE in non aqueous and aqueous media were carried by Satpati *et al* [58]. Using cyclic voltammetric technique, electrochemical parameters like diffusion coefficient, exchange current density and the anodic Tafel slopes were determined. By using an optimized differential pulse voltammetric procedure, a detection limit of 38.5 nM L^{-1} was achieved. Further, the electrochemical properties of quercetin were studied using a DNA-modified CPE using electrochemical impedance spectroscopy. From spectroelectrochemical measurements it was clear that the hydroxyl groups at the 3′ and 4′ positions were oxidized in the presence of Cu(II) and are responsible for DNA damage. Impedance measurements supported the intercalation of quercetin into the DNA strands.

The electrochemical studies of four structurally related flavonoids, quercetin, quercetin-3-*O*-rhamnose (quercetin), quercetin-3-*O*-rutinose (rutin) and luteolin were carried out [43]. These flavonoids were shown to undergo oxidation at a glassy carbon electrode (GCE) followed by homogenous chemical reactions. These reactions were studied using CV and rotating ring-disk voltammetry. They showed both first-order and zero-order processes. The substituent at the C-3 position of the flavonoid was shown to influence the rate of the zero-order process to a great extent. The rate of the first-order process was independent of substitution. Two products were observed using liquid chromatography. These products did not match with previously reported products of enzymatic oxidation. The products were not stable under conditions for isolation. The electrochemical properties of (+)-catechin, quercetin and rutin on a GCE were studied at different pH by cyclic, differential pulse and square-wave voltammetry [59]. It was noticed that in all investigated compounds, oxidation of catechol 3',4'-dihydroxy group on the B-ring (the first oxidation peak) occurred. This process was found to be reversible, pH dependent and involved the transfer of 2e− and $2H^+$s. The products of electrochemical oxidation of all the flavonoids (especially of quercetin) strongly adsorbed on the electrode surface. Peak current dependence on pH revealed the maximum at around pH = 4.5 for (+)-catechin and at pH = 5.5 for quercetin and rutin (it decreased in more acidic and alkaline media).

Koyama *et al.* have determined the oxidation potentials of 15 flavonoids in phosphate buffer at pH 7.2 by CV [60]. A good correlation was found between the oxidation potentials and the ability of flavonoids to inhibit Epstein–Barr virus early antigen (EBV-EA) activation. Further, multiple regression analysis revealed that the solvent-accessible surface area (SASA) was a useful parameter for estimating the inhibitory effects of flavonoids on EBV-EA activation. Another study, primarily aimed on investigation of the basic electrochemical behaviour, described the electrochemical technique for the determination of selected flavonoids (quercetin, quercitrin, rutin, chrysin and diosmin) in body liquids, by using SWV using CPE [61]. The optimized conditions for determination of the flavonoids were as follows: frequency 180 Hz, step potential 1.95 mV/s and phosphate buffer of pH 7 as supporting electrolyte. Their detection limits of the flavonoids were from units to tens of nM except diosmin, for which the limits were higher than μM.

The preconcentration behaviour of nine flavonoids (six flavones of fisetin, galangin, morin, quercetin, rhamnetin and rutin, and three flavanones of hesperidin, hesperitin and naringin) were studied at a CPE (nujol/graphite) and the effect of pH, accumulation potential, presence of various surfactants were studied for their subsequent differential pulse voltammetric determination [62]. Considerable signal enhancement was observed in the case of all flavones owing to their accumulation on the CPE. Determinations as low as to 10^{-8}-10^{-7} M was achieved after preconcentration for 1-4 min. Flavanones were not preconcentrated as their lower determination limits were of the order of 10^{-6} M. A voltammetric method for the assay of rutin in multivitamin preparation was also reported. Cyclic, differential pulse and square wave voltammetric techniques were employed to understand the mechanism of electrochemical oxidation of morin in aqueous electrolyte with solid, insoluble morin hydrate transferred on to a GCE surface mechanically, over a wide pH range [42]. The proposed oxidation mechanism involved the hydroxyl groups in three aromatic rings, and proceeded in sequential steps. The oxidation potentials were found to be shifted to lower values with increasing pH over part of the pH range. Oxidation of the 2', 4'-dihydroxy moiety at the B ring of morin occured first, at very low positive potentials, and was a one electron one proton reversible reaction. The hydroxyl groups oxidized at more positive potentials were shown to undergo an irreversible oxidation process.

Electrochemical studies of rutin at GCE [63] have shown that at an inactivated GCE (IGCE), the cyclic voltammograms have the characteristics of a species in solution while at an activated GCE (AGCE), the voltammetric responses are typical of a surface-immobilized redox couple. Based on these experimental results, the oxidation of rutin at an IGCE was proposed to be an ErCi mechanism (Electrochemically reversible electron transfer followed by Chemically irreversible homogeneous step). Further, the voltammetric responses demonstrated that the subsequent chemical reactions in the ErCi mechanism were dimerization and intramolecular reactions and hydroxylation reaction were slow on the characteristic time of CV method. The influence of glycosylation on the redox behavior of anthocyanins derived from different anthocyanidins and the redox behavior of anthocyanins with the same basic structure and the influence of methoxylation were investigated [64]. Malvidin-3-*O*-glucoside chloride, malvidin-3,5-di-*O*-glucoside chloride, cyanidin- 3-*O*-glucoside chloride, cyanidin-3,5-di-*O*-glucoside chloride, peonidin-3-*O*-glucoside chloride, delphinidin-3-*O*-glucoside chloride and the anthocyanidin petunidin chloride, all of them present in *Vitis vinifera* L. were used for investigation. The experimental results suggested a complex and pH dependent oxidation process, with the occurrence of adsorption and blocking the electrode surface by oxidation products.

Spectrometric and voltammetric studies of the interaction between quercetin and bovine serum albumin (BSA) using warfarin as site marker with the aid of chemometrics were carried out by Nia *et al.* [65]. Cyclic voltammograms of quercetin showed an oxidation peak at 465 mV and a reduction peak at 430 mV. The peak potential difference of 35 mV indicated a quasi-reversible, 2- e^- redox process. Upon the addition of BSA to the quercetin solution, the oxidation and the reduction peaks were found to be shifted towards high and low potentials, respectively, and a decrease of oxidation current was also observed indicating the interaction between the most hydrophobic segment of the quercetin molecule (benzopyran-4-one) [66] and the hydrophobic part of the BSA cavity. A study of the interaction between quercetin and HSA using molecular models also supported that the benzopyran-4-one moiety was located within the binding pocket of HSA, but the B-ring protruded from it and pointed towards the interface of IIB and IIIA domains [67].

Electrochemical methods have also been used to study the interaction of DNA with electroactive molecules [68-72], because they show many advantages such as direct monitoring, high sensitivity and simplicity. However, most of them are focused on small molecules with good reversibility in electrochemical reaction while many DNA-targeting drugs show irreversibility in electrochemical reaction. In many cases, irreversible electrode process must be assumed to be reversible which can arouse some errors because the electron transfer coefficients of irreversible electroactive compounds play an important role in electrochemical reaction, otherwise it can not deal with electrochemical methods. Recently,

Wang *et al.* developed a novel equation, which is suited for investigating the interaction between DNA and irreversible redox compounds using electrochemical method [73].

Voltammetric behavior of morin was studied at GCE using CV in 0.1M HAc–NaAc + 50 mM KCl (pH 3.4) buffer solution. Morin showed an irreversible oxidation peak at 0.720V in CV which involved two electrons and two protons. Interaction of morin with double-stranded calf thymus DNA (ctDNA) was also studied by CV at GCE with the use of an irreversible electrochemical equation. As a result of interaction with ctDNA, the peak potential of morin oxidation shifted and there was a decrease in peak current. These experimental results revealed that the ring B of morin intercalated between the DNA base pairs [74]. Some of the experimental values are as follows: Diffusion coefficient of free morin (Df)=1.1086×10^{-7} cm^2 s^{-1}, diffusion coefficient of DNA bound morin (Db)= 8.2544×10^{-9} cm^2 s^{-1}, binding constant (K)= 1.7765×10^7 cm^3 mol^{-1}; binding site size (s)= 0.8510.

An electrochemical investigation of the interaction between quercetin and DNA was reported by Zhu *et al.* [75]. Upon the addition of DNA to the quercetin solution, both the reduction and oxidation currents of quercetin decreased with few changes in the peak potentials. In order to evaluate the occurrence of DNA damage caused by oxidized quercetin, its interaction with double-stranded DNA (dsDNA) was investigated electrochemically using two types of DNA biosensors [76]. The results showed that the quercetin bound to dsDNA. The radicals formed during quercetin oxidation caused cleavage of the hydrogen bonds in the dsDNA finally giving rise to 8-oxoguanine since the DNA guanosine and adenosine nucleotides in contact with the electrode surface can easily be oxidized. A mechanism for damage to dsDNA immobilized onto a glassy carbon electrode (GCE) surface by oxidized quercetin was proposed and the formation of 8-oxoguanine was explained.

Since quercetin induced DNA damage via reaction with Cu(II) has been reported, an electrochemical study of the DNA–Cu(II)–quercetin system in solution was undertaken [77]. The product of DNA interaction with quercetin–Cu(II) complex was observed. Damages to DNA were electrochemically recognized through the increased anodic peaks corresponding to the oxidation of guanosine and adenosine bases and spectrophotometrically *via* increased absorption band at 260 nm. It was also observed that the dsDNA damage produced by the quercetin–Cu(II) complex occurred with time. Control experiments with different mixtures of Cu(II), quercetin, ssDNA, dsDNA or poly[A] were carried out in order to establish a possible mechanism of interaction between DNA and quercetin *via* Cu(II).

Isolation of an anti-HIV drug baicalin from traditional Chinese medicinal plant, *Scutellaria baicalensis* Georgi was carried out and its interaction with DNA was studied by electrochemical methods on pyrolytic graphite electrodes [78]. Significant decrease in the peak currents was observed corresponding to baicalin redox reaction upon the addition of DNA. Based on UV-vis spectroscopic and electrochemical data, it was concluded that baicalin interacted with DNA via intercalation. This feature has enabled baicalin to discriminate between double-stranded DNA (dsDNA) and single-stranded DNA (ssDNA).

The interaction between double stranded (ds) calf thymus DNA and curcumin (CC) in the presence of Cu(II) was studied in solution by differential pulse adsorptive transfer voltammetry using CPE and hanging mercury drop electrode (HMDE) by Serpi *et al.* [79]. The Cu(II)–CC complex resulted changes in calf-thymus DNA. The characteristic peak of dsDNA, due to the oxidation of guanine residues, decreased. The increased DNA damage by Cu(II)–CC complex was observed in the presence of various concentrations of Cu(II). Tiffiny Chan *et al.* [80] have developed an electrochemical approach to Parkinson's disease by studying the interaction of baicalein and copper with fibrillar α-synuclein (a major Lewy body), which is the major cause of Parkinson's disease. Exposure of α-synuclein to Cu (II) causes significant acceleration in α-synuclein fibril formation. Square-wave voltammetry (SWV) of α-synuclein in the presence of baicalein and Cu(II) revealed distinct interfacial changes attributed to complex formation of α-synuclein with these species *in vitro*. The role of baicalein as a fibrillogenesis inhibitor was found to be most effective during the early stages of the fibril formation.

A poly (tetrafluroethylene)-deoxyribonucleate acid (PTFE-DNA) film-modified GCE was fabricated and it was used for the study of electrochemical oxidation behaviors of morin as well as its interaction with DNA [81]. The results were also compared with those obtained at bare GCE by electrochemical methods. The modified electrode showed enhancement of peak current towards the oxidation of morin. Multiwalled carbon nanotube (MWCNT)-modified electrode was employed for the determination of hesperidin, main component in orange peel, using adsorptive stripping voltammetry (AdSV) with accumulation at open circuit potential [82]. This method could be used as an indication of the citrus fruit juice's freshness. The oxidation mechanism involved the redox chemistry of guaiacol sub-unit within the hesperidin molecular structure. Hesperidin could be determined over a range of 0.5 to 30 μM, with a detection limit of 0.61 μM (using CV) and 7 nM (using SWV). A cheap, mass produced and disposable sensor was developed using MWCNT-modified screen-printed electrodes (MWCNT-SPEs), which is capable of measuring the concentration of hesperidin in real orange juice samples, and be applied within the citrus fruit industry.

Highly sensitive detection of quercetin using GCEs modified with multiwalled carbon nanotubes (MWCNT) dispersed in polyethylenimine and poly acrylic acid). The detection limit of 7.5 nM quercetin after 1.0 min accumulation at open circuit potential could be achieved with adsorptive stripping voltammetric technique with square wave voltammetric transduction. Another easier and simpler alternative to determine quercetin, with detection limits of 0.2 mM is also reported with amperometric detection at 0.150 V. Quercetin was successfully quantified from onion samples using the proposed method [83]. Determination of quercetin from apple juice was studied by using carbon nanotube paste electrode modified with copper nanoparticles (CNTPE-Cu). Differential pulse voltammetry (DPV) was used to optimise the content of multiwalled carbon nanotubes/copper nanoparticles and it was found to be 6% (w/w) copper nanoparticles. Scanning electron microscopy (SEM) and electrochemical impedance spectroscopy (EIS) were used to characterize the modified electrode. Voltammetric behaviour of quercetin was also carried out by DPV in phosphate buffer of pH 7.1. One peak was observed at 0.213 V (*vs.* Ag/AgCl) in contrast to the two peaks observed on CNTPE at 0.082 and 0.40 V where, peak I corresponds to the oxidation of the 3´,4´-dihydroxy substituent on ring B, while peak II represents oxidation involving the hydroxyl group substituent on ring C of quercetin. Above results indicated that on CNTPE-Cu electrode, quercetin-Cu(II) complex was formed by interaction of the group 4-carbonyl of ring C on quercetin with Cu (II), and the peak at 0.213 V corresponds to the oxidation of the quercetin- Cu(II) complex [84]. The determination of 12 flavonoids (fisetin, galangin, hesperetin, hesperidin, kaempherol, morin, myricetin, naringin, quercetin, quercitrin, rutin and rhamnetin) using adsorptive stripping voltammetry in a flow injection system was studied [85]. CPEs based on mixtures of either nujol–graphite or diphenylether–graphite were used and compared as working electrodes. The effect of experimental parameters such as pH, accumulation potential and accumulation time, presence of various compounds was examined. A simple procedure was also developed for the electrochemical determination of rutin in a multivitamin preparation.

Electrochemistry coupled with mass spectrometry (EC–MS) was used for the online study of oxidation processes of anthocyanins. Two cyanidin glycosides, cyanidin-3-galactoside (ideain) and cyanidin- 3,5-diglucoside (cyanin), were subjected to electrochemical oxidation and subsequent online mass spectrometric identification of the formed products by Jirovsky *et al.* [86]. Application of relatively low working potentials (around 100 mV *vs.* Pd/H_2) using a porous graphite coulometric electrode yielded detectable oxidation products. It was found that the monoglycosylated analogue underwent anodic oxidation easier than the diglycosylated one. Incorporation of a hydroxyl group was observed in both glycosides as a first step of the electrochemically induced oxidation. Further, an oxidative condensation of two anthocyanin molecules was noticed. The proposed oxidative condensation was further supported by consecutive fragmentation (*i.e.*, collection of MS2, MS3, and MS4 spectra) in which corresponding subsequent losses of the sugar moiety were observed.

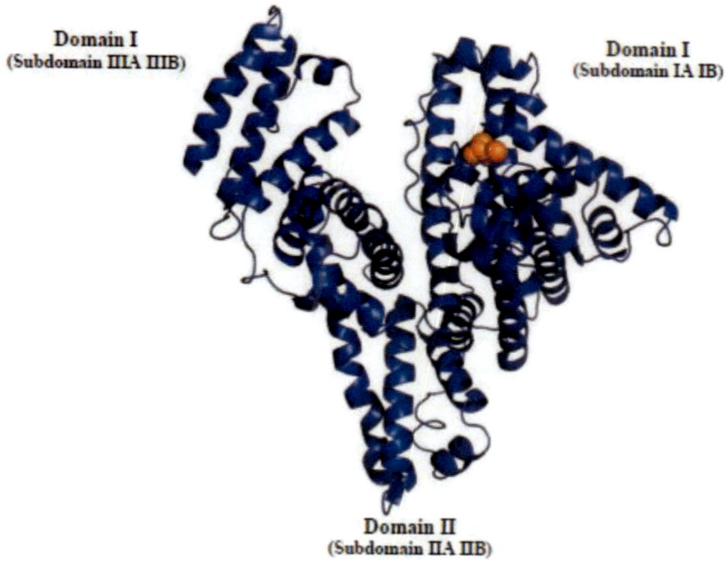

Figure 1. Structure of serum albumin.

Hesperidin, a flavanone glucoside found in the skins and juices of citrus fruits is frequently seen as a component in pharmaceutical preparations. It was determined by adsorptive stripping voltammetric method in 0.1 M NaAc-HAc solution [87]. This method was based on adsorptive accumulation of hesperidin on hanging mercury drop electrode followed by linear sweep voltammetry. Peak current response was studied with respect to accumulation time, pH effect and concentration of HAc-NaAc buffer solution. The peak potential was observed to be -1.01V (*vs.* Ag/AgCl). The peak current was directly proportional to the concentration of hesperidin with a detection limit of 3.0×10^{-7} M at an accumulation time of 120 s. The electroreduction was irreversible and showed adsorptive characteristics. An e-tongue has been developed to distinguish between wines aged in oak barrels and wines matured in steel tanks, in contact with oakwood chips based on voltammetric technique [88]. They have also been analysed by conventional chemical methods. Both the methods have been proved to be good to distinguish between wines aged by these two methods. The electrochemical signals have been succesfully utilized to determine polyphenolics or tannins.

4. Structures of Macromolecules

4.1. Serum Albumin

It is a globular protein (Figure 1) comprised of three structurally similar domains (I, II and III), each containing two subdomains (A and B), and are divided into nine loops (L1–L9) by 17 disulfide bonds [89, 90]. Albumin has two specific drug binding sites with high affinity in the IIA and IIIA subdomains. The pocket of subdomain IIA corresponds to site I, which is thought to be a binding site for salicylates, sulfonamides and a number of other drugs [91]. The inside wall of the pocket is formed by hydrophobic side chains, whereas the entrance of the pocket is surrounded by positively charged residues, *i.e.*, Arg 257, Arg 222, Lys 199, His 242, Arg 218 and Lys 195. Aromatic carboxylic acids largely ionized at physiological pH are bound to site II, the putative binding site for tryptophane, thyroxine, octanoate and other drugs, would correspond to the pocket of subdomain IIIA, which is almost the same size as site I. Various drugs are transported in the blood while bound to albumin. The remarkable binding properties of albumin accounts for the central role it can play in both the efficacy and rate of delivery of drugs. Among serum albumins, bovine and human serum albumins have been well studied.

4.2. Structures of DNA and RNA

Deoxyribonucleic acid (DNA) is a nucleic acid that contains the genetic instructions used in the development and functioning of all known living organisms and some viruses. DNA is a long polymer made from their repeating units called nucleotides. Each nucleotide is made up of a sugar called deoxyribose, a phosphate group (PO_4^{3-}) and an organic base (Figure 2). In DNA the two strands run in opposite directions to each other anti-parallel forming a double helical structure (Figure 2) [92, 93]. In RNA, the sugar present is ribose (thus ribonucleic acid). It usually exists as a single stranded polymer that is composed of a long sequence of nucleotides. Each nucleotide consists of a sugar, a PO_4^{3-} group and an organic base. DNA contains adenine (A), guanine (G), cytosine (C) and thymine (T) while RNA contains A, G, C and U as bases. A and G are double-ring bases called purines, while C, T and U are single-ring bases called pyrimidines. The DNA double helix is stabilized by hydrogen bonds between the bases attached to the two strands. The information in DNA is stored as a code made up of four chemical bases, A, G, C and T. The order or sequence of these bases determines the information available for building and maintaining an organism, similar to the way in which letters of the alphabet appear in a certain order to form words and sentences.

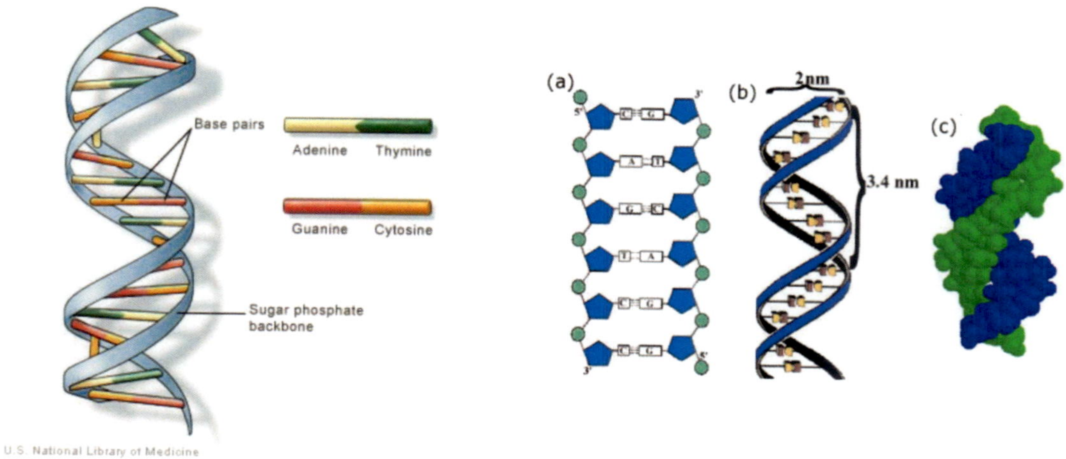

Figure 2. Structure of DNA.

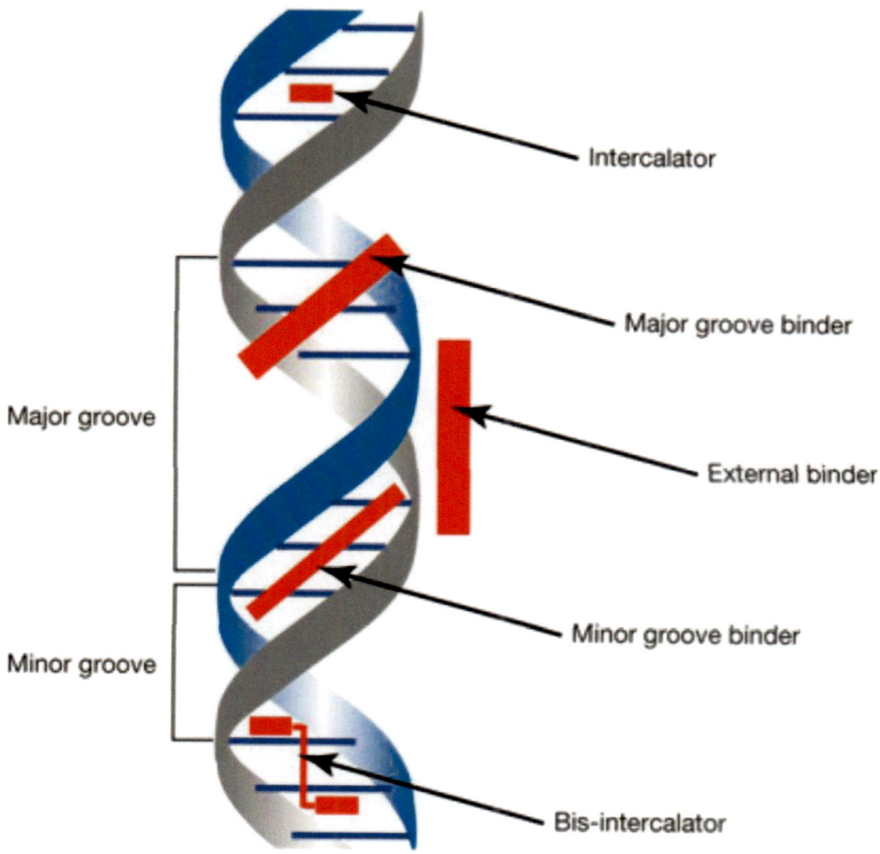

Figure 3. Modes of small molecule binding to DNA.

Due to the central role of DNA in replication and transcription, DNA has been a major target for antibiotic, anticancer and antiviral drugs. Drugs bind to DNA in different ways. These include intercalation between adjacent base pairs, intrusion into the minor groove and major groove and electrostatic interaction (Figure 3) [94]. Intercalation and minor-groove binding are the predominant DNA-binding modes of small ligands [95-97], while electrostatic binding interactions between cationic species and the negatively charged DNA phosphate backbone usually occur along the exterior of the helix.

5. Interaction of Flavonoids with Biomacromolecules

5.1. Significance of Binding Studies of Flavonoids to Macromolecules

The mechanism of binding of a bioactive compound of natural or synthetic origin to transport the blood proteins such as serum albumins and hemoglobin can give valuable information about the pharmacodynamics and pharmacokinetics of the molecule. Generally, the weak binding leads to a short lifetime or poor distribution, while strong binding decreases the concentrations of free drug in plasma. Thus, the effectiveness of a molecule as a pharmaceutical agent depends upon its binding capacity to serum albumins because they are chiefly responsible for delivering the drugs to their target organs.

Polyphenols characteristically possess a significant binding affinity to these proteins, which can lead to the formation of soluble and insoluble protein–phenol complexes [98]. With regard to the human health with flavonoid rich diet, knowledge of flavonoid action in the living system is required *viz.* how they are absorbed, metabolized, transported and delivered to specific tissues. The distribution of flavonoids in biological systems will depend on their relative hydrophilicity/hydrophobicity and on their interactions with particular macromolecules. These factors will determine the local concentration of flavonoids which will influence, among other factors, their capacity to regulate cellular events. A number of biochemical and molecular biological investigations have revealed that proteins (including enzymes) are frequently the "targets" for therapeutically active flavonoids of both natural and synthetic origin [99]. The binding of flavonoids to albumins can occur both noncovalently [100] and covalently [101] depending on conditions. Flavonoids are also capable of chelating transition and noble metal ions what may result in enhancing their anti-inflammatory and antioxidant properties [102].

DNA contains all of the genetic information related to cellular function, including DNA replication and gene expression. However, DNA molecule can be easily damaged, by reactive oxygen species and reactive nitrogen species. Therefore, in a health preventing perspective, there is a deep interest in identifying free radical scavengers or antioxidants that inhibit oxidative DNA damage. Flavonoids are strong antioxidants that prevent DNA damage. Hence, the study of interaction of natural antioxidant polyphenols with DNA can provide useful information about the mechanism of action of the polyphenol molecule. The nature and dynamics of binding of small molecules (drugs and flavonoids) to biomacromolecules like DNA represent an active area of investigation which can lead to the design of DNA-binding agents as anticancer therapeutics [103].

Different analytical methods like UV absorption, spectrofluorimetry, Fourier transform infrared spectroscopy (FT-IR), nuclear magnetic resonance spectrometry (NMR), circular dichroism (CD), thermal methods, voltammetry, high performance liquid chromatography (HPLC), liquid chromatography-mass spectrometry (LC-MS), tandem mass spectrometry (MS/MS), matrix-assisted laser desorption-ionization mass spectrometry (MALDI-TOF-MS) *etc.* have been employed for the investigation of interactions of bioactive flavonoids and anthocyanins with macromolecules and various parameters like mode of binding, binding site, thermodynamic parameters of binding, the distance between donor and acceptor, conformational changes of macromolecules etc have been determined.

5.2. Interactions of Flavonoids with Nucleic Acids

Shuyun Bi *et al.* [104] have explored the interactions of quercetin, kaempferide and luteolin with fish sperm deoxyribonucleic acid (DNA) using acridine orange (AO) as a fluorescence probe. The fluorescence intensity of DNA is very so weak that the direct use of the fluorescence emission of DNA is limited to study its properties [105]. Several small molecules have already proven to be sensitive probes of DNA. Acridine is one of such molecule that was employed as a fluorescence probe. Since, it has planar aromatic chromophore, it can insert between two adjacent base pairs in a DNA helix. The variations in the spectroscopic (UV absorption and fluorescence) characteristics of DNA-AO were used to evaluate the binding constants of flavonoids-double stranded DNA (dsDNA) or flavonoids-single stranded DNA (ssDNA). Further, the investigations on competitive bindings among these drugs interacting with DNA-AO systems, melting temperature (T_m) curves and viscosity measurements were carried out to understand the binding mechanism of these flavonoids with DNA. The binding mode of quercetin, kaempferide and luteolin with DNA was evaluated to be groove binding. The binding constants of quercetin, kaempferide and luteolin with dsDNA (or ssDNA)-AO complex were noticed to be 3.19×10^4 (or 1.19×10^4), 5.63×10^4 (or 4.58×10^4) and 2.33×10^4 (or 1.03×10^4) L mol^{-1}, respectively.

Spectroscopic studies of the interaction between quercetin and G-quadruplex DNA were carried out by Sun *et al.* [106]. In this study, they observed that the quercetin interacted with monomeric and dimeric G-quadruplexes differently exhibiting two different kinds of modes: stacking mode for monomeric G-quadruplexes and quercetin with the molecular plane being parallel with G-tetrads and groove binding mode for dimeric G-quadruplexes and quercetin with the molecular plane of quercetin being vertical with the planes of G-tetrads. Further, it was reported that the telomerase inhibition activity of drugs was strongly related to the stabilization of quadruplex structure. Therefore, quercetin was proposed to be a promising candidate for potent anti-cancer drug. Linear dichroism (LD) spectra of quercetin-DNA solutions revealed that the quercetin interacted with DNA through intercalation [107]. The DNA concentration and the planarity of the chromophore were found to be the limiting factors in the interaction. There were no induced LD signals for concentrations of DNA less than 7.8×10^{-3} M phosphate. The interaction was most probably of a hydrophobic nature between the most hydrophobic segment of the quercetin (benzopyran-4-one) and the intercalation site, which allowed the chromophore to penetrate the DNA helix and to arrange its planar structure more or less parallel to the adjacent planes of the nitrogenous bases. A comparison between the planar and hydrophobic flavonol quercetin, and the non-planar and hydrophilic flavanone dihydroquercetin showed that the interaction of the latter with DNA was strongly limited.

Binding of baicalein, wogonin and baicalin to fish sperm DNA was studied by using a well known intercalator, ethidium bromide dye as a fluorescence probe [108]. In this study, the absorption, fluorescence, melting temperature and viscosity measurements were carried out. These studies indicated that the planar flavonoids intercalated to the DNA helix. Hyperchromicity and blue shift were noticed in the absorption spectra and fluorescence quenching (>50%) in the fluorescence spectra. Further, the intercalative binding mode was consistent with the red shift in the position of λ_{max} in the fluorescence spectra. Stern–Volmer plots at 298 K and 310 K have revealed that the quenching of fluorescence by flavonoids was a combined quenching process. The binding constant, binding site number and thermodynamic parameters were determined from spectral results. Flow linear dichroism technique was employed by Solimani *et al.* [109] to investigate the interaction between DNA and flavonoids *viz.*, quercetin, rutin, morin and naringin. The results indicated that these flavonoids bound to DNA through intercalation and that their affinity for DNA increased following the same sequence of their biological activity.

The interaction of copper complexes of (-)-epicatechin gallate (ECG) and (-)-epigallocatechin gallate (EGCG) with calf thymus DNA (ct-DNA) was investigated by absorption, fluorescence and CD along with melting studies [110]. It was observed that both copper complexes quenched the fluorescence intensity of ethidium bromide bound ct-DNA upon binding, resulting in a ground state complex formation by a static quenching process. The binding constants as evaluated from fluorescence data were found to be in agreement with those obtained by absorption study. The values ranged from 0.84 to 1.07×10^5 M^{-1} and from 1.14 to 1.04×10^5 M^{-1} for Cu(II)-ECG and Cu(II)-EGCG, respectively for the temperature range of 21-42 ^{0}C with two binding sites. Thermodynamic parameters obtained were suggestive of the involvement of different modes of interaction during binding for each complex although both were found to bind through intercalation. The CD studies and variations in the melting temperature have revealed unwinding of the ct-DNA helix with conformational changes due to binding.

Binding mode of morin, quercetin and rutin with chicken blood dsDNA (ck.DNA) has been investigated spectrophotometrically at different temperatures under two physiological pH values, 7.4 (human blood pH) and 4.7 (stomach pH) [111]. The binding constants indicated that the flavonoids bound effectively through intercalation at both pH values and body temperature. Quercetin exhibited greater binding affinity for DNA compared to others. The free energies of flavonoid–DNA complexes indicated the spontaneity of their binding. The binding affinities of genistein and quercetin to DNA oligomers were investigated employing NMR spectroscopy by studying the diffusion coefficients of these compounds with and without the presence of DNA [112]. Genistein was found to bind very weakly ($K=1.54 \times 10^2$ M^{-1}) as compared to quercetin ($K=5.75 \times 10^3$ M^{-1}) and luteolin ($K=2.17 \times 10^4$ M^{-1}). In the case of genistein, a hydrogen bond between the NHF, B cytidine protons of the edge base pair and a genistein hydroxyl was proposed based on the shape of the diffusion ordered spectroscopy (DOSY) spectrum. MP2 and DFT calculations of genistein showed a 9 kcal mol^{-1} excess energy of the planar conformation as compared with a twisted one. Due to this, intercalation of the planar genistein into the base pairs was proposed to be unlikely.

Interaction of fisetin (3,7,3′,4′-tetrahydroxyflavone), a bioactive plant flavonoid, with ct-DNA was studied using fluorescence technique by Sengupta *et al.* [113]. Significant changes were noticed in the intrinsic fluorescence behaviour of fisetin in the presence of DNA. These, along with other relevant supporting spectroscopic data suggested that the fisetin intercalatively bound between the base pairs of ct-DNA. Gyrases are DNA topology modifying enzymes present only in prokaryotes which make them an attractive target for antibacterial drugs. Quercetin was found to be inhibited the supercoiling activity of bacterial gyrase and induced DNA cleavage [114]. It was observed that the quercetin bound to the 24 kDa fragment of gyrase B of *Escherichia coli* with a K_D value of 15 μM and inhibited ATPase activity of gyrase B. Its binding site overlapped with ATP binding pocket and could be competitively replaced by either ATP or novobiocin. Our research group has investigated the mechanism of binding of hesperitin or naringenin with ct-DNA by employing the different spectroscopic techniques (un published results).

The interaction of quercetin–Cu(II) complex with ct-DNA was investigated by employing neutral red (NR) dye as a spectral probe by the application of UV–vis spectrophotometry, CV and synchronous fluorescence spectroscopy[115]. The results showed that both quercetin–Cu(II) complex and the NR molecule intercalated into the double helix of the ct-DNA. The 2:1 quercetin-Cu(II) complex (estimated binding constant=2.85×10^9) was stabilized by intercalation in the DNA (binding constant, K [quercetin–Cu(II)–DNA] = $(1.82 \pm 0.20) \times 10^5 M^{-1}$), and displaced the NR dye from the NR–DNA complex in a competitive reaction. CV studies also confirmed the intercalation reaction and showed that the ratio (K_R/K_O) of binding constants for the reduced and oxidized forms of the metal complex to be 2.05. Furthermore, the alternative least squares (ALS) method was applied to resolve a complex two-way array of the absorption spectra data. This yielded the equilibrium concentration profiles of each component in the reaction [NR, NR–DNA and quercetin–Cu(II)] as well as the corresponding pure spectra. The extracted profiles showed that at equilibrium the [NR–DNA] and [NR] trends decreased and increased symmetrically, respectively, with approximately linear behaviour being observed below 10×10^{-6} mol L^{-1} of the added quercetin–Cu^{2+} complex. Thereafter, these trends converged asymptotically. The free [quercetin–Cu(II)] trend-line at equilibrium was linear over the whole range of the complex added. It was possible to estimate the approximate value of the equilibrium constant of the exchange process (approximately 5×10^{-1}) involving the intercalation of the quercetin–Cu(II) complex. It was also found that about 35% of the bound complex was unaccounted by the intercalation reaction, presumably being stabilized at an alternative site.

The interaction between DNA and Cu(II)–naringin complex was evaluated employing electrochemical ssDNA- and dsDNA-based biosensors and the results were supported by UV, CD and ^1H NMR data [116]. The interaction was evident from the changes in the oxidation peak of the guanine and adenine bases (as obtained by square wave voltammetry (SWV)). The variations of the spectroscopic characteristics of DNA and Cu(II)–naringin complex in aqueous medium demonstrated the intercalative mode of binding predominantly wherein Cu(II)–naringin complex interacted with dsDNA probably *via* N (7) of guanine site. The *in vitro* binding of flavonoids with DNA *in vitro* and interaction of transfer RNA with quercetin, kaempferol and delphinidin at physiological conditions and a comparison with the corresponding pigment–DNA adducts was examined by Kanakis *et al.* [117]. Spectroscopic methods *viz.*, FTIR and UV-vis difference were employed to determine the drug binding mode, the binding constants, and the effects of drug complexation on the stability and conformation of tRNA duplex. Both intercalative and external binding modes were proposed. Structural analysis revealed that all the flavonoids intercalated tRNA duplex with minor external binding to the major or minor groove and the backbone phosphate group. Further, the binding constants were observed to be in the order of 10^4. A partial B to A–DNA transition occurred at a high drug concentration, while tRNA remained in the A-family structure.

Interactions of morin, apigenin and naringin with yeast RNA in aqueous solution at physiological conditions were carried by Nafisi *et al.* employing FTIR, UV–vis absorption methods [118]. The binding constants were noticed to be $9.150x10^3$, $4.967x10^4$ and $1.144x10^4$ M^{-1} for Mor-RNA, Api-RNA and Nar-RNA, respectively. No biopolymer secondary structural changes were observed upon flavonoid interaction and RNA remained in the A-family structure in these pigment complexes. Quercetin could be used as the selective fluorescent probe for ds-RNA as it exhibits increased fluorescence intensity in the presence of ds-RNA compared to that with ds-DNA [119]. Further, the addition of poly G yielded more than order of magnitude stronger changes in absorption and fluorescence spectra of quercetin compared to the changes upon addition of poly A and poly U revealing possible usage of quercetin as a powerful spectroscopic probe for poly G sequences. Stability and stoichiometry of lanthane (III)/quercetin complexes in physiologically relevant aqueous media were determined. The interactions of $[LaQ]^{3+}$ with ds-DNA and ds-RNA were noticed to be significantly different compared to the free quercetin, revealing increase of complex stability and thus significant impact of La(III) in binding of $(LaQ)^{3+}$ to polynucleotides. Similar results were observed for interactions of $(LaQ)^{3+}$ with single stranded RNA.

5.3. Mechanism of Interaction of Flavonoids with Proteins

Fluorescence technique was employed by Sengupta *et al.* [113] to examine interactions of bioactive plant flavonoid fisetin (3,7,3′,4′-tetrahydroxyflavone) with the physiologically important circulatory protein, human serum albumin (HSA). From the interaction studies on fisetin–HSA, the existence of two distinct binding sites was proposed. Further, the authors have presented evidence for the occurrence of efficient Forster type fluorescence resonance energy transfer from tryptophan to fisetin, indicating that both binding sites of fisetin in HSA are proximal to the unique Trp–214 residue present in the interdomain (between IIA and IIIA domains) loop region of the protein. Quercetin (3,3′,4′,5,7-pentahydroxyflavone), a ubiquitous, bioactive plant flavonoid, is known to possess anti-cancer, antitumor, and other important therapeutic activities of significant potency and low systemic toxicity. Sengupta *et al.* have studied the interaction of quercetin with the plasma protein, HSA, exploiting the intrinsic fluorescence emission properties of quercetin as a probe [120]. Quercetin is weakly fluorescent in aqueous buffer medium with an emission maximum at 538 nm. Binding of quercetin to HSA led to dramatic enhancement in the fluorescence emission intensity and anisotropy (r), along with significant changes in the fluorescence excitation and emission profiles. The excitation spectrum suggested occurrence of efficient Forster type resonance energy transfer (FRET) from the single Trp-214 residue of HSA to the protein bound quercetin. Further more, the binding constant ($K=1.9x10^5\ M^{-1}$) and Gibb's free energy change ($\Delta G^0 = -30.12$ kJ/mol) for quercetin–HSA interaction have been calculated from the relevant anisotropy data.

5,7,4′-Trihydroxy-6,3′,5′-trimethoxyflavone is one of the bioactive components isolated from *Artemisia* plants possessing antitumor therapeutic activities [121]. The binding properties and binding sites located on HSA have been studied using UV absorption, fluorescence and FT-IR spectroscopy. The binding constants were noticed to be in the range of 10^5. The study suggested that the hydrophobic interaction was the predominant intermolecular force stabilizing the complex and the flavonoid bound to the site I of HSA. Interaction of resveratrol, a polyphenolic compound found largely in the skin of red grape and wine that exhibits a wide range of pharmaceutical properties and plays an important role in the prevention of human cardiovascular diseases, with HSA was reported [122]. Structural analysis showed that resveratrol binds non-specifically (H-bonding) *via* polypeptide polar groups with overall binding constant of 2.56×10^5 M^{-1}. The protein secondary structure as analyzed by CD spectroscopy showed no major alterations at low resveratrol concentrations (0.125 mM), whereas at high pigment content (1 mM) major increase of α-helix from 57% (free HSA) to 62% and a decrease of β-sheet from 10% (free HSA) to 7% occurred in resveratrol-HSA complexes. These results indicated a partial stabilization of protein secondary structure at high resveratrol content.

The binding of scutellarin with HSA was investigated by Tian *et al.* by fluorescence, CD, FT-IR and molecular modeling [123]. Quenching of the fluorescence of HSA in the presence of scutellarin was observed. The binding parameters were obtained from Scatchard's plot which is approximately in agreement with the results of Stern-Volmer analysis. The values of $\Delta H°$ and $\Delta S°$ were calculated to be -8.55 kJ/mol and 65.15 J/mol K, respectively. Data obtained by fluorescence spectroscopy, CD experiment, FT-IR experiment and molecular modeling method suggested that the scutellarin was strongly bound to HSA and the primary binding site of scutellarin was located in site I of HSA. It was proposed that the scutellarin bound to site I (subdomain II) mainly by a hydrophobic interaction and there were hydrogen bond interactions between the scutellarin and the residues, Arg 222 and Arg 257. Mandeville *et al.* examined the interactions of curcumin and genistein with HSA at physiological conditions employing FTIR, UV–vis, CD and fluorescence spectroscopic methods to understand the drug binding mode, the binding constant and the effects of pigment complexation on HSA stability and conformation [124]. The structural analysis showed that the curcumin and genistein bound HSA *via* polypeptide polar groups with overall binding constants of $K_{curcumin}$=5.5 (±0.8)x10^4 M^{-1} and $K_{genistein}$=2.4 (±0.40)x10^4 M^{-1} and the number of bound pigment (n) to be 1.33 for curcumin and 1.49 for genistein. The HSA conformation was altered by pigment complexation with reduction of α-helix and increase of random coil and turn structures suggesting a partial protein unfolding.

Liu Yuan *et al.* have made use of fluorescence spectroscopic, CD, UV–vis absorption spectroscopic and molecular modelling studies [125] to characterize the interaction of apigenin with HSA. The results indicated that the apigenin strongly quenched the fluorescence of HSA through static quenching mechanism. Hydrophobic and electrostatic interaction were found to be the predominant forces in the apigenin–HSA complex, and the process of apigenin binding HSA was driven by enthalpy (ΔH=-17.497 kJ mol^{-1}) and entropy (ΔS = 37.04 J mol^{-1}). The results from synchronous fluorescence showed that the microenvironment around tyrosine (Tyr) had a slight trend (about blue shift of 1 nm) of polarity decreasing. CD results have revealed that the formation of the complex has not resulted any changes in the secondary structure of HSA. The binding distance (r) between HSA and apigenin was noticed to be 3.21 nm as per the theory of Forster resonance energy transfer. Molecular modelling studies indicated that the apigenin was located to the entrance of site I by hydrophobic and electrostatic forces, which matched exactly the corresponding experimental results.

Fluorescence anisotropy, FT-IR spectroscopy, CD spectroscopy and molecular modeling methods were employed by Li *et al.* to characterize optical properties of the daidzein (an isoflavone)-HSA complex and to gain information on the binding mechanism at molecular level [126]. Daidzein is weakly fluorescent having an emission maximum at 466 nm in aqueous buffer medium. Binding of daidzein with HSA led to dramatic enhancement in the fluorescence intensity and anisotropy (r), along with significant changes in the fluorescence excitation and emission profiles. The binding constant [K= (5.9 ± 0.6) x 10^4 M^{-1}] and the standard free energy change (ΔG=27.5 kJ/mol) for daidzein–HSA interaction were calculated from the fluorescence data. The secondary structure compositions of free HSA and its daidzein complexes were estimated by the FT-IR spectra and the curve-fitted results of amide I band, which are in good agreement with the analyses of CD spectra. Furthermore, the displacement experiments with site probes indicated that the daidzein bound to the site I of HSA which is also in agreement with the result of molecule modeling study.

Kanakis *et al.* [127] have studied the interaction of HSA with quercetin, kaempferol and delphinidin in aqueous solution at physiological conditions. FTIR and UV–vis spectroscopic methods were used to determine the polyphenolic binding mode, the binding constant and the effects of flavonoid complexation on protein secondary structure. The spectroscopic results showed that flavonoids were located along the polypeptide chains through H-bonding interactions with overall affinity constant of K_{que}=1.4x10^4 M^{-1}, K_{kae}=2.6x10^5 M^{-1} and K_{del}=4.71x10^5 M^{-1}. The protein secondary structure showed no alterations at low pigment concentration (1 μM), whereas at high flavonoid content (1 mM), major reduction of α-helix from 55% (free HSA) to 42–46% and increase of β-sheet from 15% (free HSA) to 17–19% and β-anti from 7% (free HSA) to 10–20% occurred in the flavonoid–HSA adducts. The decreased α-helicity indicated the partial protein unfolding upon flavonoid interaction.

Steady-state and time-resolved fluorescence spectroscopic techniques were employed by Pastukhov *et al.* to study the interaction between rutin and HSA [128]. The binding constant as determined from the quenching of the Trp214 fluorescence by rutin was found to be (6.87 ± 0.22) x 10^4 M^{-1} and that obtained from the fluorescence of HSA-bound rutin was (3.8 ± 0.4) x 10^4 M^{-1}. The composition of HSA-rutin complex was determined to be 1:1. The efficient quenching of the Trp214 fluorescence by rutin, fluorescence resonance energy transfer from excited Trp214 to rutin, and competitive binding of warfarin indicated that the binding site for the flavonoid was located within subdomain IIA of HSA. Further, it was observed that the presence of sugar moiety in the flavonoid molecule reduced affinity of rutin for binding to HSA but did not affect the binding stoichiometry and location of the binding site compared with aglycone analogs.

The interaction of hesperetin with HSA was investigated by UV absorption, fluorescence and FT-IR spectrometry [129]. Fluorescence results showed that one molecule of protein interacted with one molecule of drug at the molar ratio of drug to HSA ranging from 0.3 to 7 and the binding affinity (K_A) was 8.11x10^4 M^{-1}. The binding ability of the drug to protein decreased from pH 6.4 to 8.4 in the drug to protein molar ratio of 1 and the drug was found to bind at subdomain IIA. Upon the addition of increased concentrations of drug, the protein α-helicity decreased. The reduction of α-helix structure of protein was found about 7–9% after the protein interacted with hesperetin in D_2O and H_2O buffer solution at pH 7.4 when the drug to protein molar ratio was 10. From the spectral results, the main mechanism of protein fluorescence quenching was proposed to be static quenching process and the hydroxyl groups of the drug in its neutral part played an important role in the binding process.

Bi *et al.* [130] have demonstrated that the quercetin, rutin, hyperin and baicalin quenched the fluorescence of HSA. The binding constants and the number of binding sites of the flavonoids with HSA were obtained by three different calculation methods and the results were compared. The effects of various metal ions on the binding constants of these flavonoids with HSA were also investigated. Based on the mechanism of the Forster energy transfer, the energy transfer efficiency between the acceptors and HSA were found. The relationship between the structure characteristics of these compounds and their binding properties with HSA were explored. Cao *et al.* have reported that the methoxylation of tangeretin improved the affinity for HSA by 100 times in citrus flavanones [131]. The 2,3-double bond in conjugation with a 4-oxo group played an important role for the affinity for HSA. Further, the affinity of apigenin for HSA was observed to be about 10,000-times higher than that of naringenin. Hydroxylation on position 30 of flavonol significantly improved the binding affinity for HSA. The affinity of quercetin for HSA was about 100-times higher than that of kaempferol. The hydroxylation on position 30 of flavone also slightly improved the binding affinity for HSA. The affinity of luteolin for HSA was about 1.38-times higher than that of apigenin.

The relationship between the structural properties of selected dietary flavanoids and their affinities for HSA were investigated by fluorescence analysis by Xiao *et al.* and found that the binding process with HSA was strongly influenced by the structural differences of the compounds [132]. Further, methylation of hydroxyl groups improved the affinities for HSA by 2 to 16-fold. Hydroxylation on rings A, B and C also affected the affinity for HSA significantly. Glycosylation decreased the affinities for HSA by 1 to 3 orders of magnitude depending on the conjugation site and the class of sugar moiety. Hydrogenation of the C_2=C_3 double bond also decreased the binding affinity. Galloylated catechins and pyrogallol-type catechins exhibited higher binding affinities for HSA than non-galloylated and catechol-type catechins, respectively. The affinities for HSA increased with increasing partition coefficients and decreased with increasing hydrogen bond donor and acceptor numbers of flavonoids, which suggested that the binding interaction was mainly caused by hydrophobic forces.

Sengupta *et al.* have reported a detailed study on the interactions of 7-hydroxyflavone (7HF) with HSA employing electronic absorption, fluorescence (steady state and time resolved) and induced circular dichroism (ICD) spectroscopy [133]. The spectral data indicated that in the protein matrix, the neutral 7HF molecule was predominantly transformed to a conjugate anion (7HFA) by a proton abstraction in the ground state. Further, the protein environment induced dramatic enhancements in the fluorescence emission intensity, anisotropy and lifetime (s) values, as well as pronounced changes in the fluorescence excitation and emission profiles of the fluorophore. They have reported that the 7HF in HSA is proximal (estimated distance, R = 23.6 Å) to Trp-214 residue present in the inter-domain (between IIA and IIIA domains) loop region of the protein. The values of binding constant and the Gibb's free energy change were found to be 9.44×10^4 M^{-1} and -28.33 kJ/mol for 7HFA–HSA interaction, respectively. CD studies indicated that the electronic transitions of 7HF were strongly perturbed on binding to the chiral host (HSA), leading to the appearance of ICD bands.

The binding of luteolin (LUT) to HSA was investigated to determine the binding mode and parameters of the interaction by Jurasekova *et al* employing CD, UV-vis absorption, fluorescence, Raman and surface-enhanced Raman scattering spectroscopy combined with molecular modeling [134]. LUT was found to bind in subdomain IIA where an interaction with Trp-214 was established. Hydrophobic and electrostatic interactions were found to be involved in the binding of LUT to HSA. The HSA conformations were slightly altered by the drug complexation with reduction of α-helix and increase of β-turns structures suggesting a partial protein unfolding.

The interaction between quercetin and bovine serum albumin (BSA) was investigated using fluorescence and UV absorption methods [135]. The apparent binding constants of quercetin-BSA were noticed to be 2.8×10^8 and 3.1×10^8 M^{-1} at 26 °C and 36 °C, respectively, while the binding sites (n) were 1.7± 0. 02. According to the Forster's theory of non-radiation energy transfer, the binding distances (r) were also calculated. The experimental results showed that the quercetin could insert into the BSA quenching the intrinsic fluorescence by forming the quercetin-BSA complex. Both static quenching and non-radiation energy transfer were proposed to be the main reasons for the fluorescence quenching. The process of binding was a spontaneous molecular interaction in which entropy increased while Gibbs free energy decreased, indicating that the interaction between quercetin and BSA was driven mainly by hydrophobic force.

Tian *et al.* have studied the binding of kaempferol to BSA at 296, 310 and 318 K employing the fluorescence, CD and FT-IR techniques [136]. The CD and FT-IR studies indicated that the kaempferol bound strongly to BSA. The association constant, K was determined based on the quenching results. The values of ΔH^0 and ΔS^0 were found to be -21.694 kJ/mol and 88.814 J/mol K, respectively. From the displacement experiments and thermodynamic results, it was proposed that the kaempferol bound at binding site II (subdomain III) mainly by hydrophobic interaction. The results obtained by FT-IR and CD experiments indicated the changes in secondary structure of the protein upon interaction with kaempferol. The distance between the tryptophan residues in BSA and kaempferol bound to site II was estimated to be 2.78 nm as per fluorescence resonance energy transfer theory.

The interaction of BSA with isoxazolcurcumin (IOC) and diacetylcurcumin (DAC) was investigated by Sahoo *et al.* [137]. Their binding constants were found to be in the range of 10^5 M^{-1}. Minor conformational changes of BSA were observed from CD and FT-IR studies upon binding. Based on the FÖrster theory of non-radiation energy transfer, the average binding distance (r) between BSA and acceptors, IOC and DAC was found to be 3.79 and 4.27 nm, respectively. Molecular docking of IOC and DAC with BSA indicated that they docked close to Trp 213, which is within the hydrophobic subdomain. The mechanism of interaction of a non-glycosidic citrus flavonoid, hesperitin with BSA was studied by Seetharamappa *et al.* [138] employing UV–vis absorption, fluorescence, FT-IR, CD, fluorescence anisotropy and synchronous fluorescence spectroscopy in phosphate buffer of pH 7.4. Fluorescence data have revealed that the fluorescence quenching of BSA by hesperitin was the result of the formed complex of HES–BSA. The binding and thermodynamic characteristics were evaluated based on fluorescence results obtained at four different temperatures. Displacement experiments were carried out using different site probes to locate the binding site in the protein. The interactions involving hydrogen bonds were found to stabilize the complex. The alterations in the protein secondary structure upon complexation with HES were evident from the gradual decrease in α-helicity. The distance between BSA and flavonoid was calculated from the fluorescence resonance energy transfer and found to be 1.978 nm. Common ions viz., Zn^{2+}, K^+, Cu^{2+}, Ni^{2+}, Mn^{2+} and Co^{2+} were found to influence the binding of flavonoid to protein.

The fluorogenic properties of quercetin and similar flavonoids commonly present in plants were exploited to analyze their interactions with target proteins, BSA and insulin [139]. Quercetin strongly quenched the fluorescence intensity of BSA and insulin. The fluorescent signal showed saturation kinetics with increasing flavonoid concentrations indicating the presence of defined peptide binding motifs. The interaction between four flavonoids, catechin, epicatechin, rutin, and quercetin and BSA was investigated using quenching of intrinsic fluorescence of tryptophan [140]. Quenching constants were determined using the Stern-Volmer equation to provide a measure of the binding affinity between the flavonoids and BSA. The order of binding affinity was found to be as follows: quercetin > rutin > epicatechin > catechin. Quercetin and rutin exhibited a total quenching effect on BSA tryptophan fluorescence at a molar ratio of 10:1 and 25:1, respectively. However, epicatechin and catechin did not fully quench the protein fluorescence over the concentration range studied. Furthermore, the spectral data suggested that the association between flavonoids and BSA did not result any changes in molecular conformation of BSA and that hydrogen bonding, ionic and hydrophobic interactions were equally important driving forces for protein-flavonoid association.

The relationship between the molecular structure of dietary flavonoids and their affinities for BSA was fully investigated by fluorescence titration analysis by Shi and Cao [141]. The methylation of hydroxyl group in flavonoids enhanced their binding affinities for BSA by 1 to 794 times. Hydroxylation on rings A, B and C also strongly influenced the affinity for BSA. The glycosylation weakened the affinities for BSA by 1-2 orders of magnitude depending on the conjugation site and the class of sugar moiety. The hydrogenation of the $C_2= C_3$ double bond slightly increased the binding affinity. It was observed that the galloylated catechins and pyrogallol-type catechins exhibited higher binding affinities for BSA than non-galloylated and catechol type catechins. Further, the affinities for BSA increased with increasing partition coefficients and decreased with increasing hydrogen bond donor and acceptor numbers of flavonoids, which suggested that the binding interaction was mainly caused by hydrophobic forces.

The interaction of five flavonoids *viz.*, formononetin-7-*O*-β-D-glucoside, calycosin- 7-*O*-β-D-glucoside, calycosin, rutin, and quercetin, with BSA was studied by fluorescence and UV-vis absorbance spectroscopy [142]. It was observed that the fluorescence quenching of BSA by flavonoids was a result of the formation of a flavonoid-BSA complex. The binding constants ranked in the order quercetin > rutin > calycosin > calycosin-7-*O*-β-D-glucoside ≈ formononetin-7-*O*-β-D-glucoside. The values of ΔG, ΔH, and ΔS at different temperatures indicated that the hydrophobic interaction played a major role in flavonoid-BSA association. The values of r between BSA and acceptor flavonoids were 4.12, 3.85, 3.01, 5.72 and 4.75 for formononetin-7-*O*-β-D-glucoside, calycosin-7-*O*-β-D-glucoside, calycosin, rutin and quercetin, respectively.

Spectrometric and voltammetric studies of the interaction between quercetin and BSA using warfarin as site marker with the aid of chemometrics were carried out by Nia *et al.* [143]. The results indicated a cooperative interaction between the quercetin–BSA complex and warfarin, which produced a ternary complex, quercetin–BSA–warfarin. It was found that both quercetin and warfarin were located in site I. This observation was consistent with the view that the quercetin–BSA interaction occurred between the most hydrophobic segment of the quercetin molecule (benzopyran-4-one) [144] and the hydrophobic part of the BSA cavity. A study of the interaction between quercetin and HSA using molecular models found that the benzopyran-4-one moiety was located within the binding pocket of HSA, but the B-ring protruded from it and pointed towards the interface of IIB and IIIA domains [145].

The binding of structurally different flavonoids to HSA and BSA was investigated by Dufour and Dangles using three methods: the quenching of the albumin fluorescence, the enhancement of the flavonoid fluorescence, the quenching of the fluorescence of the quercetin–albumin complex by a second flavonoid [146]. The latter method was extended to probes to locate the binding site in two major subdomains, subdomain IIA and subdomain IIIA. Generally, flavonoids displayed moderate affinities for albumins (binding constants being in the range of $1\text{-}15 \times 10^4$ M^{-1}, flavones and flavonols being most tightly bound). It was found that the glycosidation and sulfation lowered the affinity to albumin by one order of magnitude depending on the conjugation site.

The interaction between hesperidin and BSA was investigated by fluorescence and UV-vis absorption spectroscopy. The spectral results revealed that hesperidin caused the fluorescence quenching of BSA through a static quenching procedure. Both the hydrophobic and electrostatic interactions played a major role in stabilizing the complex. The binding site number and apparent binding constant and thermodynamic parameters at different temperatures were calculated. The effects of Cu^{2+}, Zn^{2+}, Ni^{2+}, Co^{2+} and Mn^{2+} on the binding constants of hesperidin-BSA were studied. The effect of hesperidin on the conformation of BSA was also studied. Binding ability of two flavonoid aglycones (baicalein, quercetin) and their glycosides (baicalin, quercitrin) was studied based on quenching of the protein intrinsic fluorescence [147]. From spectral results, the bimolecular quenching constants, the apparent static binding constants and binding sites were calculated. The glycosylation of flavonoids decreased the binding affinity with protein. The binding constants of quercetin-BSA and quercitrin-BSA were found to be 3.65×10^7 and 6.47×10^3 L mol^{-1}, respectively while for baicalein-BSA and baicalin-BSA, the binding constants were observed to be 4.54×10^8 and 1.63×10^6 L mol^{-1}, respectively.

The interaction between flavonoids (hesperidin and naringin) and proteins was investigated by fluorescence and absorption spectroscopy [148]. It was found that the van der Waals or hydrogen bond, electrostatic force and hydrophobic forces played a role in the interaction of said drugs with proteins. The binding distance (r) between the protein and drug was evaluated and found to be less than 3 nm. Ma *et al.* have employed, warfarin (site I) and ibuprofen or diazepam (site II) as site markers to locate the binding sites on biomacromolecules [149]. The structural alterations of the protein or flavonoids before and after the combinations were probed by FTIR, CD or UV absorption methods. Investigation of the interaction between flavonoids and biomacromolecules yielded the useful knowledge for optimizing the molecular structure of the drugs, prescriptions and route of administration, and also the information for their bioavailability and bioactivity. Our research group has employed spectroscopic techniques to understand the binding mechanism of diosmin to proteins (BSA and HSA) and ct-DNA (un published results).

Alpinetin and cardamonin, isolated from the seeds of *Alpinia katsumadai* Hayata, have been used in traditional herbs as antibacterial, anti-inflammatory, and other important therapeutic activities of significant potency and low systemic toxicity. The interactions between two flavonoids analogs and lysozyme have been studied by He *et al.* employing FT-IR spectroscopy, CD, UV-absorption spectroscopy and fluorescence quenching study [150]. Both molecules showed high affinities to lysozyme with drug concentrations from 3.33×10^{-6} to 2.67×10^{-5} mol L^{-1} for alpinetin and 1.67×10^{-6} to 13.33×10^{-6} mol L^{-1} for cardamonin. The alterations of protein secondary structure in the presence of drugs in aqueous solution were quantitatively determined from CD and FT-IR spectral results. The thermodynamic parameters obtained and the spectroscopic results suggested that the hydrophobic and electrostatic interactions stabilized the interactions. The quenching mechanism and the number of binding site ($n \approx 1$) were obtained by fluorescence titration data. The efficiency of energy transfer provided the binding distances to be 4.04 and 5.90 nm for alpinetin-LYSO and cardamonin-LYSO systems, respectively.

Pinto *et al.* have reported the interaction between quercetin and lipoxygenase by fluorescence spectroscopy [151]. The analysis of the emission quenching at different temperatures revealed the presence of static quenching mechanism as a consequence of the formation of a complex, quercetin-lipoxygenase. The thermodynamic parameters, ΔG, ΔH and ΔS were calculated to be -32.57 $kJmol^{-1}$, -3.21 $kJmol^{-1}$ and 87.14 $Jmol^{-1}K^{-1}$ respectively, which suggested that the hydrophobic forces played a major role in the stabilization of quercetin-lipoxygenase complex. The distance (r) between lipoxygenase and quercetin was calculated to be 3.84 nm. The results obtained from the evaluation of three dimensional florescence spectra suggested a conformational modification of the protein in the region of the coupling with quercetin.

The binding of quercetin to ovalbumin in aqueous solution was investigated employing fluorescence, synchronous fluorescence and UV-absorption spectroscopies and modelling at pH 7.4 by Lu *et al.* [152]. Quercetin binding was characterized by one high affinity binding site with the association constant of the order of 10^5. Based on Forster theory, the distance between ovalbumin and quercetin was estimated. Further, molecular docking showed that the quercetin can bind to the active site of ovalbumin. The binding dynamics as proposed based on thermodynamic parameters, molecular modelling and accessible surface area calculation revealed that the hydrophobic interactions, hydrogen bonding and electrostatic forces stabilized the interaction.

The interaction of curcumin with intravenous immunoglobulin (IVIG) mainly composed of immune γ-globulin (IgG) was studied *in vitro* by Liu *et al.* using spectroscopic methods [153]. Docking was used to calculate the interaction mode and force involved between curcumin and IVIG. The results showed presence of hydrogen bond and hydrophobic force. The binding parameters for the reaction were calculated according to the Sips equation, which suggested that the binding of IVIG to curcumin was characterized by two binding sites with the average affinity constant K_0 at $1.170 x 10^4 M^{-1}$ (296 K), and it was a non-specific and weak drug-protein interaction. The secondary structure compositions of free IVIG and its curcumin complexes were calculated by the FTIR difference spectra, self deconvolution, second derivative resolution enhancement and the curve-fitting procedures of amide I band. The observed spectral changes revealed a partial unfolding of the protein structure, but the typical b structural conformation of IVIG was still retained. The average binding distance between curcumin and the chromophore of IVIG was observed to be 5.57 nm.

Zhao and Ren have investigated the influence of hydroxylation and glycosylation of soybean isoflavones in ring A on the interaction with BSA [154]. The hydroxylation and glycosylation of soybean isoflavones in ring A significantly affected the binding/quenching process; in general, the hydroxylation increased the binding affinity, while the glycosylation decreased the binding affinity. The binding constants of daidzein-BSA and daidzin-BSA were found to be $5.2 x 10^4$ and $5.58 x 10^3$ L mol^{-1}, respectively, while for genistein-BSA and genistin-BSA, the binding constants were $8.40 x 10^5$ and $1.44 x 10^5$ L mol^{-1}, respectively.

5.4. Binding Characteristics of Flavonoid(s) –Hemoglobin

Steady state and time resolved fluorescence spectroscopy combined with molecular modeling computations, have been used by Chaudhuri *et al.* [155] to explore the interactions of fisetin (3,7,3′,4′-OH flavone) and 3-hydroxyflavone (3-HF), with normal human hemoglobin (HbA). Distinctive 'two color' fluorescence signatures and fairly high fluorescence anisotropy (r=0.12–0.28) of fisetin and 3-HF revealed their specific interactions with HbA. The specific interactions with HbA were further confirmed from flavonoid-induced static quenching of the protein tryptophan fluorescence as indicated by bimolecular quenching constant, K_q; closely matched values of Stern–Volmer quenching constant and binding constant, $\tau_o/\tau\approx 1$ (where τ_o and τ are the unquenched and quenched tryptophans fluorescence lifetimes respectively).

Sentchouk and Bondaryuk have studied the fluorescence properties of flavonols, quercetin, fisetin, morin and rutin with a view to understand their possible interaction with hemoglobin (Hb) and BSA [156]. They observed the increased intensity of intrinsic fluorescence for all the flavonols except rutin in the presence of BSA. They have also investigated the interconnection between the structure of related flavonols and their fluorescent properties on interaction with proteins. The binding of quercetin and rutin to Hb has been investigated by employing the fluorescence, absorption and CD spectroscopy by Xi and Guo [157]. The binding parameters and binding mode between flavonoids and Hb were determined. The results of CD and synchronous fluorescence spectra indicated the conformational changes in Hb upon the addition of flavonoids. The effects of lecithin liposomes on the binding parameter of quercetin and rutin to Hb (binding constants were 6.02×10^5 and 2.34×10^5 M^{-1}, respectively) were also investigated. The difference of the structure characteristics between quercetin and rutin exhibited a significant effect on their binding affinity for Hb.

Chaudhuri *et al.* have reported the binding of robinetin (a bioflavonoid with remarkable 'two color' intrinsic fluorescence properties), with egg yolk phosphatidylcholine (EYPC) liposomes and normal human hemoglobin A (HbA) using steady state and time resolved fluorescence spectroscopy [158]. Distinctive fluorescence signatures obtained for robinetin indicated its partitioning (K_p= 8.65×10^4) into the hydrophobic core of the membrane lipid bilayer. The HbA–robinetin interaction was examined by using both robinetin fluorescence and flavonoid induced quenching of the protein fluorescence. Specific interaction with HbA was confirmed from higher bimolecular quenching constant, K_q compared to diffusion controlled limit; closely matched values of Stern-Volmer quenching constant and binding constant and $\tau_0/\tau=1$ (where τ_0 and τ are the unquenched and quenched tryptophan fluorescence lifetimes, respectively).

Binding characteristics of apigenin and genistein to Hb were analyzed by fluorescence spectroscopy, CD and UV–vis absorption spectroscopy by Yuan *et al.* [159]. The spectral data revealed that both apigenin and genistein exhibited the strong fluorescence quenching of Hb by static quenching mechanism. The binding site, n close to unity suggested that there was a single binding site in Hb for apigenin and genistein. The results of synchronous fluorescence showed that the microenvironment around Tyr residues of Hb had a slight trend of polarity decreasing, but the polarity around Trp residues increased upon the addition of apigenin. The CD results indicated that the apigenin and genistein did not cause any changes in secondary structure of Hb. The binding distance r between Trp37, and apigenin or genistein was calculated to be 3.4/3.32 nm according to the FÖrster theory of resonance energy transfer. The binding affinity of genistein was observed to higher when compared to that of apigenin for Hb.

5.5. Binding of Anthocyanins to Proteins and DNA

Wiese *et al.* have studied protein interactions with cyanidin-3-glucoside (Cy3glc) and its influence on α-amylase activity [160]. Cy3glc with HSA, α-amylase, saliva, pepsin, casein and α-lactoglobulin HSA and casein gave relatively high association constants, K_A, of around 77×10^3 L mol^{-1} (7.0 pH, 37^0C), whereas other proteins displayed the lower affinity [(26 – 37)x10^3 L mol^{-1}] towards Cy3glc. There was a slight change in the α-helicity and β-strand and β-turn (CD). Anthocyanins are structurally very closely-related to several potent intercalators. Anthocyanins such as malvidin, cyanidin, peonidin, petunidin and delphinidin have been reported [161] to be useful to stabilize DNA triple-helical complexes, although relatively weak. Anthocyanins were identified from five purple sweet potato (*Ipomoea batatas* L.) varieties by high-performance liquid chromatography-tandem mass spectrometry (HPLC-MS/MS) to assess their calf thymus DNA-binding ability *in vitro* [162]. The acylated groups influenced the ability of the interaction with DNA. It was found that the anthocyanins from purple sweet potato with more acylated groups in sorphorose exhibited a stronger binding ability with DNA.

Webb *et al.* observed that anthocyanidins such as cyanidin chloride, cyanidin 3-*O*-glucoside, cyanidin 3,5-*O*-diglucoside, malvidin 3-*O*-glucoside and luteolinidin chloride were capable of inducing significant oxidative DNA strand cleavage (nicking) in the presence of 1 mM dithiothreito [163]. The interaction of gliadins (wheat protein) with some anthocyanins (*e.g.* myrtillin, malvin, keracyanin, callistephin) and anthocyanidins (*e.g.* delphinidin, pelargonidin, cyanidin) was studied by Mazzaracchio *et al.* at pH condition of the stomach employing NMR, FT-IR and UV–vis spectroscopic methods [164]. Binding mode and binding parameters were determined. From these results, it was proposed that the anthocyanins were located along the polypeptide chains of gliadins (minors protein along with gluten and glutelin in wheat) in a generical molecular interaction between the two moieties.

A study of the inter-reaction of anthocyanins and DNA was carried out by Sarma and Sharma [165]. They proposed the formation of a cyanidin–DNA copigmentation complex that displayed a 15–20 nm bathochromic shift due to interaction between cyanin and calf thymus DNA (ct-DNA). Exposure of either cyanidin or ct-DNA individually to hydroxyl radicals (OH) obtained in a Fenton reaction caused severe oxidative damage. However, formation of the cyanin-DNA complex prior to exposure to OH protected both the cyanidin and ct-DNA from oxidative damage. Based on the above results, they suggested that cyanidin-DNA copigmentation might be a possible defense mechanism against the oxidative damage of DNA and might have their *in vivo* physiological functions attributable to the antioxidant ability of anthocyanins.

The proanthocyanidins or condensed tannins, phenolic polymers which are synthesized by many plants, characteristically bind and precipitate proteins. The specificity of the interaction was investigated using a competitive binding assay to compare directly the affinities of various proteins and synthetic polymers for the tannin obtained from *Sorghum bicolor* (Linn.) [166]. At pH 4.9, the relative affinities range over more than 4 orders of magnitude, indicating that this proanthocyanidin interacted quite selectively with protein and protein-like polymer. The affinity for tannins was found to be an inverse function of the size of the polymer, and peptides with less than six residues interacted very weakly with tannin. Proteins were precipitated by proanthocyanidins most efficiently at pH values near their isoelectric points. Proline-rich proteins and polymers have very high affinities for tannin. Tightly coiled globular proteins have much lower affinities for tannin.

Table 4. Identification of anthocyanins by HPLC–MS in red wine extract

Comp no.	t_R (min)	Peak area (%)	Mole-cular ion, *m/z*	Fragment ions (*m/z*)	Compound name	Ref.
1	19.6 - 20.4	5.3	465 781	303 619	delp-3-*O*-gluc, mal-delp-3-*O*-gluc-cat	181
2	21.3 - 21.6	4.0	579	---	peon-3-*O* -gluc-4-vinylphen	
3	23.8 - 25.2	4.5	971	---	mal-3-*O*-gluc-pet-3-O-gluc	
4	27.8 - 29.3	53.2	493	331	mal-3-*O*-gluc	
5	34.9 - 35.3	5.9	1104	---	mal-3-*O*-gluc-delp-6-*O*-coum-3-*O*-gluc	
6	38.1 - 38.3	1.6	805	---	mal-3-*O*-gluc-4-vinyl-epic	
7	39.4 - 40.9	9.1	535 927 985	331 --- 823	mal-6-*O*-ace-3-*O*-gluc mal-6-*O*-coum-3-*O*-gluc–cat mal-3-*O*-gluc(dimmer)	
8	46.3 - 46.9	1.5	708	---	mal-6-*O*-*p*-coum-3-*O*-gluc-pyr	
9	47.7 - 48.0	2.1	1116	---	mal-3-*O*-gluc-pet-6-*O*-coum-3-*O*-gluc(dimer)	
10	48.3 - 48.9	3.9	639	331	mal-6-*O*-*p*-coum-3-*O*-gluc	
11	50.3 - 50.6	6.6	609	447	mal-3-*O*-gluc-4-vinylphen	
12	51.2	---	847	685	mal-6-*O*-ace-3-*O*-gluc-epic	
13	51.7 - 52.0	1.6	839	677	mal-6-*O*-ace-3-*O*-gluc–gallocat	
14	53.2 - 53.4	0.7	655 926 1094	331 --- ---	mal-3-*O*-digluc mal-6-*O*-ace-3-*O*-gluc-4-vinyl-epic mal-6-*O*-ace-3-*O*-gluc-4-vinyl-diepic	

6. Analysis of Anthocyanins and Related Compounds

Paper chromatographic methods were developed for the assay of flavonoids in the 1950s and 1960s [167, 168]. These techniques were replaced by thin-layer chromatography (TLC) in the 1970s providing a low-cost and useful method for the simultaneous analysis of numerous samples [169]. TLC is still in common use for preparative separations [170] and as a rapid low-cost screening method for the determination of flavonoid class of compounds present in fruits [171] and honey [172] since they are hydrophilic in nature.

Table 5. Identification of anthocyanins in a mixture of *Vaccinium angustifolium* and *V. Myrtilloides* Berries by employing HPLC-DAD

1	16.8	6.5	465	303	del-3-gala	182
2	20.6	8.2	465	303	delp-3-gluc	
3	22.0	7.6	465	303	delp-3-hex	
4	24.3	8.8	465	303	delp-3-hex	
5	26.6	4.8	449	287	cya-3-gala	
9	28.1	5.6	335	303	delp-3-ara	
11	29.6	4.8	449	287	cya-3-gluc	
13	31.8	5.5	479	317	pet hex	
14	32.9	2.9	419	287	cya-3-ara	
16	34.9	6.7	479	317	pet hex	
17	36.7	2.6	463	301	peo-3- gala	
c	39.8	5.1	493	331	mal-3-hex	
19	41.4	3.1	463	301	peo-3-gluc	
21	42.6	5.4	493	331	mal hex	
C	45.8	8.4	493	331	mal hex	
22	50.0	2.4	493	331	mal-3-ara	
23	52.2	1.0	535	465/303	delp-oxa-hex	
28	53.5	0.8	535	465/303	delp-oxa-hex	
33	55.9	0.7	515	303	------	
34	57.0	1.2	507	449/287	cya-3-prop-gala	
35	64.3	1.4	535	449/287	cya-3-malo-gala	
38	65.3	0.5	535	479/317	pet-3-prop-gala	
41	67.2	0.8	535	479/317	pet-3-prop-hex	
43	70.2	1.0	521	----	pet-3-ace-gluc	
44	73.2	1.5	535	433/301	peo-3-succ-ara	
46	73.8	0.8	535	493/331	peo-3-oxa-gala	
48	74.6	1.6	535	493/331	mal-3-ace-hex	
49	75.2	0.7	535	493/331	mal-3-ace-hex	

Table 6. Identification of anthocyanins in herbaceous peony cultivars

1	18.6	---	611	449	cya-3,5-di-*O*-gluc	183
2	22.5	---	595	433	pel-3,5-di-*O*-gluc	
3	24.8	---	449	287	cya-3-gluc	
4	27.5	---	625	463	peo-3,5-di-*O*-gluc	
5	35.4	---	463	301	peo-3-gluc	

Gas chromatography (GC) analysis with mass spectrometric detection has been applied for the analysis of flavonols [173, 174]. However, it has limited applicability in the analysis of flavonoids and other phenolics because an extra step is required to ensure the volatility of the phenolics [170]. High-performance liquid chromatography (HPLC) has been the most widely employed chromatographic technique in anthocyanins analysis [175]. Advantages of this technique are the improved resolution of flavonoid mixtures compared to other chromatographic techniques, the ability to obtain both qualitative and accurate quantitative information in one operation, and the great rapidity of analysis [176]. The solvent systems used in HPLC usually include binary gradient elutions using solvents of aqueous acetic, formic or phosphoric acids with methanol or acetonitrile as an organic modifier. The pH and ionic strength of the mobile phase are known to influence the retention of anthocyanins or phenolics in the column depending on protonation, dissociation or a partial dissociation [177]. A change in pH which increases the ionisation of a sample could reduce the retention in a reversed-phase separation. Thus, small amounts of acetic (2– 5%), formic, phosphoric or trifluoroacetic acid (0.1%) are included in the solvent system to suppress ionisation of phenolic and carboxylic groups and hence to improve resolution and reproducibility of chromatographic runs.

Table 7. Anthocyanins profile of individual blueberry cultivars

1	12.19	---	465	303	delp-3-gala	184
2	13.87	---	465	303	delp-3-gluc	
3	15.03	---	449	287	cya-3-gala	
4	16.45	---	435	303	delp-3-ara	
5	17.69	---	449	287	cya-3-gluc	
6	19.21	---	479	317	pet-3-gala	
7	20.03	---	419	287	cya-3-ara	
8	22.14	---	479	317	pet-3-gluc	
9	23.28	---	463	301	peo-3-gala	
10	25.36	---	449	317	pet-3-ara	
11	27.15	---	463 493	301 331	peo-3-gluc mal-3-gala	
12	30.17	---	433 493	331 331	peo-3-ara mal-3-gluc	
13	34.12	---	463	331	mal-3-ara	
14	41.12	---	521	303	Unidentified	

Table 8. Identification of anthocyanins isolated from black rice

1	20.27	---	611.4	287.2	cya-3,5-di-*O*-gluc	185
2	23.08	---	449.3	287.2	cya-3-gluc	
3	24.69	---	595.4	287.2	cya-3-rut	
4	29.48	---	493.0	301.0	peo-3-gluc	

Table 9. Identification of flavanol-anthocyanins adducts in pressure extracted Pomegranate (*Punica granatum* L.) Juice by HPLC-DAD-ESI/MS^n

1	7.10	---	931	769,607	(epi)gcat-delp -3,5-dihex	186
5	15.39	---	915	753,591	(epi)cat-delp-3,5-dihex	
3	13.57	---	769	607	(epi)gcat-delp-3-hex	
7	17.27	---	931	769,607	(epi)gcat-delp-3,5-dihex	
11	18.50	---	769	607	(epi)gcat-delp-3-hex	
12	18.70	---	899	737,575	(epi)afz-delp-3,5-dihex	
14	19.10	---	753	591	(epi)cat-delp-3-hex	
18	20.86	---	915	753,591	(epi)cat-delp-3,5-dihex	
21	21.81	---	753	591	(epi)cat-delp-3-hex	
22	22.52	---	737	575	(epi)afz-delp-3-hex	
28	24.86	---	899	737,575	(epi)afz-delp-3,5-dihe	
31	25.23	---	737	575	(epi)afz-delp-3-hex	
2	10.97	---	915	573, 591	(epi)gcat-cya-3,5-dihex	
6	15.86	---	753	591	(epi)gcat-cya-3-hex	
9	17.77	---	899	737,557	(epi)cat-cyd-3,5-dihex	
10	18.20	---	915	753,591	(epi)gcat-cya-3,5-dihex	
13	18.84	---	753	591	(epi)gcat-cya-3-hex	
19	20.91	---	737	575	(epi)cat-cya-3-hex	
20	21.43	---	883	721,559	(epi)afz-cya-3,5-dihex	
24	22.85	---	899	737,575	(epi)cat-cya-3,5-dihex	
25	23.31	---	737	575	(epi)cat-cya-3-hex	
29	24.88	---	721	559	(epi)afz-cya-3-hex	
33	27.35	---	883	721,559	(epi)afz-cya-3,5-dihex	
34	27.77	---	721	559	(epi)afz-cya-3-hex	
4	14.8	---	899	737,575	(epi)gcat-pel-3,5-dihex	
8	17.63	---	737	575	(epi)gcat-pel-3-hex	
15	19.13	---	899	737,575	(epi)gcat-pel-3,5-dihex	
16	19.49	---	883	721,559	(epi)cat-pel-3,5-dihex	
17	19.85	---	737	575	(epi)gcat-pel-3-hex	
23	22.70	---	721	559	(epi)cat-pel-3-hex	
26	23.69	---	867	705,543	(epi)afz-pel-3,5-dihex	
27	24.82	---	883	721,559	(epi)cat-pel-3,5-dihex	
30	25.07	---	721	559	(epi)cat-pel-3-hex	
32	27.07	---	705	543	(epi)afz-pel-3-hex	
35	29.99	---	705	543	(epi)afz-pel-3-hex	

Table 10. Identification of flavonoids in aqueous spinach extract using positive electrospray ionisation tandem quadrupole mass spectrometry

1	8.51	---	303	257,153	Quer	187
2	16.05	---	287	213,153	Kaem	
3	16.92	---	271	153	Apigenin	
4	---	---	287	161,153	Luteolin	
5	---	---	319	245, 153	Myricetin	

Table 11. Comparative study of the flavonoids of some *Verbena* species cultivated in Egypt by using high-performance liquid chromatography coupled with ultraviolet spectroscopy and atmospheric pressure chemical ionization mass spectrometry

1	----	----	449	---	Orientin	188
2	---	---	433	---	Vitexin	
3	---	---	433	---	Isovitexin	
4	---	---	449	---	Luteolin-7-*O*-glucoside	
5	---	---	443	---	Apigenin-7-*O*-glucoside	
6	---	---	287	---	Luteolin	
7	---	---	286	---	Chrysoeriol	
8	---	---	271	---	Apigenin	

Table 12. Chromatographic, spectroscopic and mass spectrometric characteristics of the anthocyanins from *Berberis boliviana* L., obtained by HPLC-PDA

1	6.9	24.1	465	303	delp-3-gluc	189
2	7.1	2.24	611	303	delp -3-rut	
3	8.0	9.49	449	287	cya-3-gluc	
4	8.4	3.45	595	287	cya-3-rut	
5	8.9	24.4	479	317	pet-3-gluc	
6	9.3	6.86	625	317	pet-3-rut	
7	10.7	0.95	463	301	peo-3-gluc	
8	11.1	0.80	609	301	peo-3-rut	
9	1.7	22.1	493	331	mal-3-gluc	
10	12.3	5.61	639	331	mal-3-rut	

Table 13. Identification of flavonoids in *Epimedium wushanese* by HPLC-DAD-ESI-MS

1	8.5	---	463	301/271	quer-3-*O*-gluc	190
2	10.6	---	447	285/255	kaem-3-*O*-gluc	
3	15.1	---	807	645/353	desmethylicar-3-*O*-di-rha-7-*O*-gluc	
4	15.4	---	661	353/298	Desmethylicar-3-*O*-rha-7-*O*-gluc	
5	20.7	---	515	353/335	desmethylicar-7-*O*-gluc	
6	21.7	---	837	675/367	icar-3-*O*-rha-gluc-7-*O*-gluc	
7	23.8	---	821	659/367	icar-3-*O*-di-rha-7-*O*-gluc	
8	25.3	---	821	659/337	icar-3-*O*-di-rha-7-*O*-gluc	
9	27.0	---	659	367/337	icar-7-*O*-di-rha	
10	28.5	---	819	657/513	icar-7-*O*-gluc-3-*O*-rha-furanoic acid	190
11	35.2	---	691	367	icar-7-*O*-di-gluc	
12	36.6	---	631	353/296	Desmethylicar-3-*O*-rha-xyl	
13	50.1	---	657	365/650	anhydroicar-3-*O*-di-rha	
14	51.4	---	821	659/367	isomer of epi	

Table 14. Polyphenolic compounds identified by chromatography and mass spectrometry in *Vitis vinifera* grapes

					Anthocyanins	
1	14.22	---	465	303	delp-3-gluc	
2	17.85	---	449	287	cya-3-gluc	
3	20.31	---	479	317	pet-3-gluc	
4	21.58	---	463	301	peo-3-gluc	
5	27.01	---	493	331	mal-3-gluc	
6	29.75	---	507	303/465	delp-3-*O*-(6-*O*-ace)-gluc	
7	33.42	---	491	287	cya-3-*O*-(6-*O*-ace)-gluc	
8	36.50	---	521	317/479	pet-3-*O*-(6-*O*-ace)-gluc	
9	39.74	---	505	301	peo-3-*O*-(6-*O*-ace)-gluc	
10	41.12	---	773	303/465	delp-3-*O*-(6-*O*-coum)-gluc	
11	42.40	---	535	331/496	mal-3-*O*-(6-*O*-ace)-gluc	
12	45.23	---	795	287/449	cya-3-*O*-(6-*O*-coum)-gluc	
13	49.98	---	625	317/479	pet-3-*O*-(6-*O*-coum)-gluc	
14	48.78	---	655	331/493	mal-3-*O*-(6-*O*-caff)-gluc	
15	51.35	---	609	301/463	peo-3-*O*-(6-*O*-coum)-gluc	191
16	52.50	---	639	331/493	mal-3-*O*-(6-*O*-coum)-gluc	
					Flavanols	
17	6.95	---	578	577	Procyanidin B1	
18	9.63	---	290	290	Catechin	
19	10.34	---	578	577	Procyanidin B2	
20	16.34	---	290	289	Epicatechin	
21	23.30	---	442	290	Epicatechin gallate	
22	35.96	---	316	315	Isorhamnetin	
					Flavonols	
23	2360	---	480	318	myricetin 3-*O*-glucoside	
24	29.42	---	610	---	rutin	
25	30.65	---	478	302	quercetin 3-*O*-glucuronide	
26	31.78	---	464	302	quercetin 3-*O*-glucoside	
27	38.34	---	449	287	kaempferol 3-*O*-glucoside	
28	39.95	---	479	317	isorhamnetin 3-*O*-glucoside	

Table 15. Detection of anthocyanins/phenolics pigmented potatoes by HPLC/DAD/MS

1	---	---	801	331	mal-3-*O*-rut-5-*O*-gluc	
2	---	---	949	317	pet-3-*O*-caf-rut-5-*O*-gluc	
3	---	---	919	303	delp-3-*O*-*p*-coum rut-5-*O*-gluc	
4	---	---	963	331	mal-3-*O*-caf-rut-5-*O*-gluc	
5	---	---	933	371	pet-3-*O*-*p*-coum-rut-5-*O*-gluc	
6	---	---	963	317	pet 3-*O*-ferul-rut-5-*O*-gluc	192
7	---	---	947	331	mal 3-*O*-*p*-coum-rut-5-*O*-gluc	
8	---	---	977	331	mal 3-*O*-ferul-rut-5-*O*-gluc	
9	---	---	741	271	pel 3-*O*-rut-5-*O*-gluc	
10	---	---	579	271	pel 3-*O*-rut	
11	---	---	---	271	pel derivative	
12	---	---	903	271	pel 3-*O*-caf-rut-5-*O*-glu	

Table 15. (Continued)

13	---	---	887	271	pel-3-*O*-*cis*-*p*-coum-rut-5-*O*-gluc	
14	---	---	---	301	pel derivative	
15	---	---	887	271	pel-3-*O*-*p*-coum-rut-5-*O*-gluc	
16	---	---	917	301	peo 3-*O*-*p*-coum-rut-5-*O*-glu	
17	---	---	917	271	pel 3-*O*-ferul rut-5-*O*-gluc	
18	---	---	947	301	peo 3-*O*-ferul-rut-5-*O*-gluc	
19	---	---	725	271	pel 3-*O*-*p*-coum-rut	
20	---	---	755	271	pel 3-*O*-ferul-rut	
21	---	---	609	301	peo 3-*O*-rut	
22	---	---	---	301	peo derivative	
23	---	---	771	301	peo 3-*O*-rut-5-*O*-gluc	
24	---	---	787	317	pet 3-*O*-rut-5-*O*-gluc	

Table 16. Anthocyanins found in the red petals of Korean edible rose (*Rosa hybrida cv.* Noblered) by HPLC-DAD-ESI/MS

1	7.8	85	611	449/287	cya-3,5-di-*O*-gluc	193
2	9.5	11	595	443/271	pel-3,5-di-*O*-gluc	

Table 17. Anthocyanins identified in young and aged red wine extracts by HPLC–MS/MS

1	17.5-17.8	---	465	303/275	delp-3-*O*-gluc	
2	22.3-23.3	---	479	317	pet-3-*O*-gluc	
3	25.2-26.2	---	463	301/286	peo-3-*O*-gluc	
4	26.4-27.7	---	493	331/299	mal-3-*O*-gluc	
5	29.9-30.8	---	507	303	delp-3-*O*-ace-gluc	
6	32.1-32.4	---	535	331	mal-3-*O*-ace-gluc	
7	32.5-32.7	---	611	303	delp-3-*O*-coum-gluc	
8	32.8-32.9	---	595	287	cya-3-*O*-coum-gluc	194
9	33.1-33.3	---	625	317	pet-3-*O*-coum-gluc	
10	33.3-33.4	---	609	301	peo-3-*O*-coum-gluc	
11	33.4-33.5	---	639	331	mal-3-*O*-coum-gluc	
12	33.6-33.8	---	627	303	delp 3-*O*-caf-gluc	
13	33.9	---	547	385	Visitin A of pet-3-*O*-gluc	
14	33.9-34.0	---	531	369	Visitin A of peo-3-*O*-gluc	
15	34.1-34.8	---	561	399	Visitin A	
16	34.9-35.0	---	707	399	Visitin A of mal-3-*O*-coum-gluc	
17	35.1-35.6	---	581	419	Adduct 4-vinylphenol-delp-3-*O*-gluc	
18	35.8-35.9	---	579	417	Adduct 4-vinylphenol-peo-3-*O*-gluc	
19	36.0-36.3	---	609	447	Adduct 4-vinylphenol-mal-3-*O*-gluc	
20	---	---	651	447	Adduct 4-vinylphenol-mal-3-*O*-ace-gluc	
21	40.0-55.0	---	727	419	Adduct 4-vinylphenol-delp-3-*O*-ace-gluc	
22	40.0-55.0	---	741	433	Adduct 4-vinylphenol-pet-3-*O*-coum-gluc	
23	40.0-55.0	---	625	301	peo-3,7-*O*-di-gluc	
24	40.0-55.0	---	551	317	pet-3-*O*-gluc + lactic acid	194
25	40.0-55.0	---	565	331	mal-3-*O*-gluc + lactic acid	
26	40.0-55.0	---	753	591	Dimer delp-3-*O*-gluc	
27	40.0-55.0	---	783	621	Adduct galloc-pet-3-*O*-gluc	
28	40.0-55.0	---	767	605	Adduct galloc-peo-3-*O*-gluc	
29	40.0-55.0	---	797	635	Adduct epic-mal-3-*O*-gluc	
30	40.0-55.0	---	943	---	Adduct galloc-mal-3-*O*-gluc	
31	40.0-55.0	---	751	589	Adduct epic-peo-3-*O*-gluc	
32	40.0-55.0	---	781	619	Adduct cat-mal-3-*O*-gluc	

Table 18. Identified anthocyanins in blue-berried honeysuckle juices by HPLC

7	12.9	---	287	---	cya-3,5-di-gluc	195
9	16.5	---	449/287	---	cya-3-gala	
10	17.9	---	449/287	---	cya-3-gluc	
11	19.4	---	419/287	---	cya-3-arabinoside	
12	19.5	---	595/287	---	cya3-rut	
13	21.9	---	463/301	---	peo-3-gluc	
14	24.6	---	419/287	---	cya-3-xyloside	
15	27.1	---	465/303/487	----	quercetin-3-gala	
16	27.9		611		quercetin-3-rut	
17	32.1		303		Quercetin	

Table 19. Determination of anthocyanins and related compounds in berry samples

Berry type/ Source	Method used for Analysis	Stationary Phase	Mobile Phase	Anthocynin Concentration	Ref
Lingonberry Cranberry	HPLC-DAD HPLC-EC	ODS-3 (4.0 X 150 mm, 3 μm)	A: 50 mM H_3P $_4$, pH- 10 B: ACN	**Quercetin:** 10 mg 100 g^{-1} FW **Myricetin:** 6.9 mg 100 g^{-1} FW **Quercetin:** 10. 4 mg 100 g^{-1} FW	196
Strawberry (cvs. Earliglow and Kent/Allstar and Honeoye)	HPLC-DAD (510 nm)	Novapak C18 (150 X 3.9 mm, 4μm)	A: 2.5% FA in H_20 B: ACN	**cya-3-gluc:** 4.5–34 μg g^{-1} FW, 561–1153 μg g^{-1} DW, 18.5 - 45.8 μg g^{-1} juice **pel-3-gluc:** 275.6–847.1 μg g^{-1} FW, 2124–3669 μg g^{-1} DW, 291.3– 945.1 μg g^{-1} juice **cya-3-gluc-succinate:** 1.31– 5.62 μg g^{-1} FW, 37–75 μg g^{-1} DW, 6.1–19.8 μg g^{-1} juice **pel-3-rut:** 24.7–50.9 μg g^{-1} juice **pel-3-gluc-succ:** 45.6– 95.9 μg g^{-1} FW, 323–681 μg g^{-1} DW, 62.2–224.5 μg g^{-1} juice. according to cultivar and soil supplement or growing temperature	197-199
Strawberry (cvs. Camarosa, Dorit, Chandler, Osmanli)	HPLC-DAD	Nucleosil C18 Analytical column (150 X 4.6 mm, 5μm)	A: 2.5% FA in H_20 B: 2.5% FA in CAN	**cya-3-gluc:** 4.5–17.5 9 μg g^{-1} FW **pel-3-gluc:** 53.3–441.0 9 μg g^{-1} FW	200
Strawberry (cvs. Honeoye, Jonsok and Polka)	HPLC	Li ChroCART Purospher RP-18e (125 X 3 mm, 5μm)	A: 5% FA in H_20 B: ACN	**cya-3-gluc:** 22 μg g^{-1} FW **pel-3-gluc:** 248 μg g^{-1} FW; **pel-3-rut:** 12 μg g^{-1} FW; **pel-3-mal-gluc:** 25 μg g^{-1} FW **pel-3-succ:** 45 μg g^{-1} FW	201
Bilberry	HPLC-DAD	Beckman Ultrasphere ODS (250 X 4.6 mm, 5 μm)	A: 0.4% TFA in H_2O B: 0.4% TFA in can	**delp:** 6.93 % **cya:** 7.65 % **pet :** 2.3 % **peo :** 1.81 % **mal:** 2.89 %	202
Strawberry (cvs. Senga Sengana, Dukat, Elkat, Selva, Elsanta and Kent)	HPLC-DAD	Li ChroCART Purospher RP-18e (125 X 3 mm, 5μm)	A: 4.5% FA B: 80% ACN and 20% solvent A	**cya-3-gluc:** 6.3–27.2 μg g^{-1} FW **pel-3-gluc:** 202.8–331.9 μg g^{-1} FW **pel-3-ara:** 2.3–8.4 μg g^{-1} FW	203

Table 19. (Continued)

Berry type/ Source	Method used for Analysis	Stationary Phase	Mobile Phase	Anthocynin Concentration	Ref
Black Sorghum	HPLC	ODS (250 X 4.6 mm, 2.5 μm)	A: 10% FA in H_2O B: (5:4:1) ACN:H_2O:FA	**TA:** 4.0 - 9.8 mg g^{-1} Black sorghum brans were a very good source of anthocyanins (4.0 - 9.8 mg g^{-1}) relative to the commercially available fruits and vegitables (mainly contains berries) (0.2 - 10 mg g^{-1}).	204
Pigmented potato (cvs. Hermanns Blaue, Highland Burgundy Red, Shetland Black, and Vitelotte	HPLC-DAD	Synergi MAX-RP (250 X 4.6 mm)	H_2O: ACN : FA A: 87:3:10 B: 40:50:10 (v/v/v)	Hermanns Blaue : **pet-3-rut-5-glc:** 30 mg : **pet-3-coum-rut-5-glc:** 38 mg Highland : **pel-3-rut-5-glc:** 35 mg Burgundy Red **: pel-3-coum-rut-5-glc:** 52mg Shetland Black: **peo-3-rut-5-glc:** 46 mg **pet-3-coum-rut-5-glc:**31 mg, **pet-3-coum-rut-5-glc:**5 mg, **peo-3-coum-rut-5-glc:**0.6 mg Vitelotte: **mal-3-rut-5-glc:** 57 mg, **pet-3-coum-rut-5-glc:** 5 mg **mal-3-coum-rut-5-glc:**11 mg	205
Ruscus *aculeatus* *(Red Berry)*	HPLC-DAD	SS Wakosil C18 (150 X 4.6 mm, 5 μm)	A: 0.1 % TFA in H_2O B: 0.1 % TFA in ACN	**pel- 3-*O*- rut**: 46 mg g^{-1} **pel-3-*O*- gluc**: 8.6 mg g^{-1} **pel-3-*O*- *p*-cougluc**: 6.0 mg g^{-1}	206
Blackcurrant (cv. Öjebyn)	HPLC-DAD	Hypersil ODS (60 X 4.6 mm, 3μm)	A: 5% FA in H_2O (v/v) B: ACN	**delp-3-gluc:** 303–391 μg g^{-1} FW **delp-3-rut:** 808.69–1174.39 μg g^{-1} FW **cya-3-gluc:** 151.31– 267.77 μg g^{-1} FW **cya-3-rut:** 844.69– 947.24 μg g^{-1} FW according to the cultivation system	207
Strawberry (cvs. Totem, and Puget)	HPLC	Prodigy ODS-3 (250 X 4.6 mm, 5 μm)	A: ACN B: 70 mM KH_2PO_4 with H_3PO_4	**Flesh** (freeze-dried) in % | **cya-3-gluc** | **pg-3-gluc** | **pg-3-rut** | **pg 3-*O*-gluc-mal** | **cyd 3-*O*-gluc-mal** Totem | 0.9 | 86.4 | 2.5 | 9.5 | -- Puget | 4.2 | 91.3 | 2.9 | -- | -- **Flesh** (Puree) in % Totem | 1.1 | 85.8 | 2.1 | 10.1 | -- Puget: | 3.8 | 90.9 | 3.5 | -- | -- **achenes**(freeze-dried) in % Totem | 39.3 | 36.1 | -- | 6.1 | 18.6 Puget: | 43.6 | 34.8 | -- | -- | 21.6 **achenes** (Puree) in % Totem | 41.1 | 37.3 | -- | 7.5 | 13.7 Puget: | 37.0 | 51.7 | -- | -- | 11.3	208
Strawberry (cv. Senga Sengana)	HPLC-DAD (510 nm)	Betasil C 18 (250 X 2.1 mm, 5 μm)	A: phosphoric acid/acetic acid/H_2O (1:10:89, v/v/v) B: ACN	**cya-3-gluc:** 19.0 μg g^{-1} FW **pel-3-gluc:** 400 μg g^{-1} FW **pel-3-rut:** 16.3 μg g^{-1} FW **5-carboxypyrano-pg-3-gluc:** 2.4 μg g^{-1} FW **pel-3-mlgluc:** 65.9 μg g^{-1} FW **pel-3-acetylgluc:** 3.9 μg g^{-1} FW	209
Strawberry (cvs. Camarosa, Ventana, Aromas, Diamante and Medina)	HPLC-DAD	C-18 Novapack (300 X 3.9 mm, 5μm)	A:FAN/FA/ H_2O(3:10:87) B: ACN/FA/ H_2O(50:10:40)	**pel- der1:** 0.18–1.0 μg g^{-1} FW **cya-3-gluc:** 0.64–5.30 μg g^{-1} FW **pel-3-gluc:** 73.69–165.66 μg g^{-1} FW **pel-3-rut:** 3.24–23.92 μg g^{-1} FW; **pel der2:** 0.18–1.30 μg g^{-1} FW; **pel-ace-gluc:** 0.55– 3.62 μg g^{-1} FW	210

Berry type/ Source	Method used for Analysis	Stationary Phase	Mobile Phase	Anthocynin Concentration	Ref
Blackcurrant & Strawberry (cvs. not specified)	HPLC-DAD	Li ChroCART Purospher RP-18e (125 X 3 mm, 5 m)	A: 8.5% FA in H_2O B:ACN/ MeOH/ (85:15, v/v)	**cya (as sum of all derivatives):** 13– 50 μg g^{-1} FW in strawberries and 993 μg g^{-1} FW in blackcurrants **pel (as sum of all derivatives):** 294–510 μg g^{-1} FW in strawberries and 1017 μg g^{-1} FW in blackcurrants	211
Strawberry (cvs. Eris, udnew,Oso Grande, arisma, & Camarosa)	HPLC-DAD (520 nm)	AQUA C18 (150 X 4.6 mm, 5 μm)	A: 0.1% TFA/ in H_2O B : ACN	**cya-3-gluc:** 10–41 μg g^{-1} FW **pel-3-gluc:** 162–468 μg g^{-1} FW	212
Strawberry (cv. Elsanta)	HPLC-DAD (520 nm)	Alltech Allosphere ODS-1 (250 X 4.6 mm, 5 μm)	A: 1% Phosphoric acid and 10% acetic acid (v/v) in H_2O B: CAN	**cya-3-gluc:** 2.15 μg g^{-1} FW **pel-gluc deriv:** 34.99 μg g^{-1} FE **pel-3-gluc:** 125.99 μg g^{-1} FW	213
Black-currant (17 UK-grown cultivars)	HPLC-DAD (520 nm)	Alltech Allosphere ODS-1 (250 X 4.6 mm, 5 μm)	A: 1% H_3PO_4 10% AC (v/v) in H_2O B: ACN	**cya-3-gluc:** 68.81 μg g^{-1} FW **cya-3-rut:** 546.21 μg g^{-1} FW **delp-3-gluc:** 133.06 μg g^{-1} FW **delp-3-rut:** 605.23 μg g^{-1} FW	214
Strawberry (cvs.Alba, Irma, Patty, Adria, and Sveva and four advance sec.)	HPLC-DAD	C18 (250 X 4.6 mm)	A: 0.1% TFA in H_2O B: 0.1% TFA in ACN	**cya-3-gluc:** 0–11.15 μg g^{-1} FW **pel-3-gluc:** 95.8–282.3 μg g^{-1} FW **pel-rut:** 0–13.8 μg g^{-1} FW **pel-mal-gluc:** 0–26.8 μg g^{-1} FW	215
Strawberry Elderberry Black carrot (from concentrated juices upon thermal treatment)	HPLC-DAD	C18 Sunfire (250 X 4.6 mm, 5μm)	A: 5% (v/v) FA in water B: ACN	**pel-3-glc:** 1.00 mg $L^{-1\ b}$ **5-Carboxypyrano-pel-gluc:** 2.01mg $L^{-1\ a}$ ***p*-Coumaroylhexose:** 2.92 mg $L^{-1\ a,b}$ **Cya-3-gluc-xyl-5-glc + Cyd-3,5-di-glc:** 2.47 mg $L^{-1\ a}$ **Cya-3-gluc-xyl:** 1.72 mg $L^{-1\ a}$ **Cya-3-gluc:** 1.76 mg $L^{-1\ a,b}$ **Cya-3-gala-xyl-gluc**: 1.74 mg $L^{-1\ b}$ **Cya-3-gala-xyl:** 1.24 mg $L^{-1\ b}$ **Cya-3-gala-xyl-gluc-sinap:** 2.62 mg $L^{-1\ b}$ **Cya-3-gal-xyl-gluc-fer:** 2.52 mg $L^{-1\ a}$ **Cya-3-gal-xyl-gluc-coum:** 2.13 mg $L^{-1\ b}$ [a]without supplemented sugar or ascorbic acid [b]without supplemented sugar or ascorbic acid	216
Strawberry juice	HPLC	Zorbax SB C18, (150 X 4.6 mm, 5 μm)	A: ACN + MeOH B: ACN	**AC:** 53.3 mg 100 ml^{-1} The analysis was done after ozonation of strawberry juice	217
Elderberry (15 Portugal-grown cultivars)	HPLC-PDA - ESI/tandem MS	Spherisorb ODS2 (150 X 2.1 mm, 3 μm)	A:2% aqueous FA B: MeOH	**TP**: 8.9 % GAE, **TF:** 7.5% ECE, **TA:** 7.6 % **cya-3-gluc:** 3.6% **cya-3-sam:** 3.1 % **cya-3- sam-5-gluc:** 1.1%	218

Table 19. (Continued)

Berry type/ Source	Method used for Analysis	Stationary Phase	Mobile Phase	Anthocynin Concentration	Ref
White strawberry (local grown cultvors)	HPLC-DAD HPLC–ESI -MS	C18-RP Luna (250 X 4.6 mm, 5 μm)	A: 1% FA B: ACN	**TP:** fruit: 2.72 g GAE 100 g^{-1} DW leaves: 1.99 g GAE 100 g^{-1} DW rhizome: 1.45g GAE 100 g^{-1} DW **TA:** fruit: 43.6 mg cya-3-gluc equivalents 100 g^{-1} DW, : 4.4 mg cya-3-gluc 100 g^{-1} FW **TF:** fruit: 2 g quercetin equivalents 100 g^{-1} DW leaves: 0.83 g quercetin equivalents 100 g^{-1} DW rhizome: 0.55 g quercetin equivalents 100 g^{-1} DW	219
Strawberry (cv. Camarosa) Bilberry (cv. Santa rosa) Red raspberry (cv. Santa rosa)	HPLC	C18 Spherisorb ODS-1 (250 X 4.6 mm, 5 μm)	A: ACN : H_2O (1:1)with TFA B:0.5 % TFA	Strawberry: **pel-3-gluc:** 100% Bilberry: **delp-3-gluc:** 18%, **cya-3-gluc:** 3%, **pet-3-gulc:** 8%, **mal-3-gulc:** 71 % Red raspberry: **cya- 3-gluc:** 100%	220
Strawberry (cvs. Antea, Clery, & Matis)	HPLC-DAD	XDB-C18 (250 X 4.6 mm, 5 μm)	A: H_2O B:ACN(mp)	**cvs. cya-3-gluc pg-3- gluc pg- der1 pg- der2** (μg g^{-1}) (μg g^{-1}) (μg g^{-1}) (μg g^{-1}) Antea 118.6 2474.5 74.5 Clery -- 3544.9 66.1 1060.9 Matis 84.9 2013.3 57.4 285.5	221

Table 20. Determination of anthocyanins in wine

					cya-3-gluc (mgl^{-1})	peo-3-gluc (mgl^{-1})	mal-3-gluc (mgl^{-1})	
Rioja Rioja Valdepenas Valdepenas Valladolid Navarra	HPLC	Ultrabase C18 (250 X 4.6 mm, 5 μm)	A: ACN B: 10% FA	Rioja	1.8	1.10	12	222
				Rioja	1.4	0.61	5.3	
				Valdepenas	2.8	0.92	9.0	
				Valdepenas	3.1	0.50	4.5	
				Valladolid	1.1	0.52	5.1	
				Navarra	0.54	0.15	3.2	

Table 21. Concentration of flavonoids, flavonols and antocyanins in the skin of four different red grape varieties

<table>
<tr><td rowspan="14">Alfrocheiro
Jaen
Tinta Roriz
Touriga
Nacional</td><td rowspan="14">HPLC-
ECD</td><td rowspan="14">Xbridg
e C18
steel
(150 X
4.6
mm, 5
μM)</td><td rowspan="14">A: (v/v/v)
H_2O:Methanol
:FA
(83:16:1)
B:
H_2O:MeOH
:FA
(68.5:30:1.5)</td><td></td><td>Alfro-
cheiro</td><td>Jaen</td><td>Tinta
Roriz</td><td>Touriga
Nacional</td><td rowspan="14">223</td></tr>
<tr><td colspan="5">Anthocyanins (μg g^{-1})</td></tr>
<tr><td>myrtillin</td><td>142.9</td><td>10.36</td><td>123.8</td><td>691.2</td></tr>
<tr><td>kuromanin</td><td>0.330</td><td>---</td><td>0.170</td><td>0.730</td></tr>
<tr><td>peo-3-gluc</td><td>617.5</td><td>95.17</td><td>418.4</td><td>1761</td></tr>
<tr><td>pet-3-gluc</td><td>139.5</td><td>61.36</td><td>338.2</td><td>542.0</td></tr>
<tr><td colspan="5">Flavan-3-ols(μg g^{-1})</td></tr>
<tr><td>catechin</td><td>11.12</td><td>3.640</td><td>2.920</td><td>7.640</td></tr>
<tr><td colspan="5">Flavonols (μg g^{-1})</td></tr>
<tr><td>rutin</td><td>46.67</td><td>29.11</td><td>78.06</td><td>214.3</td></tr>
<tr><td>myricetin</td><td>44.38</td><td>---</td><td>---</td><td>---</td></tr>
<tr><td>fisetin</td><td>---</td><td>---</td><td>3000</td><td>3.750</td></tr>
<tr><td>quercetin</td><td>---</td><td>---</td><td>---</td><td>---</td></tr>
<tr><td>morin</td><td>---</td><td>---</td><td>---</td><td>---</td></tr>
<tr><td rowspan="8">Cranberry
Pomegranate
Blueberry
Tart cherry
Grape
Black cherry
Concord</td><td rowspan="8">HPLC-
DAD</td><td rowspan="8">(250 X
4.6
mm, 5
μM)</td><td rowspan="8">A: 100% ACN
B: 10% AC:
1%H_3PO_4 in
H_2O(v:v)</td><td>juice</td><td colspan="2">TA(mg 100 ml^{-1} ece of cya)</td><td colspan="2">TA(mg 100 ml^{-1} ece of mal)</td><td rowspan="8">224</td></tr>
<tr><td>Cranberry</td><td colspan="2">1.08</td><td colspan="2">1.86</td></tr>
<tr><td>Pomegranate</td><td colspan="2">43.54</td><td colspan="2">23.63</td></tr>
<tr><td>Blueberry</td><td colspan="2">24.35</td><td colspan="2">9.97</td></tr>
<tr><td>Tart cherry</td><td colspan="2">7.16</td><td colspan="2">4.41</td></tr>
<tr><td>Grape</td><td colspan="2">8.08</td><td colspan="2">4.55</td></tr>
<tr><td>Black cherry</td><td colspan="2">21.67</td><td colspan="2">11.79</td></tr>
<tr><td>Concord</td><td colspan="2">15.95</td><td colspan="2">9.98</td></tr>
<tr><td rowspan="8">Cranberry
Pomegranate
Blueberry
Tart cherry
Grape
Black cherry
Concord</td><td rowspan="8">HPLC-
DAD</td><td rowspan="8">(150 X
2 mm,
4 μM)</td><td rowspan="8">A:AC :TFA
:CAN:H_2O
(10:0.2:5:84.8)
(v/v/v/v)
B:ACN</td><td>juice</td><td colspan="2">TA(mg 100 ml^{-1} ece of cya)</td><td colspan="2">TA(mg 100 ml^{-1} ece of mal)</td><td rowspan="8">224</td></tr>
<tr><td>Cranberry</td><td colspan="2">1.08</td><td colspan="2">1.86</td></tr>
<tr><td>Pomegranate</td><td colspan="2">43.54</td><td colspan="2">23.63</td></tr>
<tr><td>Blueberry</td><td colspan="2">24.35</td><td colspan="2">9.97</td></tr>
<tr><td>Tart cherry</td><td colspan="2">7.16</td><td colspan="2">4.41</td></tr>
<tr><td>Grape</td><td colspan="2">8.08</td><td colspan="2">4.55</td></tr>
<tr><td>Black cherry</td><td colspan="2">21.67</td><td colspan="2">11.79</td></tr>
<tr><td>Concord</td><td colspan="2">15.95</td><td colspan="2">9.98</td></tr>
<tr><td colspan="10">Quantification of Onobrychis viciifolia compounds</td></tr>
<tr><td>Onobrychis
viciifolia</td><td>HPLC-
DAD</td><td>C18
column</td><td>A: H_2O + 5%
FA
B: MeOH</td><td colspan="5">Flavones(mg g^{-1} DW)
Vitexin: 0.048; Unknown flavones: 0.127
Flavan-3-ols(mg g^{-1} DW)
Catechin: 0.100; Epicatechin: 0.261; Gallocatechin: 0.058; Epigallocatechin 0.111; Procyanidin B_2: 0.054; Procyanidin C1: 0.013; Unknown flavanols: 0.201
Flavonols(mg g^{-1} DW)
Kaempferol 3-rutinoside: 1.874
Quercetin 3-rutinoside: 6.147
Isorhamnetin 3-rutinoside: 0.375
Kaempferol 3-rhamnosylrutinoside: 0.285
Quercetin 3-rhamnosylrutinoside: 0.997
Unknown flavonol glycosides: 0.693
Unknown acylated flavonols: 0.110
Isoflavones(mg g^{-1} DW)
Unknown isoflavones 0.034</td><td>225</td></tr>
</table>

Table 22. Major anthocyanin compounds identified and quantified in sour cherry varieties (mg/100 g)

					cya	pel	mal	delp	peo	
Gerema (GE) Maliga emleke (ME) Erdi botermo (EB) Erdi jubileum (EJ) Kantorjanosi (KJ)]	HPLC-DAD	SYMMETRY C18 (4.6 X 50 mm, 5 μM)	A:H_2O:MeOH: ACN B: ACN	EJ	4.03	1.35	8.31	0.40	0.12	226
				EB	6.64	5.98	0.27	0.01	0.01	
				ME	3.41	21.14	1.41	0.16	0.02	
				KJ	4.57	15.42	0.53	0.09	0.01	
				GE	6.10	64.36	2.66	0.52	0.19	

Table 23. Quantitative Analysis of quercetin in *Euphorbia helioscopia* L

roots	HPLC	TC-C18 (250 X 4.6 mm, 5 μM)	A:0.2% MeOH B: MeOH: H_3PO_4(65:35)	1.4179 mg g^{-1}	227
stems				0.0211 mg g^{-1}	
leaves				0.0437 mg g^{-1}	

Table 24. Average amounts of free anthocyanins found in different samples

				delp	cya	pet	pel	peo	mal	
Pineapple juice	HPLC	ODS-3 18 (150 X 4.6 mm, 3 μM)	A:H_2O:FA: ACN (85:10:5) B: (35:10:55)	0.04	0.04	---	---	---	---	228
Cherry				6.42	37.77	1837.14	6.70	12.33	55.07	
Red Cabbage				49.92	1.40	6.03	44.83	69.38	2.60	
Black Carrot juice				80.08	75.73	76.90	0.79	0.29	0.60	
Apricot				0.01	0.02	---	---	0.07	---	
Sour cherry				1111.8	8.10	212.66	1.31	1.88	---	
Red rose				19.30	14.90	3.08	0.44	0.97	---	
Parsley				---	---	0.03	0.02	---	---	
Grape juice				3.37	1.83	---	1.54	3.50	5.25	
Black lettuce				---	0.01	0.02	---	---	0.05	
Whine				2.02	0.10	---	2.26	4.33	23.84	
Carrot				--	0.12	---	---	---	---	
Guelder rose juice				0.05	6.34	0.03	0.01	0.16	---	
Grape				4.18	2.28	---	1.76	4.58	6.53	
Black mulberry				2.40	53.03	108.59	9.22	8.12	---	
Sour cherry juice				2.81	1.39	13.30	0.19	0.12	0.02	
Red plum				---	0.04	0.32	0.08	1.02	---	
Pomegranate juice				0.07	3.15	2.24	0.13	0.16	---	
Cherry laurel				9.60	3.39	70.35	0.70	0.49	---	
Cherry juice				0.28	7.07	962.51	2.99	81.45	0.16	
Pomegranate				248.12	68.90	338.46	10.84	1.82	5.20	

Table 25. Determination of phenolic compounds from grape berry skins using liquid chromatography and spectrophotometry [229]

Spectrophotometric parameters

phenolic compounds	Standard used	$\lambda_{measurement}$ (nm)
Total phenols	Gallic acid in 10% ethanol	280
Flavonols	Quercetin in 95% ethanol	360
Anthocyanins	Malvidin-3-glucoside in 10% ethanol	520

Conditions

Method	Column	Solvent used
HPLC-DAD	4 U MAX-RP 80 A (450 X 4.6 µM)	A: 0.01 M L^{-1} H_3PO_4 B: MeOH

Contents of phenolic compounds (mg kg^{-1} FW) determined by HPLC

Extraction	Maturity	(+)-Catechin	(-)-Epicatechin	Oenin	Rutin	Myricetin	Kaempferol	Quercetin
Method-1	Ripe	62	25	310	2.2	0.34	1.9	3.1
	Over ripe	56	23	290	2.3	0.32	1.6	2.9
Method-2	Ripe	74	35	450	3.5	0.46	2.3	3.2
	Over ripe	68	32	440	2.9	0.42	2.3	3.6

Using spectrophotometry, an arbitrary procedure was followed for determination of phenolic compounds extracted with ethanol/formic acid/water [228].

Contents	Concentration found
Total phenols	791-127 mg kg^{-1}
Flaonols	163-167 mg kg^{-1}
Anthocyanins	2179-2219 mg kg^{-1}

Table 26. Determination of the polyphenol contents in Macedonian grapes by standardized spectrophotometric method [230]

Grape variety	Grape part	TP (mg L^{-1})	TF (mg L^{-1})	TA (mg L^{-1})	TC (mg L^{-1})
Vranec	Pulp	2.58	0.77	0.15	0.15
	Seed	166	18.9	---	25.3
	Skin	48.4	7.01	11.9	2.71
Cabernet Sauvignon	Pulp	1.65	0.65	0.02	0.08
	Seed	113	7.35	---	17.2
	Skin	31.5	23.3	5.67	2.62
Muscat Hamburg	Pulp	1.14	0.39	0.03	0.05
	Seed	135	9.4	---	24.2
	Skin	37.6	2.94	2.74	4.08
Riesling	Pulp	1.06	0.50	---	0.05
	Seed	126	15.8	---	22.1
	Skin	10.2	3.65	---	1.55

Table 27. Total anthocyanin content of pomegranate juice samples from various processing stages [231]

Pomegranate juice	Total CAN content (equivalents of cya-3-gluc, mg/kg)	
	Spectrophotometric	**HPLC**
Control (from *sacs*)	364	382
Clarified (from *sacs*)	349	355
Unclarified[a] (from *sacs*)	335	363
Clarified (from *sacs*)	229	331
Control (from *whole fruit*)	365	291
Clarified (from *whole fruit*)	296	227
Unclarified[a] (from *whole fruit*)	316	278
Clarified (from *whole fruit*)	269	254

[a] Pasteurised juice.

High performance liquid chromatography- mass spectrometry (HPLC-MS) is a rapid and reliable method for the analyses of non-volatile phenolics and anthocyanins, because better techniques (interfacing systems) have been developed for the removal of the liquid mobile phase before ionization showed that thermospray liquid chromatography (LC) MS is an excellent technique for the analysis of flavonol glycosides from medicinal plants [178, 179]. Positive ion fast atom bombardment MS and tandem MS have been used to study the glycosidic linkages in diglycosyl flavonoids [180]. Electrospray and its several variations are more recent developments in atmospheric pressure ionisation mass spectrometers [167]. HPLC electrospray ionisation mass spectrometry (ESI-MS) offers advantages in terms of sensitivity and capacity for the analysis of large, thermally labile and highly polar compounds [167, 178]. Identification of anthocyanins and related compounds are shown in Tables 4-27.

There are relatively very few standardised or validated methods for the determination of flavonoids, flavonols, flavonones and anthocyanins in plant extracts, fruits, vegetables, etc. Many liquid chromatographic methods have been reported from 2000 onwards for the assay of phenolics/ anthocyanins in berries (different cultivars) and these methods are shown in Tables 19-27.

REFERENCES

[1] Asgary S, Naderi Gh, Sarrafzadegan N, Ghassemi N, Boshtam M, Rafie M, Arefian A. *Pharm Acta Helv* 73, 223, 1999.

[2] Oteiza PI, Erlejman AG, Verstraeten SV, Keen CL, Fraga CG. *Clin Dev Immunol* 12, 25, 2005.

[3] Beecher GR. *J Nutr* 133, 3248S, 2003.

[4] Robards K, Antolovich M. *Analyst* 122, 11R, 1997.

[5] Fleschin S, Fleschin M, Nita S, Pavel E, Magearu V. *Roun Biotechnol Lett* 5, 479, 2003.

[6] Takahama UJ. *J Photochem Photobiol* 38, 363, 1983.

[7] Yanmg CS, Landau JM, Huang MT, Newmark HL. *Annu Rev Nutr* 21, 381, 2001.

[8] Middleton JE, Kandaswami C. *The impact of plant flavonoids on mammalian biology: implications for immunity, inflammation and cancer*. Chapter 15 in: Harbourne JB (ed). The flavonoids: advances in research since 1986. Chapman & Hall, London, 1993; pp 619-652.

[9] Carlo GD, Mascolo N, lzzo AA, Capasso F. *Life Sci* 65, 337, 1999.

[10] Rice-Evans CA., Miller NJ, Paganga G. *Free Radical Biol Med* 20, 933, 1996.S

[11] Pazmino-Duran AE, Giusti MM, Wrolstad RE, Gloria BA. *Food Chem* 75, 211, 2001.

[12] Konczak I, Zhang W. *J Biomed Biotech* 239, 2004.

[13] Lule SU, Xia W. *Food Rev Int* 21, 367, 2005.

[14] Stintzing FC, Carle R. *Trends Food* Sci *Tech* 15, 19, 2004.

[15] da Costa CT, Horton D, Margolis SA. *J Chromatogr A* 881, 403, 2000.

[16] Heredia FJ, Francia-Aricha EM, Rivas-Gonzalo JC, Vicario IM, Santos-Buelga C. *Food Chem* 63, 491, 1998.

[17] Kennedy JA, Waterhouse AL. *J Chromatogr A* 866, 25, 2000.

[18] Castaneda-Ovando A, Pacheco-Hernandez MDL, Paez-Hernandez ME, Rodriguez JA, Galan-Vidal CA. *Food Chem* 113, 859, 2009.

[19] Kahkonen MP, Hopia AI, Heinonen M. *J Agric Food Chem* 49, 4076, 2001.

[20] Amr A, Al-Tamimi E. *Int J Food Sci Tech* 42, 985, 2007.

[21] Awika JM, Rooney LW, Waniska RD. *Food Chem* 90, 293, 2005.

[22] Cacace JE, Mazza G. *J Food Process Eng* 59, 379, 2003.

[23] Donner H, Gao L, Mazza G. *Food Res Int* 30, 637, 1997.

[24] Fossen T, Andersen OM. *Phytochemistry* 62, 1217, 2003.

[25] Phippen WB, Simon JE. *J Agric Food Chem* 46, 1734, 1998.

[26] Kapasakalidis PG, Rastall RA, Gordon MH. *J Agric Food Chem* 54, 4016, 2006.

[27] Metivier RP, Francis FJ, Clydesdale FM. *J Food Sci* 45, 1099, 1980.

[28] Aguirre MJ, Chen YY, Isaacs M, Matsuhiro B, Mendoza L, Torres S. *Food Chem* 121, 44, 2010.

[29] Korotkova EI, Karbainov YA, Avramchik OA. *Izv Vyssh Uchebn Zaved Ser Khim Tekhnol* 45, 110, 2002.

[30] Korotkova EI. *Zh Fiz Khim* 74, 1704, 2000.

[31] Korotkova EI, Karbainov YA, Avramchik OA. Abstracts of Papers. *The 6th Int Conf "Bioantioxidant"* [in Russian], Moscow (2002), pp 298–299.

[32] Korotkova EI, Avramchik OA, Yusubov MS, Belousov MV, Andreeva TI. *Pharm Chem J* 37, 511, 2003.

[33] Cosio MS, Buratti S, Mannino S, Benedetti S. *Food Chem* 97, 725, 2006.

[34] Šeruga M, Novak I, Jakobek L. *Food Chem* 124, 1208, 2011.

[35] Lunte SM, Blankenship KD, Read SA. *Analyst* 113, 99, 1988.

[36] Lunte SM. *J Chromatogr* 384, 371, 1987.

[37] Cren-Olive C, Hapiot P, Pinson J. *J Am Chem Soc* 124, 14027, 2002.

[38] Hotta H, Sakamoto H, Nagano S, Osakai T, Tsujino Y. *Biochim Biophys Acta,* 1526, 159, 2001.

[39] Jovanovic SV, Steenken S, Tosic M, Marjanovic B, Simic MG. *J Am Chem Soc* 116, 4846, 1994.

[40] Slabbert NP. *Tetrahedron* 33, 821, 1977.

[41] Steenken S, Neta P. *J Phys Chem* 86, 3661, 1982.

[42] Janeiro P, Brett AMO. *Electroanalysis* 17, 733, 2005.

[43] Hendrickson HP, Kaufman AD, Lunte CE. *J Pharm Biomed Anal* 12, 325, 1994.

[44] Hendrickson HP, Sahafayen M, Bell MA, Kaufman AD, Hadwiger ME, Lunte CE. *J Pharm Biomed Anal* 12, 335, 1994.

[45] Zoulis NE, Efstathiou CE. *Anal Chim Acta* 320, 255, 1996.

[46] de Lima AA, Sussuchi EM, Giovani WFD. *Croat Chem Acta* 80, 29, 2007.

[47] Janeiro P, Brett AMO. *Electroanalysis* 19, 1779, 2007.

[48] Vestergaard M, Kerman K, Tamiya E. *Anal Chim Acta* 538, 273, 2005.

[49] Moncada MC, Mesquita MFD, Santos MMCD. *J Electroanal Chem* 636, 60, 2009.

[50] He JB, Yuan SJ, Du JQ, Hu XR, Wang Y. *Bioelectrochemistry* 75, 110, 2009.

[51] Masek A, Zaborski M, Chrzescijansk E. *Food Chem* 127, 699, 2011.

[52] Sokolovaa R, Degano I, Ramesova S, Bulickova J, Hromadov M, Gal M, Fiedler J, Valasek M. *Electrochim Acta* 56, 7421, 2011.

[53] Pineda LEG, Guzmán MTO, Pina-Luis G. *ECS Trans* 20, 141, 2009.

[54] Brett AMO, Ghica ME. *Electroanalysis* 15, 1745, 2003.
[55] Nematollahi D, Malakzadeh M. *J Electroanal Chem* 547, 191, 2003.
[56] Farghaly OA. *Anal Lett* 38, 2247, 2005.
[57] Jørgensen LV, Cornett C, Justesen U, Skibsted LH, Dragsted LO. *Free Radical Res* 29, 339, 1998.
[58] Satpati AK, Sahoo S, Dey MK, Reddy AVR, Mukherjee T. *Anal Methods* 3, 1344 , 2011.
[59] Medvidović-Kosanović M, Šeruga M, Jakobek L, Novak I. *Croat. Chem Acta* 83, 197, 2010.
[60] Koyama J, Morita I, Kobayashi N, Konoshima T, Takasaki M, Osakai T, Tokuda H. *Cancer Lett* 263, 61, 2008.
[61] Adam V, Mikelova R, Hubalek J, Hanustiak P, Beklova M, Hodek P, Horna A, Trnkova L, Stiborova M, Zeman L, Kizek R. *Sensors* 7, 2402, 2007.
[62] Zoulis NE, Efstathiou CE. *Anal Chim Acta* 320, 255, 1996.
[63] Zare HR, Samimi R, Ardakani MM. *Int. J Electrochem Sci.* 4, 730, 2009.
[64] Janeiro P, Brett AMO. *Electroanalysis* 19, 1779, 2007.
[65] Nia Y, Zhangb X, Kokot S. *Spectrochim Acta Part A* 71, 1865, 2009.
[66] Zhu ZW, Li C, Li NQ. *Microchem J* 71, 57, 2002.
[67] Zsila F, Bikadi Z, Simonyi M. *Biochem Pharmacol* 65, 447, 2005.
[68] Oshima H, Yoshie Y, Auriol S, Gilibert I. *Free Radical Biol Med* 25, 1057, 1998.
[69] Sergediene E, Jonsson K, Szymusiak H, Tyrakowska B, Rietjens IMCM, Cenas N. *FEBS Lett* 462, 392, 1999.
[70] Zhou J, Wang LF, Wang JY, Tang N. *J Inorg Biochem* 83, 41, 2001.
[71] Zhou J, Wang JY, Tang N. *Indian J Chem A* 40, 149, 2001.
[72] Zhou J, Wang LF, Wang JY, Tang N. *Trans Met Chem* 26, 57, 2001.
[73] Wang SF, Peng TZ, Yang CF. *J Biochem Biophys Meth* 55, 191, 2003.
[74] Kang J, Li Z, Lu X. *J Pharm Biomed Anal* 40, 1166, 2006.
[75] Zhu Z, Li C, Li NQ. *Microchem J* 71, 57, 2002.
[76] Brett AMO, Diculescu VC. *Bioelectrochemistry* 64, 143, 2004.
[77] Brett AMO, Diculescu VC. *Bioelectrochemistry* 64, 133, 2004.
[78] Sun Z, Ma Z, Zhang W, Wang X, Fan C, Li G. *Anal Bioanal Chem* 379, 283, 2004.
[79] Serpi C, Stanic Z, Girousi S. *Talanta* 81, 1731, 2010.
[80] Chan T, Chow AM, Tang DWF, Li Q, Wang X, Brown IR, Kerman K. *J Electroanal Chem* 648, 151, 2010.
[81] Wang F, Xu Y, Zhao J, Hu S. *Bioelectrochemistry* 70, 356, 2007.
[82] Sims MJ, Li Q, Kachoosangi RT, Wildgoose GG, Compton RG. *Electrochim Acta* 54, 5030, 2009.
[83] Gutierrez F, Ortega G, Cabrera JL, Rubianes MD, Rivas GA. *Electroanalysis* 22, 2650, 2010.
[84] Oliveira AC., Mascaro LH. *Int J Electrochem Sci* 6, 804, 2011.
[85] Volikakis GJ, Efstathiou CE. *Talanta* 51, 775, 2000.
[86] Jirovsky D, Bednar P, Myjavcova R, Bartosova Z, Skopalova J, vrdonova M, Lemr K. *Monatsh Chem* 142, 1211, 2011.
[87] Hua J, Lia Q, Tana X *Anal Lett* 29, 1779, 1996.
[88] Apetrei C, Apetrei IM, Nevares I, Alamo MD, Parra V, Rodrıguez-Mendez ML, Saja JAD. *Electrochim Acta* 52, 2588, 2007.

[89] Peters T. *Adv Protein Chem* 37, 161, 1985.
[90] http://www.scielo.br/img/revistas/bjmbr/v42n4/html/7566i01.htm
[91] Sudlow G, Birkett DJ, Wade DN. *Mol Pharmacol* 11, 824, 1975.
[92] http://ghr.nlm.nih.gov/handbook/basics/dna
[93] http://mjlightfoot.edublogs.org/2011/06/09/3-3-4-explain-how-a-dna-double-helix-is-formed-using-complementary-base-pairing-and-hydrogen-bonds/.
[94] http://www.invitrogen.com/site/us/en/home/References/Molecular-Probes-The- and book/Nucleic-Acid-Detection-and-Genomics-Technology/Nucleic-Acid-Stains.html.
[95] Bischoff G, Hoffmann S. *Curr Med Chem* 9, 312, 2002.
[96] Han X, Gao X. *Curr Med Chem* 8, 551, 2001.
[97] Neidle S, Nunn CM. *Nat Prod Rep* 5, 1, 1998.
[98] Papadopoulou A, Frazier RA. *Trends Food Sci Tech* 15, 186, 2004.
[99] Sugio S, Kashima A, Mochizuki S, Noda M, Kobayashi K. *Protein Eng* 12, 439, 1999.
[100] Dangles O, Dufour C, Manach C, Morand C, Remesy C. *Methods Enzymol* 335, 319, 2001.
[101] Kaldas MI, Walle UK, Woude HVD, McMillan JM, Walle T. *J Agric Food Chem* 53, 4194, 2005.
[102] Afanasev IB, Ostrakhovitch EA, Mikhalchik EV, Ibragimova GA, Korkina LG. *Biochem Pharmacol* 61, 677, 2001.
[103] Waring MJ. *J Mol Biol* 54, 247, 1970.
[104] Shuyun Bi, Qiao C, Song D, Tian Y, Gao D, Sun Y, Zhang H. *Sens Actuat B* 119, 199, 2006.
[105] Udenfriend S, Zaltzman P. *Anal Biochem* 3, 49, 1962.
[106] Sun H, Tang Y, Xiang J, Xu G, Zhang Y, Zhang H, Xu L. *Bioorg Med Chem Lett.* 16, 3586, 2006.
[107] Solimani R, *Int J Biol Macromol* 18, 287, 1996.
[108] Sun Y, Bi S, Song D, Qiao C, Mu D, Zhang H. *Sens Actuat B* 129, 799, 2008.
[109] Solimani R, Bayon F, Domini I, Pifferi PG, Todesco PE, Marconi G, Samori B. *J Agric Food Chem* 43, 876, 1995.
[110] Ghosh KS, Sahoo BK, Jana D, Dasgupta S. *J Inorg Biochem* 102, 1711, 2008.
[111] Janjuaa NK, Siddiqab A, Yaquba A, Sabahata S, Qureshia R, Haque SU. *Spectrochim Acta Part A* 74, 1135, 2009.
[112] Bocian W, Kawecki R, Bednarek E, Sitkowski J, Ulkowska A, Kozerski L. *New J Chem* 30, 467, 2006.
[113] Sengupta B, Banerjee A, Sengupta PK. *J Photochem Photobio B Biol* 80, 79, 2005.
[114] Plaper A, Golob M, Hafner I, Oblak M, Solmajer T, Jerala R. *Biochem Biophys Res Commun* 306, 530, 2003.
[115] Ni Y, Du S, Kokot S. *Anal Chim Acta* 584, 19, 2007.
[116] Mello LD, Pereira RMS, Sawaya ACHF, Eberlin MN, Kubota LT. *J Pharm Biomed Anal* 45, 706, 2007.
[117] Kanakis CD, Tarantilis PA, Polissiou MG, Tajmir-Riahi HA. *DNA Cell Biol* 25, 116, 2006.
[118] Nafisi S, Shadaloi A, Feizbakhsh A, Tajmir-Riahi HA. *J Photochem Photobiol B Biol* 94, 1, 2009.
[119] Marinic M, Piantanida I, Rusak G, Zinic M. *J Inorg Biochem* 100, 288, 2006.
[120] Sengupta B, Sengupta PK. *Biochem Biophys Res Commun.* 299, 400, 2002.

[121] Tang J, Wang W, Luan F, Chen X. *Int J Biol Macromol* 37, 85, 2005.
[122] Pendurthi UR, Williams JT, Rao LVM. *Arterioscler Thromb Vasc Biol* 19, 419, 1999.
[123] Tian J, Liu J, He W, Hu Z, Yao X, Chen X. *Biomacromolecules* 5, 1956, 2004.
[124] Mandeville JS, Froehlich E, Tajmir-Riah HA. *J Pharm Biomed Anal* 49, 468, 2009.
[125] Yuan JL, lv Z, Liu ZG, Hu Z, Zou GL. *J Photochem Photobiol A Chem* 191, 104, 2007.
[126] Li Y, He WY, Liu H, Yao X, Hu Z. *J Mol Struct* 831, 144, 2007.
[127] Kanakis CD, Tarantilis PA, Polissiou MG, Diamantoglou S, Tajmir-Riahi HA. *J Mol Struct* 798, 69, 2006.
[128] Pastukhov AV, Levchenko LA, Sadkov AP. *J Mol Struct* 842, 60, 2007.
[129] Xie MX, Xu XY, Wang YD. *Biochim Biophys Acta* 1724, 215, 2005.
[130] Bi S, Ding L, Tian Y, Song D, Zhou X, Liu X, Zhang H. *J Mol Struct* 703, 37, 2004.
[131] Cao H, Chen L, Xiao J. *Mol Biol Rep* 38, 2257, 2011.
[132] Xiao J, Chen T, Cao H, Chen L, Yang F. *Mol Nutr Food Res* 55, 310, 2011.
[133] Banerjee A, Basu K, Sengupta PK. *J Photochem Photobiol B Biol* 90, 33, 2008.
[134] Jurasekova Z, Marconi G, Sanchez-Cortes S, Torreggiani A. *Biopolymers* 91, 917, 2009.
[135] Pu G, Yu X, Pu G, Xi F. *Spec Spectral Anal* 26, 1672, 2006.
[136] Tian J, Liu J, Tian X, Hu Z, Chen X. *J Mol Struct* 691, 197, 2004.
[137] Sahoo BK, Ghosh KS, Dasgupta S. *Biophy Chem* 132, 81, 2008.
[138] Hegde AH, Sandhya B, Seetharamappa J. *Mol Biol Rep* 38, 4921, 2011.
[139] Gutzeit HO, Henker Y, Kind B, Franz A. *Biochem Biophys Res Comm* 318, 490, 2004.
[140] Papadapoulou A, Green RJ, Frazier RA. *J Agric Food Chem* 53, 158, 2005.
[141] Shi J, Cao H. *Braz J Pharmacognosy* 21, 594, 2011.
[142] Liu EH, Qi L, Li P. *Molecules* 15, 9092, 2010.
[143] Nia Y, Zhangb X, Kokot S. *Spectrochim Acta A* 71, 1865, 2009.
[144] Zhu ZW, Li C, Li NQ. *Microchem J* 71, 57, 2002.
[145] Zsila F, Bikadi Z, Simonyi M. *Biochem Pharmacol* 65, 447, 2005.
[146] Dufour C, Dangles O. *Biochim Biophys Acta* 1721, 164, 2005.
[147] Cao H, Wu D, Wang H, Xu M. *Spectrochim Acta A* 73, 972, 2009.
[148] Sun Y, Zhang H, Sun Y, Zhang Y, Liu H, Cheng J, Bi S, Zhang H. *J Lumin* 130, 270, 2010.
[149] Ma J, Yin YM, Liu HL, Xie MX. *Curr Org Chem* 15, 2627, 2011.
[150] He W, Li Y, Tang J, Luan F, Jin J, Hu Z. *Int J Biol Macromol* 39, 165, 2006.
[151] Pinto MDC, Duque AL, Macías P. *J Fluorescence* 21, 1311, 2011.
[152] Lu Y, Wang YL, Gao SH, Wang GK, Yan CL, Chen DJ. *J Luminescence* 129, 1048, 2009.
[153] Liu Y, Yang Z, Du J, Yao X, Lei R, Zheng X, Liu J, Hu H, Li H. *Immunobiology* 213, 651, 2008.
[154] Zhao J, Ren F. *Spectrochim Acta A* 72, 682, 2009.
[155] Chaudhuri S, Chakraborty S, Sengupta PK. *Biophys Chem* 154, 26, 2011.
[156] Sentchouk VV, Bondaryuk EV. *J Appl Spectrosc* 74, 731, 2007.
[157] Xi J, Guo R. *Int J Biol Macromol* 40, 305, 2007.
[158] Chaudhuri S, Pahari B, Sengupta B, Sengupta PK. *J Photochem Photobiol B Biol* 98, 12, 2009.
[159] Yuan JL, Liu H, Kang X, Lv Z, Zou GL. *J Mol Struct* 891, 333, 2008.

[160] Wiese S, Gartner S, Rawel HM, Winterhalter P, Kulling SE. *J Sci Food Agric* 89, 33, 2009.
[161] Mas T, Susperregui J, Berke B, Chèze C, Moreau S, Nuhrich A, Vercauteren J. *Phytochemistry* 53, 679, 2000.
[162] Wang D, Wang X, Zhang C, Ma Y, Zhao X. *J Agric Food Chem* 59, 7405, 2011.
[163] Webb MR, Min K, Ebeler SE. *J Food Biochem* 32, 576, 2008.
[164] Mazzaracchio P, Tozzi S, Boga C, Forlani L, Pifferi PG, Barbiroli G. *Food Chem* 129, 1100, 2011.
[165] Sarma AD, Sharma R. *Phytochemistry* 52, 1313, 1999.
[166] Hagerman AE, Butler LG. *J Biol Chem* 256, 4444, 1981.
[167] Robards K, Antolovich M. *Analyst* 122, 11R, 1997.
[168] Markham KR. *Techniques of Flavonoid Identification*. New York, USA: Academic Press, 1982.
[169] Harborne JB. Phytochemical Methods: *A Guide to Modern Techniques of Plant Analysis*. London, UK: Chapman & Hall, 1998.
[170] Lee HS, Widmer BW. Phenolic compounds. In: Nollet LML, ed. *Handbook of Food Analysis. Physical Characterization and Nutriet Analysis*. New York, USA: Marcel Dekker, Inc., 1996, Vol 1, pp 821–894.
[171] Simón BFD, Pérez-Ilzarbe J, Hernández T, Gómez-Cordovés C, Estrella I. *J Agric Food Chem* 40, 1531, 1992.
[172] Sabatier S, Amiot MJ, Tacchini M, Aubert S. *J Food Sci* 57, 773, 1992.
[173] Mouly P, Gaydou EM, Estienne J. *J Chromatogr* 634, 129, 1993.
[174] Schmidt TJ, Merfort I, Willuhn G. *J Chromatogr A* 669, 236, 1994.
[175] Merken HM, Beecher GR. *J Agric Food Chem* 48, 577, 2000.
[176] Markham KR. *Flavones, flavonols and their glycosides*. Dey PM, Harborne JB, Eds. *Methods in Plant Biochemistry*, Vol 1, Plant Phenolics. London, UK: Academic Press, 1989, pp 197–235.
[177] Marko-Varga G, Barcelo D. *Chromatographia* 34, 146, 1992.
[178] Careri M, Mangia M, Musci M. *J Chromatogr A* 794, 263, 1998.
[179] Pietta P, Facino RM, Carini M, Mauri P. *J Chromatogr A* 661, 121, 1994.
[180] Li QM, Claeys M. *Biol Mass Spectrom* 23, 406, 1994.
[181] Aguirre MJ, Chen YY, Isaacs M, Matsuhiro B, Mendoza L, Torres S. *Food Chemistry* 121, 44, 2010.
[182] Nicoueä EE, Savard S, Belkacemi K. *J Agric Food Chem* 55, 5626, 2007.
[183] Jia N, Shu Q, Wanga L, Dua H, Xu Y, Liu Z. *Scientia Horticulturae* 117, 167, 2008.
[184] Lohachoompol V, Mulholland M, Srzednicki G, Craske J. *Food Chemistry* 111, 249, 2008.
[185] Hou Z, Qin P, Zhang Y, Cu S, Ren G. *Food Res Int*. In Press, 2011.
[186] Sentandreu E, Navarro JL, Sendra JM. *J Agric Food Chem* 58, 10560, 2010.
[187] Dehkharghanian M, Adenier H, Vijayalakshmi MA. *Food Chemistry* 121, 863, 2010.
[188] El-Hela AA, Al-Amier HA., Ibrahim TA. *J Chromatogr A* 1217, 6388, 2010.
[189] Jiménez CDC, Flores CS, He J, Tian Q, Schwartz SJ, Giusti MM. *Food Chemistry* 128, 717, 2011.
[190] Li H, Guan X, Ye M, Xiang C, Lin C, Sun C, Guo D. *J Sep Sci* 34, 1437, 2011.
[191] Liang Z, Owens CL, Zhong G, Cheng L. *Food Chemistry* 129, 940, 2011.

[192] Ieri F, Innocenti M, Andrenelli L, Vecchio V, Mulinacci N. *Food Chemistry* 125, 750, 2011.
[193] Lee JH, Lee H, Choung M. *Food Chemistry* 129, 272, 2011.
[194] Aguirre MJ, Isaacs M, Matsuhiro B, Mendoza L, Santos LS, Torres S. *Food Chemistry* 129, 514, 2011.
[195] Piasek A, Kusznierewicz B, Grzybowska I, Czyk EM, Piekarska A, Azqueta A, Collins AR, Namiesnik J, Bartoszek A. *J Food Comp Anal* 24, 880, 2011.
[196] Mattila P, Astola J, Kumpulainen J. *J Agric Food Chem* 48, 5834, 2000.
[197] Wang SY, Zheng W. *J Agric Food Chem* 49, 4977, 2001.
[198] Wang SY, Lin HS. *J Agric Food Chem* 51, 6844, 2003.
[199] Wang SY, Bunce JA, Maas JL. *J Agric Food Chem* 51, 4315, 2003.
[200] Kosar M, Karkas E, Paydas S, Husnu K, Baser C. *J Agric Food Chem* 52, 1586, 2004.
[201] Määttä-Riihinen KR, Kamal-Eldin A, Mattila PH, González-Paramás AM, Törrönen AR. *J Agric Food Chem* 52, 4477, 2004.
[202] Zhang Z, Kou X, Fugal K, Mclaughlin J. *J Agric Food Chem* 52, 668, 2004.
[203] Skupien K, Oszmianski J. *Eur Food Res Tech* 219, 66, 2004.
[204] Awika JM, Rooney LW, Waniska RD. *Food Chemistry* 90, 293, 2004.
[205] Eichhorn S, Winterhalter P. *Food Res Int* 38, 943, 2005.
[206] Longo L, Vasapollo G. *J Agric Food Chem* 53, 475, 2005.
[207] Anttonen MJ, Karjalainen RO. *J Agric Food Chem* 54, 7430, 2006.
[208] Aaby K, Krede GR, Wrolstad RE. *J Agric Food Chem* 53 , 4032, 2005.
[209] Aaby K, Wrolstad R, Ekeberg D, Skrede G. *J Agric Food Chem* 55, 5156, 2007.
[210] Hernanz D, Recamales AF, Melendez-Martinez AJ, Gonzalez-Miret ML, Heredia FJ. *J Agric Food Chem* 55, 1846, 2007.
[211] Koponen JM, Happonen AM, Mattila PH, Torronen AR. *J Agric Food Chem* 55, 1612, 2007.
[212] da Silva FL, Escribano-Bailon MT, Alonso JJP, Rivas-Gonzalo JC, Santos-Buelga C. *Food Sci. Tech* 40, 374, 2004.
[213] Terry LA, Chope GA, Bordonaba JG. *J Agric Food Chem* 55, 10812, 2007.
[214] Bordonaba JG, Terry LA. *J Agric Food Chem* 56, 7422, 2008.
[215] Tulipani S, Mezzetti B, Capocasa F, Bompadre S, Beekwilder J, Vos CHRD, Capanoglu E, Bovy A, Battino M. *J Agric Food Chem* 56, 696, 2008.
[216] Sadilova E, Stintzing FC, Kammerer DR, Carle R. *Food Res. Int* 42, 1023, 2009.
[217] Tiwari BK, O'Donnell CP, Patras A, Brunton N, Cullen PJ. *Food Chemistry* 113, 1119, 2009.
[218] Seabra IJ, Brag MEM, Batist MT, Sousa HCD. *J Supercrit Fluids* 54, 145, 2010.
[219] Simirgiotis MJ, Schmeda-Hirschmann G. *J Food Comp Analysis* 23, 545, 2010.
[220] Hernandez-Herrero JA, Frutos MJ. *Int. J Food Sci Tech* 46, 2550, 2011.
[221] Bordonaba JG, Crespo P, Terry LA. *Food Chemistry* 129, 1265, 2011.
[222] Mataix E, Castro MDLD. *J Chromatogr A* 910, 255, 2001.
[223] Novak I, Janeiro P, Seruga M, Oliveira-Brett AM. *Anal Chim Acta* 630, 107, 2008.
[224] Lee J, Rennaker C, Wrolstad RE. *Food Chemistry* 110, 782, 2008.
[225] Regos I, Treutter D. *J Chromatogr A* 1217, 6169, 2010.
[226] Doka O, Ficzek G, Bicanic D, Spruijt R, Luterotti S, Toth M, Buijnsters JG, Vegvari G. *Talanta* 84, 341, 2011.

[227] Liu HP, Shi XF, Zhang YC, Li ZX, Zhang L, Wang ZY. *Cell Biochem Biophys* 61, 59, 2011.
[228] Can NO, Arli G, Atkosar Z. *Food Chemistry* 130, 1082, 2012.
[229] Rusjan D, Korosec-Koruza Z. *Acta Chim Slov* 54, 114, 2007.
[230] Ivanova V, STefova M, Chinnici F. *J Serb Chem Soc* 75, 45, 2010.
[231] Turfan O, Turkyılmaz M, Yemis O, Ozkan M. *Food Chemistry* 129, 1644, 2011.

In: Anthocyanins: Structure, Biosynthesis and Health Benefits
Editor: Noboru Motohashi
ISBN: 978-1-62257-329-5

Chapter 4

ANTHOCYANINS IN RICES: BIOSYNTHESIS AND HEALTH BENEFITS

Yafang Shao and Jinsong Bao*
Institute of Nuclear Agricultural Sciences, College of Agriculture and Biotechnology, Zhejiang University, Hua Jiachi Campus, Hangzhou, PR China

ABSTRACT

Anthocyanins are a kind of reddish-purple water-soluble flavonoids that are present in the red and black pigmented rice (*Oryza sativa* L.) grains as well as in colored plants. Anthocyanins show potent antioxidative effects through scavenging reactive oxygen generated in the body, proving that they are benefit for human health. Many studies indicate that consumption of black rice have a number of beneficial properties including antioxidant, α-glucosidase inhibitory, hepatoprotective, antidiabetic, anti-inflammatory, and anti-carcinogenic effects. Understanding of the biosynthetic pathway of anthocyanins in rices could provide possible strategy to improve anthocyanins and other flavonoid contents in rice grains by genetic engineering. Black rice accumulating anthocyanins in the endosperm are strongly expected. This chapter summarizes the knowledge on the genetics, biosynthesis and health benefits of anthocyanins in rice grains.

Keywords: Anthocyanin, biosynthesis, black rice, genetics, health benefits

1. INTRODUCTION

Rice (*Oryza sativa* L.) is one of the most important staple foods, feeding about half the world's people. The commonly consumed rice is white, also called polished rice or milled rice, in which the bran layer of rice grain has been removed by polishing. There are specialty rice with different grain colors, which are called pigmented rice in common or more specially,

* Corresponding author. Tel.: +86 571 86971932; fax: +86 571 86971421.
E-mail address: jsbao@zju.edu.cn.

red rice and black (purple) rice (Figure 1). In addition to the black rice grain, some rice plants have a dark-purple appearance all over the body; the plants appear greenish-dull-red in the early stages, but gradually become full purple as the pigment extends all over the leaf blades and other parts of the plants [1]. In addition to the purple plant, many rices also accumulate pigments in different tissues [2].

Figure 1. Different color of rice grain (up left: brown rice; up right: middle: red rice; down: black rice).

The reddish to purple appearance of rice grain or plant is derived from their accumulation of anthocyanins. The kernel of red rice is high in the proanthocyanidins, whereas black rice is high in anthocyanins [3]. Few black rices may accumulate both proanthocyanidins and anthocyanins [4]. After removing the bran layers of pigmented rice, the milled is also white because of no anthocyanins accumulated in the endosperm of rice grain. In the purple plant, the anthocyanin pigment commences to develop as soon as the seedling emerges [1]. The genetic mechanism and biosynthetic pathway underlying the accumulation of anthocynidins need to be clarified.

In the black rice grain, the existence of anthocyanin is mainly in free form, and the percentage of free anthocyanin to total anthocyanin ranged from 99.5-99.9% [5]. Two major anthocyanins have been identified in black rice: cyanidin 3-glucoside (C-3-G) and peonidin 3-

glucoside (P-3-G) in the black rice (Figure 2) [5, 6]. The content of C-3-G is significantly higher than that of P-3-G [5-11].

Besides the anthocyanins, black rice grain also contains gallic, hydroxybenzoic, and protocatechuic acids, carotenoids, γ-tocols, other flavonoids, *etc*., which shows higher than red rice or normal rice [11, 12]. As far as nutritional quality of rice grain is concerned, the colored rice becomes much more popular because of high density of phytochemicals presented in the bran layer. Recently, the health benefits of anthocyanins have been received increasing attentions.

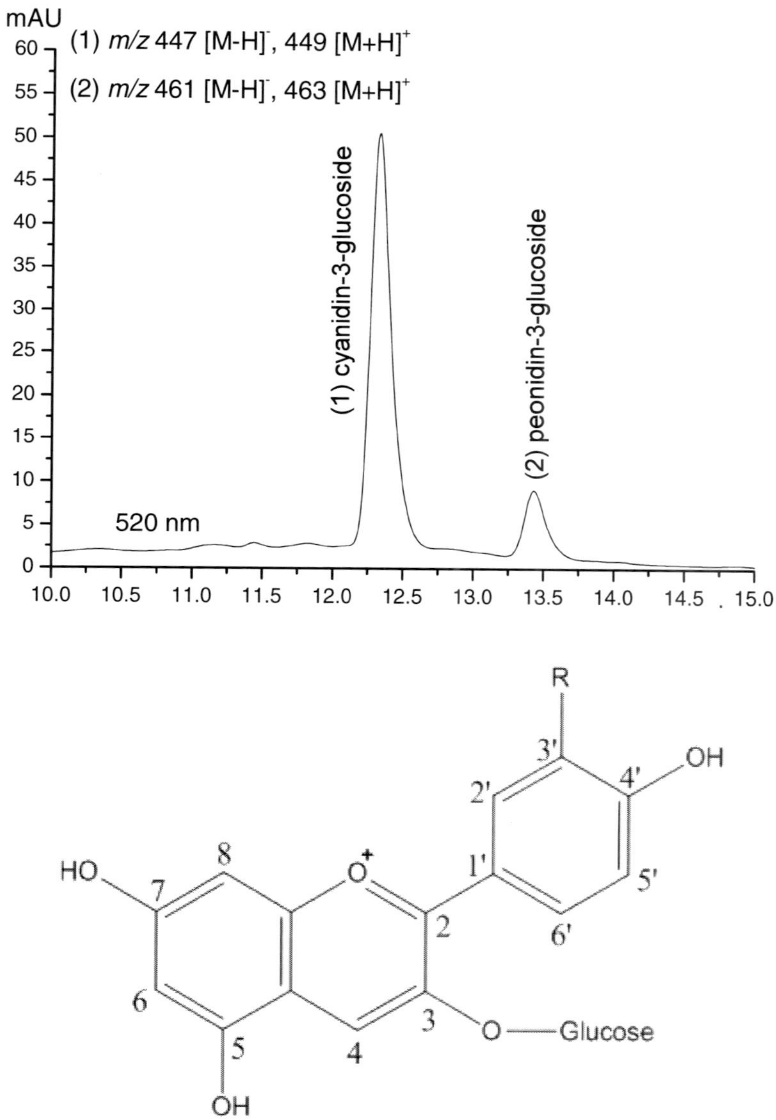

Figure 2. Identification of anthocyanin components in black rice. Up: Typical HPLC chromatogram and ESI-MS data of anthocyanins in black rice detected at 520 nm. (1) cyanidin-3-glucoside (C-3-G) and (2) peonidin-3-glucoside (P-3-G). Down: the structure of anthocyanidin in black rice grain extracts, C-3-G: R=OH; P-3-G: R=OCH_3

In this chapter, current understanding of the genetics, biosynthetic pathway and regulation of anthocyanins in black rice are reviewed, and the health benefits of consuming the whole black rice grain are summarized.

2. Biosynthesis of Anthocyanins

The genetic study of the anthocyanin pigment in various parts of the rice plant could be dated back to about 80 years ago [1]. It is found that the color of rice plant was controlled by an anthocyanin gene (*A*), which made the plant entirely purple, and a partial inhibiting gene (*I*), which suppressed the color of leaves but allowed pigment accumulation in other parts [1]. From then on, many genetic studies have been carried out about the pigment of different parts in rice. The C-A-P genic system has been applied to explain the genetics of pigment accumulation in rice plant. *C* and *A* are basic genes that associated with anthocyanin pigments, because they exist in all colored tissues of rice [13]. The gene C governs the production of chromogen, the gene A converts or oxidizes chromogen into anthocyanin. The gene P stands for pleiotropic complementary genes [14]. Genetically, the pericarp color of red rice is controlled by two complementary genes, *Rc* (brown pericarp) on chromosome 7 and *Rd* (red pericarp) on chromosome 1. When present together, these loci produce red seed color. *Rc* in the absence of *Rd* produces brown seeds, whereas *Rd* alone has no phenotype [15]. The color of dark purple pericarp is also controlled by two complementary genes, *Pb* and *Pp*, located on chromosome 4 and 1, respectively [16]. Wang and Shu mapped *Pb* gene and suggested that it might be the *Ra* gene [17]. Nowadays, the understanding of the anthocyanin biosynthetic pathway has become much clearer than ever, though some regulator genes have not been elucidated.

2.1. The Pathway of Anthocyanins

The biosynthetic pathway of anthocyanins derives from phenylpropanoid pathway and involves many enzymes (Figure 3). The first enzyme is phenylalanine ammonia lyase (PAL), which converts L-phenylalanine to cinnamate. Under the effects of cinnamate-4-hydroxylase (C4H) and 4-coumarate-CoA ligase (4CL), cinnmate transforms to 4-coumaroyl-CoA. At the same time, acetyl-CoA-corboxylase (ACC) transforms acetyl-CoA to malonyl-CoA. Then, chalcone synthase (CHS) catalyzes 4-coumaroyl-CoA and malonyl-CoA to naringenin chalcone, which is a precursor for various phenylpropanoid biosynthetic derivatives, including flavonoid and anthocyanin. Next, chalcone isomerase (CHI) produces narigenin, and flavonoid 3'-hydroxylase (F3'H) produces eriodictyol. On the other hand, narigenin can be converted into flavones including tricin and isovitexin [18] by the action of flavone synthases I (FNS I) or FNS II, or converted into dihydrokaempferol by the action of flavanone 3-hydroxylase (F3H). F3Hs also use flavanone, naringenin and eriodictyol as substrates to produce dihydrokaempferol and taxifolin [19]. Under the action of F3'H or F3H, dihydrokaempferol or eriodictyol is converted into dihydroquercetin. Dihydrokaempferol and dihydroquercetin can be catalyzed into flavonols by flavonol synthase (FLS) or catalyzed into leucoanthocyanidins by dihydroflavonol 4-reductase (DFR). Afterward, leucoanthocyanidin reductase (LAR) catalyzes the conversion of leucoanthocyanidins to catechin, which is finally converted to proanthocyanidins by condensing enzyme or enzyme complex. Another enzyme, anthocyanidin synthease (ANS), catalyzes leucoanthocyanidins to anthocyanidins, which is subsequently catalyzed to cyaniding-3-glucoside by flavonoid 3-*O*-glycosyltransferase (FGT) in black rice. The peonidin-3-*O*-glucoside comes from cyaniding-3-*O*-glucoside by anthocyanin-3'-*O*-methyl transferase (A3'M), which is responsible for the higher content of cyaniding-3-*O*-glucoside than that of peonidin-3-*O*-glucoside in the black rice grain.

2.2. Associated Genes, Enzymes and Their Functions

The biosynthesis of anthocyanins requires complex interactions genes with both structural and regulatory roles. Many structural genes in the anthocyanin biosynthetic pathway have been clarified, and most genes are located on chromosome 1, 3, 4 and 11 (Table 1).

PAL, as the first gene in the phenylpropanoid pathway, is expressed in almost all parts of rice plant, but are greater in the roots and stems [23]. Its activity can be induced by UV-B radiations [21-22].

Table 1. Associated structural genes in the biosynthesis of anthocyanins*

Gene **	TIGR locus	RAP locus	Genomic sequence length (bp)	CDS length (bp)	Protein length (aa)
PAL	LOC_Os05g35290	Os05g0427400	2514	2151	716
CHS	LOC_Os11g32650	Os11g0530600	3460	1197	398
CHI1	LOC_Os11g02440	Os11g0116300	1196	651	216
CHI2	LOC_Os03g60509.1	Os03g0819600	1172	888	295
F3'H	LOC_Os10g17260	Os10g0320100	1975	1581	526
F3H	LOC_Os04g56700	Os04g0662600	2954	1134	377
DFR	LOC_Os01g44260	Os01g0633500	1981	855	284
LODX	LOC_Os01g27490	Os01g0372500	1593	1128	375
ANR	LOC_Os04g53850	Os04g0630800	2802	1035	344
LAR	LOC_Os03g15360	Os03g0259400	1444	1077	358
ANS	LOC_Os01g27490	Os01g0372500	1593	1128	375

* The information was found on http://rice.plantbiology.msu.edu.

** Genes are abbreviated as: *PAL:* phenylalanine ammonia lyase; *CHS:* chalcone synthase; *CHI:* chalcone isomerase; *F3'H:* flavonoid 3'-hydroxylase; *F3H:* flavonoid 3-hydroxylase; *DFR:* dihydroflavonol 4-reductase; *LODX:* leucoanthocyanidin dioxygenase; *ANR*: anthocyanidin reductase; *LAR*: leucoanthocyanidin reductase; *ANS:* anthocyanidin synthase.

The chalcone synthase (CHS) is a key enzyme that catalyses the first dedicated reaction of the flavonoid pathway in higher plants. It expresses in all parts of rice plant, and has high levels in the stems or seeds. Its transcript in seeds at 13 day after flowering was significantly higher than that at 2 and 7 day after flowering [23]. It has been mapped between RFLP markers RG2 and RG103 of chromosome 11 of rice [24]. When transforms this gene from rice to *Arabidopsis*, the seed coat color of transgenic plant turned from yellow to purple [18], suggesting that this gene functions in the pigment accumulation in rice. Yeast two hybrid analyses suggested that a novel organization of an rice chalcone synthase (OsCHS) homodimer interacting in specific orientations with rice Flavonoid 3-hydroxylase (OsF3H), Flavonoid 3'-hydroxylase (OsF3′H), Dihydroflavonol 4-reductase (OsDFR), and Anthocyanidin synthase (OsANS) in a macromolecular complex for anthocyanin biosynthesis [18]. Moreover, it is found that the superfamily of *CHS* is probably diverged into four branches, and members of each branch may perform specific functions [25].

Chalcone isomerase (CHI) or chalcone flavanone isomerase (CFI) is one of the key enzymes in the flavonoid biosynthesis pathway catalyzing conversion of chalcone to flavanone. Rice has only one copy of this gene whereas barely and maize has one and three copies in the genome [26]. The seed coat color of transgenic *Arabidopsis* with *OsCHI* turned from yellow to purple, but OsCHI was not involved in the interaction with other enzymes by the study of yeast two-hybrid analysis [18].

Flavonoid 3'-hydroxylase (F3′H) acts on the flavonols, flavones, and flavanones biosynthesis [27]. When transforms the *OsF3'H* to *Arabidopsis* mutant, the seed coat color turned from yellow to purple [18], indicating that this gene may have some relations to the pigment accumulation in rice. It is also found that this enzyme had some interactions with OsF3H, OsDFR and OsANS by yeast two-hybrid analysis [18].

Flavonoid 3-hydroxylase (F3H) is a key gene in the synthesis of flavonol and anthocyanin. It is expressed at extremely high level in the seeds of black rice at 13 day after flowering. The transcripts in seeds growing at high temperature (27°C) was about 3 times more than that growing at low temperature (24°C), but it had only a little expression at 21°C [23]. Kim *et al.* cloned three F3H cDNAs from rice by reverse transcription polymerase chain reaction (RT-PCR), and found that OsF3H-1 had the highest enzymatic activity, whereas the overall expression of *OsF3H-2* was highest in all tissues except seeds [19]. When transforms the *OsF3H* to *Arabidopsis*, the seed coat color turned from yellow to purple, suggested that this gene is associated with the formation of pigmentation in rice.

Dihydroflavonol 4-reductase (*DFR*) is a tissue-specific expression gene in the anthocyanin biosynthetic pathway. It is expressed mostly in roots and seed. In seeds, highest expression was found at 13 days after flowering [23]. When introducing the *DFR* into *Rcrd* mutant, the rice color is changed from brown to red, suggesting that the *Rd* locus encodes *DFR* proteins [28]. When *OsDFR* was transform to *Arabidopsis* mutant, the seed coat color turned from yellow to purple [18].

Anthocyanidin reductase (ANR) catalyses the synthesis of epicatechin, but its function in rice has not been characterized.

Leucoanthocyanidin reductase (LAR) belongs to the superfamily of isoflavone reductases, and catalyzes the synthesis of catechin from leucoanthocyanidin. It is the first committed step in proanthocyanidin biosynthesis. However, its function in rice has not been characterized.

Anthocyanidin synthase (ANS) or leucoanthocyanidin dioxygenase (LODX) is one of the four dioxygenases (DOXs) of the flavonoid biosynthetic pathway that catalyzes the formation of anthocyanidins from leucoanthocyanidins. It contains a double stranded beta helix and shows high sequence similarity with other DOXs of the pathway including flavonol synthase, flavonone-3-β-hydroxylase and flavone synthase I [29]. It is expressed in roots, stems, leaves, and especially high in the seeds of black grain rice at 13 day after flowering [23]. Overexpression of the *ANS* in a rice mutant leads to its accumulation of anthocyanin and a mixture of flavonoids with a decrease of the proanthocyanin [29]. Interestingly, the color of rice grain do not depend on whether this gene is expressed or not, but on the expression amounts of this gene, because it is also slightly expressed in white-grained rice [23]. When *OsANS* was transformed to *Arabidopsis*, the seed coat color turned from yellow to purple [18].

2.3. Regulation of Anthocyanidin Biosynthesis

2.3.1. Regulatory Genes

Besides the structural genes controlling the biosynthesis of anthocyanins, there are transcription factors (TFs) regulating the expression at the same time. The regulatory genes belong to the two gene families: *R/B* and *C1/Pl*; they can activate or inhibit the expression of structural genes in different tissues of rice plant. The *R/B* genes encode the proteins homology to the basic helix-loop-helix (bHLH) DNA-binding protein dimerization domain of Myc proteins [30]. *R* gene family is the same homologous in maize and rice [31]. The *C1/Pl* genes encode the proteins homology to the DNA-binding domain of Myb proteins [32]. Zhu *et al.* showed that a Myc protein, encoded by *OsBP-5*, exhibited 50% amino acid sequence identity with Rb protein, regulated the expression of genes related to anthocyanin biosynthesis [33].

OsC1 on the short arm of chromosome 6, is a homolog of maize *C1* on chromosome 9 [34]. Reddy *et al.* (1998) cloned this gene [35], which belongs to *Myb* family TF. It is responsible for the color of apiculus. A mutation in the third exon causes the frameshift that leads to the divergent phenotypes [36]. Fan *et al.* mapped the *C* locus between SSR markers RM111 and RM253, with genetic distances of 0.7 cM to RM111 and 0.4 cM to RM253 [37]. Gao *et al.* found that it had a 34 bp longer in the purple sheath somaclonal mutant than that in the green sheath rice [38].

The Purple leaf (Pl) locus, located on chromosome 4, is similar to maize *Rb* loci, and plays a role in the anthocyanin pathway. Takagashi thought that it was a controller of anthocyanin biosynthetic genes [13]. Kinoshita and Maekawa considered it as a transcriptional regulator, and containing three alleles---*Pl*, Pl^{w}, Pl^{i}, with each displaying tissue specificity for anthocyanin pigmentation [39]. Wataru *et al.* found that Pl^{w} allele may be complex and composed of at least two genes encoding bHLH proteins [40].

OsB2, isolated from the *Purple leaf (Pl)* locus, is a rice *myc*-type transcription factor. It encodes a transcription factor for anthocyanin synthesis in rice leaves. Its genomic sequence of 5'-portion existed about 10kb downstream of *OsB1* coding region, so *OsB2* lacks a conserved C-terminal domain [40]. When the transgenic rice plant overexpressing the *OsB2* was treated with methyl jasmonate and 2,6-dichloroisonicotinic acid, the color of shoots turned to red with the increased anthocyanin content to approximately 5-fold [41]. Hirose *et al.* also found that the leaf color of transgenic rice plants turned red, and anthocyanin content was increased approximately twice [42]. It is suggested that anthocyanin accumulation is consistent with the increase of the expression of *OsB2* and *anthocyanidin synthase (ANS)*.

The Ra gene, on chromosome 4, can activate the anthocyanin pathway. Black rice has a 2-bp (GT) deletion in exon 7 compared with white rice. It might be the *Pb* locus that controls the color of black rice [17].

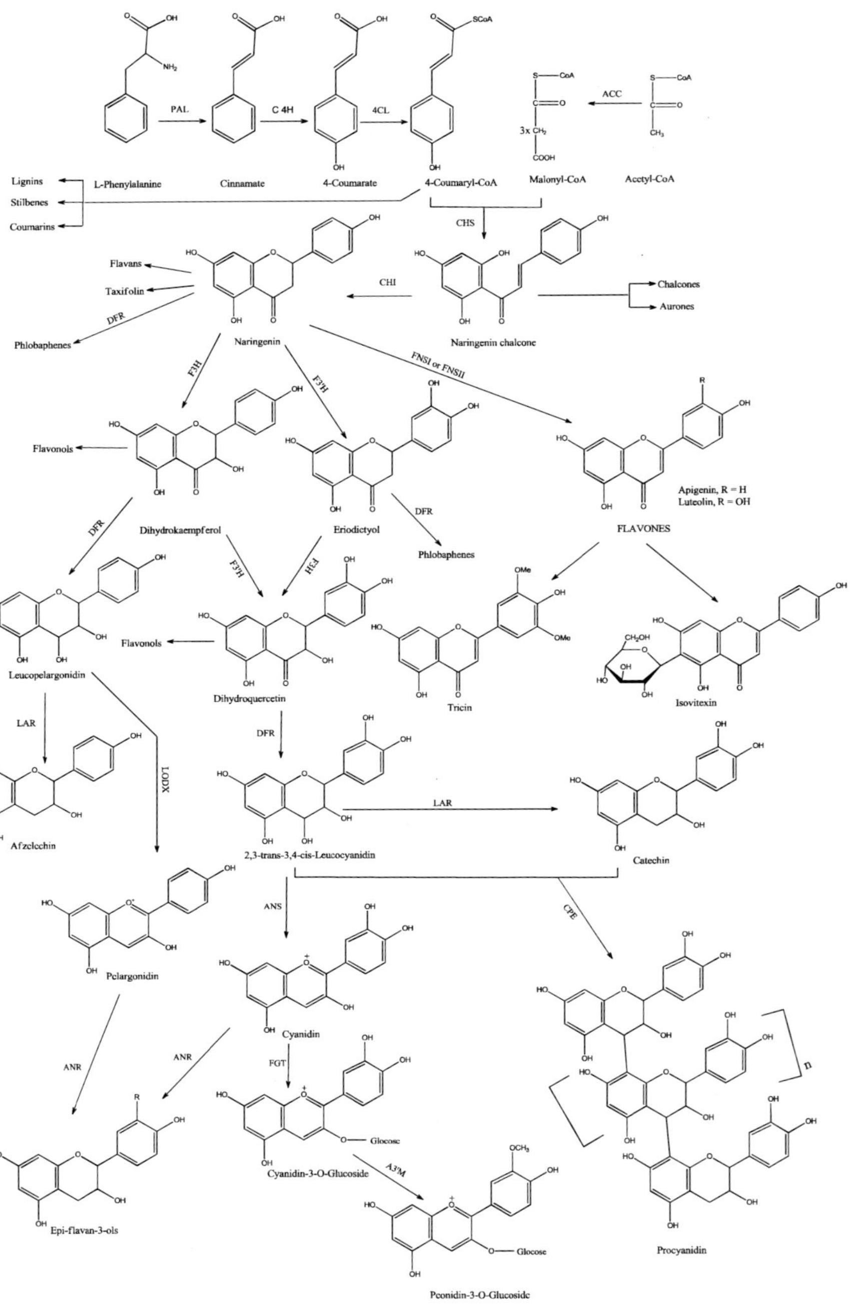

Figure 3. The biosynthesis of flavonoid pathway in rice plant. Modified from Lepiniec *et al.* [20]. Enzymes are abbreviated as follows: ACC: acetyl-CoA-corboxylase; ANR (anthocyanidin reductase); ANS: anthocyanidin synthase; A3'M: *S*-adenosyl-L-methionine; CHI: chalcone isomerase; CHS: chalcone synthase; CPE: condensing enzyme and/or chainpolymerising enzyme; C4H: cinnamate-4-hydroxylase; 4CL: 4-coumarate-CoA ligase; DFR: dihydroflavonol 4-reductase; FNS I or FNS II: flavone synthases I of flavone synthases II; FGT: flavonoid 3-*O*-glycosyltransferase; F3H: flavonoid 3-hydroxylase; F3'H: flavonoid 3'-hydroxylase; LAR: leucoanthocyanidin reductase; LODX: leucoanthocyanidin dioxygenase; PAL: phenylalanine ammonia lyase

Rb gene, as a member of *R* gene family, locates on chromosome 1 in rice, and can induce pigmentation in maize suspension cells [43].

Rc, encoding a basic helix-loop-helix (bHLH) protein, is a positive regulator of proanthocyanidin required for red pericarp in rice. The existing of a 14-bp deletion within exon 6 leads to the dominant red pericarp different from the recessive white pericarp, because of knocking out the bHLH domain of the protein. However, it is expressed in both red- and white-grained rice, yet only a little expression in white varieties by RT-PCR detection [15].

Other Transcription Factors (TFs)

Besides all above TFs, Kim *et al.* identified more TFs that had potential anthocyanin-specific functions using *hypergeometric* analysis for transcription factor *function*, such as the *Myb* and *GT* families, *PBP, PBF, RAV*, and *STF* factors [44]. To better understand the functional characterization of anthocyanin gene expression in black rice, Kim *et al.* found that a total of 8,517 genes exhibited greater than 2.5-fold up- or down regulation in three rice groups and three seed developmental stages, and identified 12 unknown and hypothetical genes that were involved in anthocyanin biosynthesis by microarray [45]. By RT-PCR analysis, expressions of 15 out of 17 unknown and hypothetical genes were verified [46]. The function of these regulatory factors or genes that related to anthocyanin metabolism in rice needs further researches.

2.3.2. Effects of Environment

In rice plant, the accumulation of anthocyanins is not only regulated by the interaction between structural gene and transcription factors, but also is affected by some environmental factors. Treatment of rice with abscisic acid (ABA) [47, 48], salts [49, 50], drought [51, 52] and cold [53] could induce accumulation of anthocyanins in the leaves to defense the stress and survive in the adverse circumstance. The adverse environment may also induce expressions of genes in anthocyanin biosynthesis, and increase the specific enzyme activities, such as PAL [47].

3. HEALTH BENEFITS

Pigmented rice is a good source of food for human health, because the bran layer contains high levels of phytochemicals that show high antioxidant capacity such as anthocyanins, γ-oryzanol, and vitamin E homologues, *etc.* The black pigment components, cyanidin 3-glucoside and peonidin 3-*O*-cyanidin have higher scavenging activities than others [54]. Therefore, regular long-term consumption of black rice is benefit for health. The functions of the extracts from black rice grain had been widely studied *in vivo* and *in vitro* models (Table 2). Oral administration is shown more effective than intraperitoneal administration [55]. However, the dose of anthocyanin is limited in a certain range, because high dose induces cytotoxicity [56].

3.1. Decrease the Serum Lipid Level

The extracts from black rice (EBR) are rich in anthocyanins. It could enhance low-density lipoprotein clearance, regulate low-density lipoprotein receptor production on the cell surface membrane and maintain lipid homeostasis at low levels [56]. When old apoE-deficient mice were fed by black rice grain extracts chronically, their serum triglyceride, total cholesterol and low-density lipoprotein cholesterol decreased, and plaque stabilization enhanced [57]. Adding EBR to the diet of hypercholesterolemic rats could effectively decrease the serum lipid levels [59]. It also showed suppressive activity on thrombin time that was inhibited on a dose-dependent way [82].

3.2. Scavenge the Free Radicals

Extracts from black rice (EBR) can eliminate 1,1-diphenyl-2-picrylhydrazyl (DPPH) radical, superoxide anion radical (•O^{2-}), and hydroxyl radical (•OH) induced by light, and possess the ability to protect retina from radical damages [61]. It also can prevent the supercoiled DNA strand scission that was induced by reactive oxygen species, specifically, peroxyl radical and hydroxyl radicals [60]. For rats, when they were fed by EBR, their oxidative stress and blood oxidized glutathione were lower than the fructose-fed controls [62].

3.3. Prevent Cardiovascular and Heart Diseases

When add 10 g of EBR to the diets of the patients with coronary heart disease, their plasma total antioxidant capacity and the sensitive C-reactive protein (hs-CRP) were higher than the patients who supplied by white rice fraction, but the plasma levels of soluble vascular cell adhesion molecule-1 (sVCAM-1) and soluble CD40 ligand (sCD40L) were lower [66]. When human umbilical vein endothelial cells were treated by EBR, their proliferation and migration can be inhibited, so it had the ability of antiangiogenesis [69].

3.4. Anti-Inflammatory

The inflammation response is one of the body's oldest defense mechanisms. Wang *et al.* showed that anthocyanins can exert a significant anti-inflammatory activity with almost no change to the antioxidative status *in vivo* [66]. Min *et al.* indicated that C-3-G and its metabolites potently inhibited many inflammatory mediators *in vivo* and *in vitro* [79]. If EBR is orally administered, its main constituent, C-3-G, may be metabolized to cyanidin and/or protocatechuic acid, which express potent anti-inflammatory effects by regulating NF-κB and MAPK activation [79].

Table 2. The Functions of Anthocyanin Extracted from Black Rice

Component	Function	Model species	Reference
Black rice grain extracts	Decrease the serum lipid level	Mice & rats	[56-59]
Extraction of black rice, mainly C-3-G and P-3-G	Scavenge free radicals	Human & rats	[60-62]
Anthocyanin extract from black rice	Protect endothelial cells	Endothelia cells	[63]
Anthocyanin-rich extract from black rice	Inhibit atherosclerotic	Mice	[64]
Anthocyanin-rich extract from black rice	Ameliorate obesity	Rats	[58, 65]
Black rice fraction	Prevent heart disease and cardiovascular	Patients	[66-69]
Rice bran fractions	Good to diabetes	Mice & rats & humans	[70-73]
P-3-G C-3-G	Anticancer	Mice & human	[74-77]
C-3-G	Antiallergic	mice	[55,78]
C3G from black rice	Anti-inflammatory	Mice	[79]
Anthocyanin compounds from black rice	Antifatigue	Human	[80]
Purple rice bran extract	Antiangiogenesis	human umbilical vein endothelial cells	[69]
Anthocyanin extracts from black rice	Reduce the adverse effect of alcohol	mice	[81]
Anthocyanin-rich extract from black rice	Improve metabolic abnormalities	Rats	[62]

C-3-G: cyanidin 3-glucoside; P-3-G: Peonidin 3-glucoside.

3.5. Improve Metabolic Abnormalities

The EBR can improve certain metabolic abnormalities. When rats were fed by EBR, the concentrations of plasma thiobarbituric acid reactive substances were lower than the fructose-fed controls, but failed to reverse the fructose induced hyperinsulinemia [62]. However, in some way, EBR can reduce the body weight gain of rats significantly, because the increasing expression of hepatic CPT-1 mRNA can lead to fatty acid oxidizing and serum triglyceride decreasing [58].

3.6. Anticancer

Malignancies are fatal diseases that threaten the people all around the world. EBR have anticancer effects. When mice were fed by red glutinous rice, they could survive longer, because of the effects on the suppression of tumor growth by elevating S phase and blocking in the step from S phase to G2 phase [74]. Peonidin 3-glucoside and cyanidin 3-glucoside could induce cancer cells die by inhibiting the DNA binding activity and exerting an inhibitory effect of cell invasion on various cancer cells (SCC-4, Huh-7, and HeLa) [75]. Hui *et al.* [76] and Leardkamolkarn *et al.* [77] also found that EBR could induce different degrees of apoptosis, and suppress angiogenesis to against human breast cancer cells.

3.7. Anti-Allergic and Dispel the Effects of Alcohol

Cyanidin 3-glucoside (C-3-G) could inhibit the expression of allergic cytokines stimulated with IgE–antigen, and inhibit the effects against histamine- or compound 48/80-induced scratching behaviors in mice [55]. Moreover, it has a beneficial effect on reducing the adverse effect of alcohol [81].

3.8. Inhibit the α-Glucosidase Activity, and Ameliorate Hyperglycemia

Yao *et al.* found that black rice had the highest α-glucosidase inhibitory activity among the colored grains, and a significant positive correlation of the antioxidant activity and α-glucosidase inhibitory activity with total anthocyanin content and total phenolic content [83]. The result has implication in the management of diabetes mellitus. Guo *et al.* reported that EBR is able to eliminate the glucose intolerance and hyperlipidemia initiated by 4 weeks of a high-fructose diet fed to rats [62]. Further studies show that C-3-G exerts a protective role against H_2O_2- or tumor necrosis factor α- induced insulin resistance in 3T3-L1 adipocytes by inhibiting the c-Jun NH_2-terminal kinase signal pathway [84]. These results indicate that anthocyanins themselves are directly beneficial for the improvement of insulin sensitivity.

4. Prospects

Understanding of the pathways of anthocyanin biosynthesis and its regulation can contribute to production of black rice high in anthocyanins. With the development of biotechnology, more and more advanced technologies have been applied to improve the functional compounds of rice. Genetic engineering of rice crops to enhance the antioxidants is highly beneficial to human health. Expression of maize C1 and R-S regulatory genes in rice grain resulted in a dark brown pericarp of the C1/R-S homozygous lines, and the major flavonoids, dihydroquercetin (taxifolin), dihydroisorhamnetin (3′-*O*-methyl taxifolin) and 3′-*O*-methyl quercetin have been identified in the rice grain [18]. Overexpression of the ANS in a rice mutant results in the ANS transcript and enzyme activity approximately ten and 4-fold increase, respectively [29]. Though only few successful cases in breeding rice with high anthocyanins, new transgenic black rice that accumulates anthocyanin in the endosperm are expected to be produced in the future when the anthocyanin synthesis pathway and its regulation mechanisms are well understood.

In food industry, the anthocyanin-rich black rice extracts can be widely used as food additive. The black rice grain is not only high in antioxidant, but also rich in other phytochemicals. Thus, black rice is strongly recommended as a good resource of food additive. There are some successful cases using red rice [85] and black rice [86-87] in wine or low-alcohol beverage production. The products show higher antioxidant activities and benefits to human health. New products are expected to produce using the black rice as food additives to improve human health.

ACKNOWLEDGMENTS

Financial support for this work was provided in part by the Zhejiang Provincial Natural Science Foundation of China (R3080016), the Program for the New Century Excellent Talents in University (NCET-08-0484), and the Science and Technology Department of Zhejiang Province (2010C32082).

REFERENCES

[1] Kadam BS. An anthocyanin inhibitor in rice. *J Heredity* 27, 405-408, 1936.

[2] Reddy VS, Dash S, Reddy AR. Anthocyanin pathway in rice (*Oryza sativa* L.): identification of a mutant showing dominant inhibition of anthocyanins in leaf and accumulation of proanthocyanidins in pericarp. *Theor Appl Genet* 91, 301-312, 1994.

[3] Min B, McClung AM, Chen MH. Phytochemicals and antioxidant capacities in rice brans of different color. *J Food Sci* 76, 117-126, 2011.

[4] Finocchiaro F, Ferrari B, Gianinetti A. A study of biodiversity of flavonoid content in the rice caryopsis evidencing simultaneous accumulation of anthocyanins and proanthocyanidins in a black-grained genotype. *J Cereal Sci* 51, 28–34, 2010.

[5] Zhang MW, Zhang RF, Zhang FX, Liu RH. Phenolic profiles and antioxidant activity of black rice bran of different commercially available varieties. *J Agric Food Chem* 58, 7580-7587, 2010.

[6] Zhu F, Cai YZ, Bao JS, Corke H. Effect of γ-irradiation on phenolic compounds in rice grain. *Food Chem* 120, 74-77, 2010.

[7] Ryu SN, Park SZ, Ho CT. High performance liquid chromatographic determination of anthocyanin pigments in some varieties of black rice. *J Food Drug Anal 6*, 729-736, 1998.

[8] Sompong R, Siebenhandl-Ehn S, Linsberger-Martin G, Berghofer E. Physicochemical and antioxidative properties of red and black rice varieties from Thailand, China and Sri Lanka. *Food Chem* 124, 132-140, 2011.

[9] Lee JH. Identification and quantification of anthocyanins from the grains of black rice (*Oryza sativa* L.) varieties. *Food Sci Biotechol* 19, 391-397, 2010.

[10] Ichikawa H, Ichiyanagi T, Xu B, Yoshii Y, Nakajima M, Konishi T. Antioxidant activity of anthocyanin extract from purple black rice. *J Med Food* 4, 211-218, 2001.

[11] Thunnop L, Charles FS, Sakda J, Vanna T. Antioxidants and antioxidant activity of several pigmented rice brans. *The J Agric Food Chem* 59, 193-199, 2011.

[12] Kim JK, Lee SY, Chu SM, Lim SH, Suh SC, Lee YT, Cho HS, Ha SH. Variation and correlation analysis of favonoids and carotenoids in Korean pigmented rice (*Oryza sativa* L.) cultivars. *J Agric Food Chem* 58, 12804-12809, 2010.

[13] Takahashi ME. Analysis on apiculus color genes essential to anthocyanin coloration in rice. *J Fac Agr Hokkaido Univ* 50, 266-362, 1957.

[14] Hagagal BN, Manjunath A, Goud JV. Linkage of genes for anthocyanin pigmentation in rice (Oryza sativa L.). Euphytica 30, 747-754, 1980.

[15] Sweeney MT, Thomson MJ, Pfeil BE, McCouch S. Caught red-handed: Rc encodes a basic helix-loop-helix protein conditioning red pericarp in rice. *The Plant Cell* 18, 283-294, 2006.

[16] Wang XG, Ji ZJ, Cai J, Ma LY, Li XM, Yang CD. Construction of near isogenic lines for pericarp color and evaluation on their near isogenicity in rice. *Chin J Rice Sci* 23, 135-140, 2009.

[17] Wang CX, Shu QY. Fine mapping and candidate gene analysis of purple pericarp gene Pb in rice (*Oryza sativa L.*). *Chinese Sci Bull* 52, 3097-3104, 2007.

[18] Shih CH, Chu H, Tang LK, Sakamoto W, Maekawa M, Chu IK., Wang MF, Lo C. Functional characterization of key structural genes in rice flavonoid biosynthesis. *Planta* 228, 1043–1054, 2008.

[19] Kim JH, Lee YJ, Kim BG, Lim Y, Ahn JH. Flavanone 3beta-hydroxylases from rice: key enzymes for favonol and anthocyanin biosynthesis. *Mol Cell* 25, 312-316, 2008.

[20] Lepiniec L, Debeaujon I, Routaboul JM, Baudry A, Pourcel L, Nesi N, Caboche M. Genatics and biochemistry of seed flavonoids. *Ann Rev Plant Biology* 57, 405-430, 2006.

[21] Reddy VS, Goud KV, Sharma R, Reddy AR. Ultraviolet-B responsive anthocyanin production in a rice cultivar is associated with a specific phase of phenylalanine ammonialyase biosynthesis. *Plant Phisiol* 105, 1059-1066, 1994.

[22] Annamraju DS, Rameshwar S. Purification and characterization of UV-B induced phenylalanine ammonialyase from rice seedlings. *Phytochemistry* 50:729-737, 1999.

[23] Kim BG, Kim JH, Min SY, Shin KH, Kim JH, Kim HY, Ryu SN, Ahn JH. Anthocyanin content in rice is related to expression levels of anthocyanin biosynthetic genes. *J Plant Biol* 50, 156-160, 2007.

[24] Reddy AR, Scheddler B, Madhuri G, Srivastava MN, Kumar A, Sathyanarayanan PV, Nair S, Mohan M. Chalcone synthase in rice (*Oryza sativa L.*): Detection of the CHS protein in seedlings and molecular mapping of the chs locus. *Plant Mol Biol* 32, 735-743, 1996.

[25] Han YY, Wang JW, Han N, Liu QJ, Liu TM, Guan FM, Ming F. Duplication and sequence divergence of rice chalcone synthase genes. *Russ J Plant Physiol* 56, 417, 2009b.

[26] Druka A, Kudrna D, Rostoks N, Brueggeman R, Wettstein DV, Kleinhofs A. Chalcone isomerase gene from rice (*Oryza sativa*) and barley (*Hordeum vulgare*): physical, genetic and mutation mapping. *Gene* 302, 171-178, 2003.

[27] Larson RL, Bussard JB. Microsomal flavonoid 3′-monooxygenase from maize seedlings. *Plant Physiol* 80(2), 483-486, 1986.

[28] Furukawa T, Maekawa M, Oki T, Sudal I, Iida S, Shimada H, Takamure I, Kadowaki KI. The Rc and Rd genes are involved in proanthocyanidin synthesis in rice pericarp. *Plant J* 49, 91-102, 2007.

[29] Reddy AM, Reddya VS, Scheffler BE, Wienandb U, Reddy AR. Novel transgenic rice overexpressing anthocyanidin synthase accumulates a mixture of flavonoids leading to an increased antioxidant potential. *Metab Eng* 9, 95-111, 2007.

[30] Chandler VL, Radicella JP, Robbins TP, Chen J, Turks D. Two regulatory genes of the maize anthocyanin pathway are homologous: isolation of B utilizing R genomic sequences. *Plant Cell* 1, 1175–1183, 1989.

[31] Hu JP, Anderson B, Wessler SR. Isolation and characterization of rice R genes: evidence for distinct evolutionary paths in rice and maize. *Genetics* 142, 1021-1031, 1996.

[32] Paz-Ares JD, Wien GU, Peterson PA, Saedler H. The regulatory c1 locus of Zea mays encodes a protein with homology to myb proto-oncogene products and with structural similarities to transcriptional activators. *EMBO J* 6, 3553-3558, 1987.

[33] Zhu Y, Cai XL, Wang ZY, Hong MM. An interaction between a MYC protein and an EREBP protein is involved in transcriptional regulation of the rice *Wx* gene. *J Biol Chem* 278, 47803-47811, 2003.

[34] Ahn S, Tanksley SD. Comparative linkage maps of the rice and maize genomes. *Proc Natl Acad Sci USA* 90, 7980-7984, 1993.

[35] Reddy VS, Scheffler BE, Wienand U, Wessler SR, Reddy AR. Cloning and characterization of the rice homologue of the maize C1 anthocyanin regulatory gene. *Plant Mol Biol* 36, 497-498, 1998.

[36] Saitoh K, Onishi K, Mikami I, Thidar K, Sano Y. Allelic Diversification at the C (OsC1) locus of wild and cultivated rice: Nucleotide changes associated with phenotypes. *Genetics* 168, 997-1007, 2004.

[37] Fan FJ, Fan YY, Du JH, Zhuang JY. Fine mapping of C (Chromogen for anthocyanin) gene in rice. *Rice Sci* 15, 1-6, 2008.

[38] Gao DY, He B, Zhou YH, Sun LH. Genetic and molecular analysis of a purple sheath somaclonal mutant in japonica rice. *Plant Cell Rep* 30, 901–911, 2011.

[39] Kinoshita T, Maekawa M. Inheritance of purple leaf color found in indica rice. Genetical studies on rice plant, XCIV. *J Fac Agr Hokkaido Univ.* 62, 453-466, 1986.

[40] Wataru S, Taku O, Keisuke K, Chikara M, Akira S, Minoru M, Kazuhiko N, Masahiko M. The *Purple leaf* (*Pl*) locus of rice: the Pl^w allele has a complex organization and includes two genes encoding basic helix-loop-helix proteins involved in anthocyanin biosynthesis. *Plant Cell Physiol* 42(9), 982-991, 2001.

[41] Kawahigashi H, Hirose S, Iwai T, Ohashi Y, Sakamoto W, Maekawa M, Ohkawa Y. Chemically induced expression of rice OsB2 under the control of the OsPR1.1 promoter confers increased anthocyanin accumulation in transgenic rice. *J Agric Food Chem* 55, 1241-1247, 2007.

[42] Hirose S, Kawahigashi H, Tagiri A, Ohkawa Y. Herbicide-induced anthocyanin accumulation in transgenic rice by expression of rice OsB2 under the control of rice CYP72A21 promoter. *J Agric Food Chem* 56, 1259-1263, 2008.

[43] Hu JP, Reddy VS, Wessler SR. The rice R gene family: two distinct subfamilies containing several miniature inverted-repeat transposable elements. *Plant Mol Biol* 42, 667–678, 2000.

[44] Kim C, Kikuchi S, Satoh K, Kim J, Kim D, Kim Y, Park S, Lee J, Yoon U. Genetic analysis of seed-specific gene expression for pigmentation in colored rice. *Biochip J* 3(2), 125-129, 2009.

[45] Kim CK, Kikuchi S, Kim YK, Park SH, Yoon UH, Lee GS, Choi JW, Kim YH, Park SC. Computational identification of seed-specific transcription factors involved in anthocyanin production in black rice. *BioChip J* 4, 247-255, 2010.

[46] Kim CK, Cho MA, Choi YH, Kim JA, Kim YH, Kim YK, Park SH. Identification and characterization of seed-specific transcription factors regulating anthocyanin biosynthesis in black rice. *J Appl Genet* 52, 161–169, 2011.

[47] Kuo TH, Deng GC, Yi TH, Ching HK. Abscisic acid-induced hydrogen peroxide is required for anthocyanin accumulation in leaves of rice seedlings. *J Plant Physiol* 165, 1280-1287, 2008.

[48] Roychoudhury A, Basu S, Sengupta DN. Effects of exogenous abscisic acid on some physiological responses in a popular aromatic indica rice compared with those from two traditional non-aromatic indica rice cultivars. *Acta physiol plant* 31, 915-926, 2009.

[49] Roychoudhury A, Basu S, Sarkar SN, Sengupta DN. Comparative physiological and molecular responses of a common aromatic indica rice cultivar to high salinity with non-aromatic indica rice cultivars. *Plant Cell Rep* 27, 1395–1410, 2008.

[50] Daiponmaka W, Theerakulpisut P, Thanonkaoc P, Vanavichit A, Prathephaa P. Changes of anthocyanin cyaniding-3-glucoside content and antioxidant activity in Thai rice varieties under salinity stress. *Sci Asia* 36, 286-291, 2010.

[51] Farooq M, Wahid A, Lee DJ. Exogenously applied polyamines increase drought tolerance of rice by improving leaf water status, photosynthesis and membrane properties. *Acta Physiol Plant* 31, 937–945, 2009a.

[52] Farooq M, Wahid A, Basra SMA , Din IU. Improving water relations and gas exchange with brassinosteroids in rice under drought stress. *J Agron Crop Sci* 195, 262-269, 2009b.

[53] Pramod K. Photosynthetic and yield performance of rice (*Oryza sativa*) genotypes under low temperature condition in hills. *Indian Council Agric Res* 72, 383-388, 2002.

[54] Kaneda I, Kubo F, Sakurai H. Antioxidantive components in the extracts of black rice brans. *J Health Sci* 52, 495-511, 2006.

[55] Han SJ, Ryu SN, Trinh HT, Joh EH, Jang SY, Han MJ, Kim DH. Metabolism of cyanidin-3-*O*-β-D-glucoside isolated from black colored rice and its antiscratching behavioral effect in mice. *J Food Sci* 74, 253-258, 2009a.

[56] Sangkitikomol W, Tencomnao T, Rocejanasaroj A. Effects of Thai black sticky rice extract on oxidative stress and lipid metabolism gene expression in HepG2 cells. *Genet Mol Res* 9, 2086-2095, 2010.

[57] Xia XD, Ling WD, Ma J, Xia M, Hou MJ, Wang Q, Zhu HL, Tang ZH, An anthocyanin-rich extract from back rice enhances atherosclerotic plaque stabilization in apolipoprotein E–deficient mice. *J Nutr* 136, 2220-2225, 2006.

[58] Yang Y, Andrews MC, Hu Y, Wang DL, Qin Y, Zhu Y, Ni H, Ling WH. Anthocyanin extract from black rice significantly ameliorates platelet hyperactivity and hypertriglyceridemia in dyslipidemic rats induced by high fat diets. *J Agric Food Chem* 59, 6759-6764, 2011.

[59] Zawistowski J, Kopec A, Kitts DD. Effects of a black rice extract (*Oryza sativa* L. *indica*) on cholesterol levels and plasma lipid parameters in Wistar Kyoto rats. *J Funct Food I,* 50-56, 2009.

[60] Hu C, Zawistowski J, Ling W, Kitts DD. Black rice (*Oryza sativa* L. indica) pigmented fraction suppresses both reactive oxygen species and nitric oxide in chemical and biological model systems. *J Agric Food Chem* 51, 5271–5277, 2003.
[61] Tanaka J, Nakanishi T, Ogawa K, Tsuruma K, Shimazawa M, Shimoda H, Hara H. Purple rice extract and anthocyanidins of the constituents protect against light-induced retinal damage in Vitro and in Vivo. *J Agric Food Chem* 59, 528-536, 2011b.
[62] Guo HH, Ling WH, Wang Q, Liu C, Hu Y, Xia M, Feng X, Xia XD. Effect of Anthocyanin-Rich Extract from Black Rice (*Oryza sativa* L. *indica*) on Hyperlipidemia and Insulin Resistance in Fructose-Fed Rats. *Plant Food Hum Nutr* 62, 1–6, 2007.
[63] Zhang MW, Zhang RF, Guo BJ, Chi JW, Wei ZC, Xu ZH. The protective effects of anthocyanidin extracted from black rice fraction on endothelia cells injured by oxidative stress. *Acta Nutrimenta Sinica* 28, 216-220, 2006.
[64] Yu XP, Xia XD, Xia M, Wang Q, Zhu HL, Hu Y, Chi DS, Lin WH. Effect of anthocyanin-rich extract from black rice on stability of atherosclerotic vulnerable plaque. *Chin J Public Health* 22, 155-156, 2006.
[65] Hu Y, Guo HH, Wang Q, Feng X, Liu C, Ling WH. Effects of anthocyanin-rich extract from black rice on rats with high fat diets-induced obesity. *Food Sci* 29, 376-379, 2008.
[66] Wang Q, Han P, Zhang M, Xia M, Zhu H, Ma J, Hou M, Tang Z, Ling W. Supplementation of black rice pigment fraction improves antioxidant and anti-inflammatory status in patients with coronary heart disease. *Asia Pac J Clin Nutr* 16, 295-301, 2007.
[67] Kris-Etherton PM, Hecker KD, Bonanome A, Coval SM, Binkoski AE, Hilpert KF, Griel AE, Etherton TD. Bioactive compounds in foods: Their role in the prevention of cardiovascular disease and cancer. *Am J Med* 113, 71-88, 2002.
[68] Dell'Agli M, Busciala A, Bosisio E. Vascular effects of wine polyphenols. *Cardiovasc Res* 63, 593-602, 2004.
[69] Tanaka J, Nakamura S, Tsuruma K, Shimazawa M, Shimoda H, Hara H Purple rice (*Oryza sativa* L.) extract and its constituents inhibit VEGF - induced angiogenesis. *Phytother Res DOI:* 10.1002/ptr.3533, 2011a.
[70] Son MJ, Rico CW, Nam SH, Kang MY. Effect of oryzanol and ferulic acid on the glucose metabolism of mice fed with a high-fat diet. *J Food Sci* 76, H7-H10, 2011.
[71] Kim SM, Rico CW, Lee SC, Kang MY. Modulatory effect of rice bran and phytic acid on glucose metabolism in high fat-fed C57BL/6N mice. *J Clin Biochem Nutr* 47, 12-17, 2010c.
[72] Ardiansyah, Hitoshi S, Takuya K, Kousaku O, Katsumi H, Michio K. Rice bran fractions improve blood pressure, lipid profile, and glucose metabolism in stroke-prone spontaneously hypertensive rats. *J Agric Food Chem* 54, 1914-1920, 2006.
[73] Qureshi AA, Sami SA, Khan FA. Effects of stabilized rice bran, its soluble and fiber fractions on blood glucose levels and serum lipid parameters in humans with diabetes mellitus Type I and II. *J Nutr Biochem* 13(3), 175-187, 2002.
[74] Koide T, Kamei H, Hashimoto Y, Kojima T, Haseqawa M. Antitumor effect of hydrolyzed anthocyanin from grape rinds and red rice. *Cancer Biother Radiopharm* 11, 273-7, 1996.

[75] Chen PN, Kuo WH, Chiang CL, Chiou HL, Hsieh YS, Chu SC. Black rice anthocyanins inhibit cancer cells invasion via repressions of MMPs and u-PA expression. *Chem-Biol Interact* 163, 218-229, 2006.

[76] Leardkamolkarn V, Thongthep W, Suttiarporn P, Kongkachuichai R, Wongpornchai S, Wanavijitr A. Chemopreventive properties of the bran extracted from a newly-developed Thai rice: The Riceberry. *Food Chem* 125, 978-985, 2011.

[77] Hui C, Bin Y, Xiaoping Y, Chunye C, Mantian M, Wenhua L. Anticancer activities of an anthocyanin-rich extract from black rice against breast cancer cells in vitro and in vivo. *Nutr Cancer* 62(8), 1128-38, 2010.

[78] Li LC, Yan GH, Qin XZ, Shen GH. Anti-allergic effects of anthocyanin-rich extract from black rice. *Chin. J. Public Health* 27(5), 613-615, 2011.

[79] Min SW, Ryu SN, Kim DH. Anti-inflammatory effects of black rice, cyanidin-3-*O*-β-D-glycoside, and its metabolites, cyanidin and protocatechuic acid. *Intl Immunopharmacol* 10, 959-966, 2010.

[80] Deng WH, Li XS. Anti-fatigue biochemical mechanism of black rice pigment and its research prospect. *Sci Technol Food Ind* 31(6), 372-374, 2010.

[81] Hou ZH, Qin PY, Ren GX. Effect of anthocyanin-rich extract from black rice (*Oryza sativa* L. Japonica) on chronically alcohol-Induced liver damage in rats. *J Agric Food Chem* 58, 3191-3196, 2010.

[82] Han S, Kim JS, Chae SW, Kang SS. Biological screening of extracts from the colored rice cultivars. *Korean J Pharmacogn* 35(4), 346-349, 2004.

[83] Yao Y, Sang W, Zhou MJ, Ren GX. Antioxidant and alpha-glucosidase inhibitory activity of colored grains in Chian. *J Agric Food Chem* 58(2), 770-774, 2010.

[84] Guo H, Ling W, Wang Q, Liu C, Hu Y, Xia M. Cyanidin 3-glucoside protects 3T3-L1 adipocytes against H2O2- or TNF-α-induced insulin resistance by inhibiting c-Jun NH2-terminal kinase activation. *Biochem Pharmacol* 75, 1393-1401, 2008.

[85] Ueda S, Ueki T, Ohba R, Teramoto Y, Yoshizawa K. Ethanol fermentation of aromatic red rice without cooking, studies on red rice wine brewing (part 1). *J Biosci Bioeng* 70, 326-328, 1990.

[86] Koguchi M, Saigusa N, Teramoto Y. Production and antioxidative activity of mead made from honey and black rice (*Oryza sativa* var. *Indica* cv. *Shiun*). *J Inst Brew* 115, 238-242, 2009.

[87] Koguchi M, Saigusa N, Teramoto Y. Antioxidative activity of alcoholic beverages made from purple rice (*Oryza sativa* var. *Indica* cv. *Shiun*). *Food Sci Technol Res* 16, 157-162, 2010.

In: Anthocyanins: Structure, Biosynthesis and Health Benefits ISBN: 978-1-62257-329-5
Editor: Noboru Motohashi

Chapter 5

EFFECT OF ANTHOCYANIN EXTRACTS FROM *JUSTICIA MATAMMENSIS* AND *JUSTICIA LAXA* ON SICKLE CELLS

P. T. Mpiana[1,*], M. T. Bokota[2], J. P. Mbula[2], K. N. Ngbolua[3], D. S. T. Tshibangu[1], E. K. Atibu [1], J. T. K. Kwembe[2] and M. K. Kakule[4]

[1]Département de Chimie, Faculté des Sciences B.P. Kinshasa XI, D R Congo
[2]Département de Chimie, Faculté des Sciences, Université de Kisangani, B.P. Kisangani, D R Congo
[3]Département de Biologie, Faculté des Sciences B.P. Kinshasa XI, D R Congo
[4]Département de Physique, Faculté des Sciences B.P. Kinshasa XI, D R Congo

ABSTRACT

Background

Sickle cell disease is due to the aggregation of deoxy haemoglobin S under hypoxic conditions conducing to erythrocytes shape modification and anaemia. In Sub-Saharan Africa, area mainly affected by this genetic disease, medicinal plants are intensely used to manage sickle anaemia. Some Congolese plants have shown an *in vitro* and *in vivo* antisickling activity. This biological activity seems to be due mainly to anthocyanin extracts. Anthocyanin extracts of some species of *Justicia* had displayed interesting antisickling activity *in vitro*. In this work biological activity of anthocyanin extracts from *Justicia matammensis* and *Justicia laxa* are tested using three different biological tests.

[*] Corresponding Author: Prof Dr PT Mpiana, e-mail: ptmpiana@yahoo.fr; tel. +243 81 811 6019

Materials and Methods

Emmel, osmotic fragility and hypoxic induced sickle erythrocyte haemolysis bioassays are used to evaluate the effect of anthocyanin extracts from Justicia matammensis and Justicia laxa on sickle cells .

Results

As for other Justicia species, anthocyanin extracts from Justicia matammensis and Justicia laxa were found to possess antisickling activity. The treated sickle erythrocyte showed the re-appearance of the normal biconcave form of red blood cells for anthocyanins from the two Justicia species. Justicia laxa that possess the highest anthcyanins ratio showed the highest activity with 85% of normalized cells for a minimal concentration of 0,63 μg/mL of anthocyanins. The rate of inhibition of hypoxic induced haemolysis and the osmotic fragility of drepanocytes increased upon treatment with anthocyanin extracts for both J. matammensis and J. laxa ¶

Conclusion

These results show that anthocyanins might act not only on hemoglobin S polymerization but also on cell membranes. This confirms earlier reports on the antisickling properties of anthocyanins of Congolese plants justifying their use in the management of sickle cell disease by Congolese traditional healers.

Keywords: Antisickling activity, anthocyanin extracts, polymerisation, haemolysis, osmotic fragility

INTRODUCTION

Sickle cell disease (SCD) is one of the most prevalent hematologic genetic disorders in the world with prominent morbidity and mortality. While the disease may affect various ethnic groups such as the people of the Hispanic and Middle East descent, it affects mostly those of African descent [1-4]. SCD result from a single point mutation of β^{Glu6} in Hb to β^{val6} in Hb S. The hydrophobic interactions, which occur only in the deoxy Hb S, induce polymerization of the molecules into fibers. The formation of Hb S polymers causes the normally flexible red blood cells (RBC) to adopt rigid, sicklelike shapes that block small capillaries, causing both local tissue damage and severe pain [2,5]. The most clinical manifestations of this disease are largely due to hemolytic processes leading to severe anemia and vaso-occlusive crises [2,4].

Several therapies have been proposed and many chemical substances investigated for their possible role in the management of SCD. Among the potential agents employed to prevent or reverse sickling include: hydroxyurea, erythropoietin, tucaresol, ciklervit. Unfortunately some of these products have yielded many pronounced side effects. Proposed medullar transplantation is not only too expensive but with many incompatibility [1, 3].

In the search for effective and affordable chemotherapeutic agents with less adverse effects on SCD patients, many researchers have shown the antisickling effectiveness of medicinal plants extracts [6-12]. In fact, the use of natural products in attempts at inhibiting sickling could be as old as the sickle cell Disease. Some African ethnomedicinal remedies for the treatment of SCD have been reported in the literature [13].

In Democratic Republic of Congo (DRC), more than fifty medicinal plants used in Congolese folk medicine for the management of SCD have shown antisickling activity among which *Justicia secunda* Valh [14-32]. The biological activity of some of these plants could be due to anthocyanins [14, 15, 17-23]. Anthocyanins from three others species of *Justicia* not mentioned by traditional practitioners among plants used in the management of SCD showed interesting *in vitro* antisickling activity [14].

Anthocyanins, one of the major groups of natural pigments belonging to the secondary metabolite group of flavonoids, are known as responsible of the orange, red, and blue colors in fruits, vegetables, flowers, and other storage tissues in plants. Thus, anthocyanins have become important as food additives. However, interest in anthocyanins has recently intensified because of their possible health benefits. Epidemiological investigations have indicated that benefits include protection against liver injuries, significant reduction of blood pressure, improvement of eyesight, strong anti-inflammatory and antimicrobial activities, inhibition of mutations caused by mutagens from cooked food, and suppression of proliferation of human cancer cells , decreasing capillary permeability and fragility, membrane strengthening etc [33-38].

In this chapter, antisickling activity of anthocyanin extracts from leaves of two species of *Justicia*, J. *matammensis* and *J. laxa*, are studied by using three different biological tests.

MATERIALS AND METHODS

Plant material

Leaves of *Justicia matammensis* and *Justicia laxa* were collected from plants growing in Kisangani and its surroundings, DRC. Botanical identification was made by M. Bola, Mabay and Kombozi in Herbarium of the Science Faculty of the University of Kisangani. Voucher specimen are on deposit at the INERA (Institut National d`Etudes et de Recherches Agronomiques) Herbarium in Yangambi.

Extraction and Chemical Screening

The dried powdered plant material (10 g) was extracted by cold percolation with 95 % EtOH and water (100 mL x 1) for 48 hrs. Extracts were filtered and the solvent was evaporated under reduced pressure using a rotary evaporator. A chemical screening has been done on the two plants using an aqueous or organic fraction following an established protocol [39]. Extraction of anthocyanins was then done by maceration of 100 g of dried powdered plant material in 200 mL of acidified methanol (0, 01% v/v HCl) according to the universal procedures [34,39].

Biological Material

The sodium citrate suspension of blood samples used in the evaluation of the antisickling activity of the plant extracts in this study were taken from known sickle cell children and adolescent patients attending the "Centre médical Gracia Fondation", in Kisangani area, DRC. None of the patients had been transfused recently with AA blood. All antisickling experiments were carried out with freshly collected blood. In order to confirm their SS nature, the above-mentioned blood samples were first characterized by hemoglobin electrophoresis on cellulose acetate gel at pH 8.5. They were found to be SS blood and were then stored at +4°C in a refrigerator. Ethical clearance on the use of SS blood was strictly observed according to international rules.

Bioassays

Test of Emmel

Sickle cell blood was diluted with 150 mM phosphate buffered saline (NaH_2PO_4 30mM, Na_2HPO_4 120mM, NaCl 150mM) and mixed with an equivalent volume of 2% sodium metabisulfite. A drop from the mixture was spotted on a microscope slide in the presence or absence of anthocyanin extracts and covered with a cover slip. Paraffin was applied to seal the edges of the cover completely to exclude air (hypoxia).

Normalization, defines as the passage of the sickled shape into the circular and normal biconcave form after treatments of red blood cells with anthocyanin extracts, was quantitatively evaluated by measuring various cellular parameters such as area, perimeter and radius using Motic Images package software. A zoom 6x Cannon-type digital camera was used to convert the photonic micrograph image into a digital image, which was then digitalized using a Motic image 2000 1.3; Motic Chine Group Co Ltd. Statistical data analysis were processed using Microcal Origin 6.1 package software.

Hypoxic Induced Haemolysis Assay

RBC were washed twice in physiological saline (NaCl 0,9 %, 1:5 v/v) by centrifugation at 3000 rpm for 10 min, re-suspended in phosphate buffer (150 mM, pH 7,4) containing 2 % sodium metabisulfite and incubated in the absence (control) or presence of anthocyanin extracts (50 µg/mL of NaCl 0,9 %) at 37° C for 60 min. At fixed time points, aliquots of the blood samples were removed and centrifuged at 3000 rpm at ambient temperature for 5 min. The absorbance of the supernatant was measured at 540 nm and was expressed as the degree of haemolysis [9, 15].

The rate of haemolysis inhibition (% HI) versus time was calculated from the absorbance of sickle erythrocyte (SS RBC) suspension by the relation:

$$HI(\%) = \frac{A_0}{A_t} x100 \tag{1}$$

With A_0 as absorbance of untreated SS RBC suspension and A_t, the absorbance of treated SS RBC suspension.

Osmotic Fragility Test

The fragility of RBC was determined by placing the cells in graded series of hypotonic saline solutions buffered at pH 7.4 with 150 mM phosphate. Concentrations ranging from 0.2% to 0.8% NaCl were made up in a final volume of 10 mL. A 10 μL sample of washed SS RBC was added to 1990 μL of each hypotonic saline solution and immediately mixed by inverting several times. The tubes were allowed to stand for 150 min at room temperature. To determine the effect of the anthocyanin extracts, 10 μL of extract (30 mg/mL) were added to 1980 μL of each hypotonic saline solution, then 10 μL of RBC added and the mixture treated as described earlier. The number of RBC not lysed at each saline concentration was determined using a photonic microscope (Motic AE 21) and a haemacytometer (Neubauer's cell). Haemolysis was calculated using the following equation: number of RBC after 150 min×100/number of RBC inoculated (0 min). The mean corpuscular fragility (determined from the concentration of saline causing 50% haemolysis of the RBC) can be obtained from a plot of lysis (%) *versus* NaCl concentration [15].

pH and Optical Absorption Measurements

The pH values were determined with Metrohm E 604 pH-meter equipped with a glass electrode. This electrode was kept soaked in 3 mol/L KCl solution and calibrated with aqueous standard buffers. A Perkin Elmer Lambda 2 UV-Visible spectrophotometer was used.

The *in vitro* bioassays were performed in triplicate and the number of observed erythrocytes was determined using Neubauer's cell counter as previously reported [15]. All results presented are mean ±SD.

RESULTS AND DISCUSSION

Phytochemical Screening and Anthocyanins Extraction

The phytochemical screening revealed the presence of alkaloids, saponins, sterols, terpenes flavonoids, tanins and anthocyanins in the two plants. This indicates a chemical similarity of these two species of *Justicia*. As for other species of *Justicia* previously studied [14, 15], both *J.laxa* and *J. matammensis* contain anthocyanins. The yield of extraction of the anthocyanins on leaves powders of the studied plants are respectively 6.36% for *Justicia laxa* and 4.62% for *Justicia matammensis*, respectively.

Effect of Anthocyanin Extracts on Sickle Cell Morphology

Figure 1 shows micrography of SS blood alone in a NaCl 0.9 % solution (negative control). Figures 2 and 3 show micrographies of the drepanocytes in the presence of anthocyanins extracted from these two species of Justicia genus.

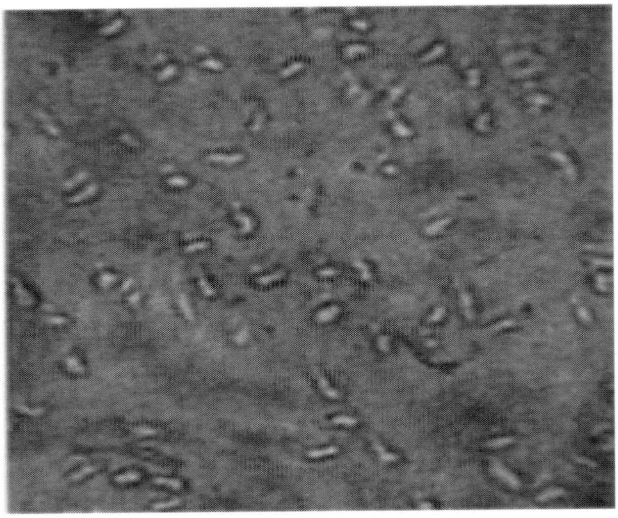

Figure.1. Morphology of RBC of untreated *SS* blood (negative control) (X500), [NaCl 0,9% ; $Na_2S_2O_4$ 2%].

Figure 1 showed that the majority of RBC are elongated (sickled), this confirms that the used blood is SS one. Figures 2 and 3 show that in the presence of anthocyanins extracts of *J. laxa* and *J. matammensis*, RBC have circular (biconcave) and normal shape, indicating the antisickling effect of anthocyanins. It should be noted that in the antisickling bioassays there was no established standard molecule that could be used as a positive control but this morphology could be compared to that of normal cell blood [15].

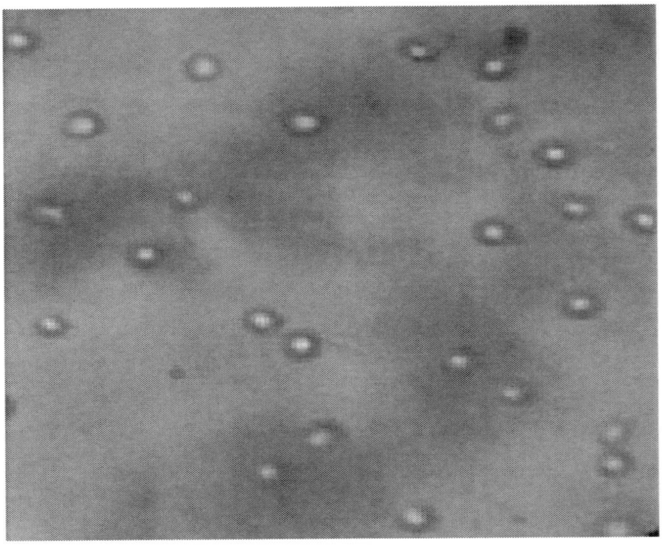

Figure 2. Morphology of SS RBC treated with anthocyanins crude extracts from *J laxa* (500 X).

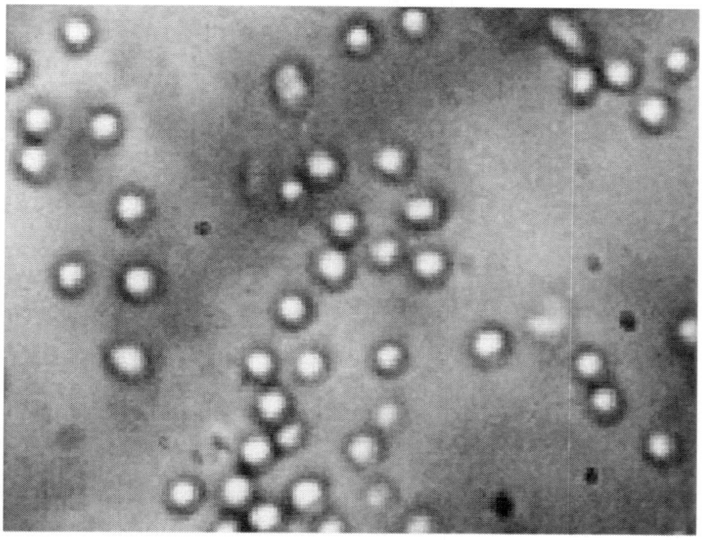

Figure 3. Morphology of SS RBC treated with anthocyanins crude extracts from *J. matammensis* (500 X).

Table 1. Average values of the perimeters, surfaces and the radius of untreated and treated sickle RBC anthocyanin extracts

Samples	Cellular perimeters (μm)	Cellular surfaces (μm^2)	Cellular radius (μm)
Control	33.12 ±2.59	21.30 ± 1.98	-
Anthocyanin extracts of *J.laxa*	19.39 ±2.43	34.02±3.32	3.32 ±0.54
Anthocyanin extracts of *J. matammensis*	20.22 ±2.18	33.13 ±3.22	3.16 ±0.51

The calculated values of perimeter, surface and radius of untreated and treated sickle RBC with the anthocyanin extracts are given in table1.

Table 1 shows that the average radius for the RBC of the sickle blood could not be calculated; because sickled RBC of untreated blood are not circular. The average radius appeared after treatment of sickle RBC with anthocyanin extracts indicating the re-appearance of the normal form of RBC.

Statistical treatment (Student test applied with a probability threshold of 0.05) [40] enabled the determination of a significant difference between the average values of both the perimeter and the surface of blood cells on the micrography, thus confirming the modification RBC morphology in the presence of anthocyanin extracts.

Effect of Anthocyanin Extracts Concentration on Sickle Cells Normalization rate

The effect of anthocyanin extract concentration of these two plants on the rate of normalization of drepanocyte forms were studied and the minimal concentration necessary to have a maximum of normalization, called minimal concentration of normalization (MCN) was evaluated. These values could be compared with those of other plants.

Figure 4 gives the evolution of the rates of normalization of the form of the drepanocytes according to the concentration of anthocyanins extracted from the two species of *Justicia*.

These curves show that the normalization rate *i.e.* the percentage of the drepanocytes which regain the normal shape, which increases with the concentration of anthocyanins for the extracts of the two species of *Justicia*. At a certain concentration, this rate reached a maximum value which remained constant although the increase of the anthocyanins concentration. Anthocyanin extracts from *J. matammensis* gave a maximal RBC normalization rate of 72 % for a MCN of 1.2 µg/mL while anthocyanin extracts from *J. laxa* reached a maximal RBC normalization rate of 85% for a MCN of only 0.63 µ g/mL. This indicates that anthocyanin extracts from *J. laxa* are more active than that from *J. matammensis*.

Figure 5 gives the values of MCN of anthocyanin extracts from *J. laxa* and *J. matammensis* compared with that of other plants.

This figure shows that a part from anthocyanin extracts of *A. cordifolia* and *F. capensis,* those from *J. laxa* had the lowest MCN. So, anthocyanins from *J. laxa* have the highest antisickling activity of anthocyanin extracts from Congolese *Justicia* species already studied [14,15].

Effect of Anthocyanin Extracts on the Haemolysis of Sickle Cells

Effect of anthocyanins on hypoxic induced membrane damage of RBC could be evaluated by comparing % of haemolysis of untreated and treated SS RBC in isotonic medium (NaCl 0,9%) by monitoring the optical density of released Hb S at 540 nm at different times.

Figure 6 shows the evolution of the haemolysis inhibition rate (HIR) of the RBC with time for the SS blood alone (control) and in presence of anthocyanin extracts of *J. laxa* and *J. matammensis*.

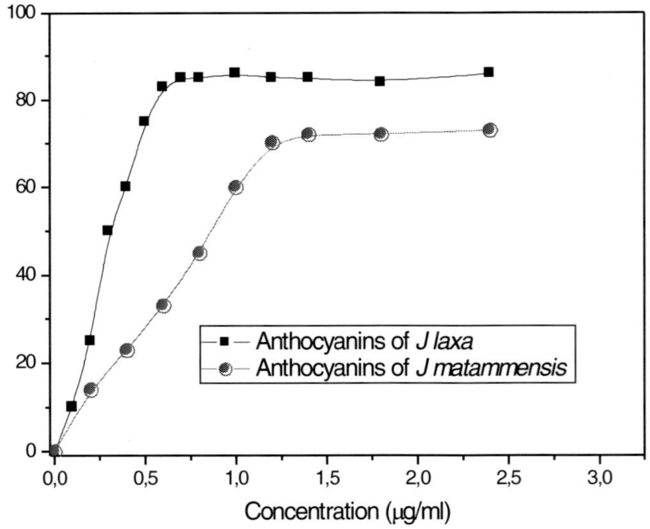

Figure 4. Concentration-dependent antisickling effect of anthocyanin extracts of *J. laxa* and *J. matammensis*.

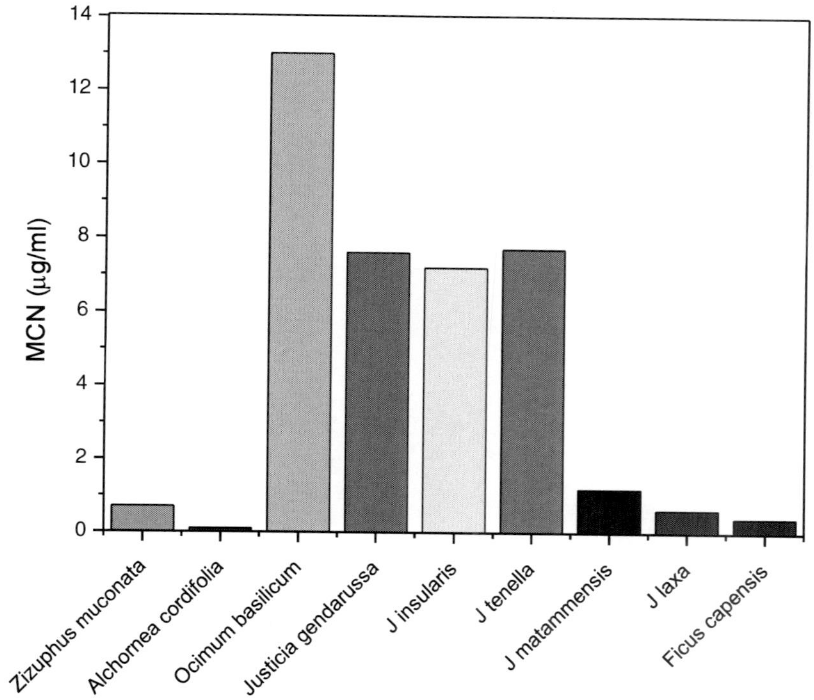

Figure 5. MCN values of *J. laxa* and *J. matammensis* compared with that of other plants.

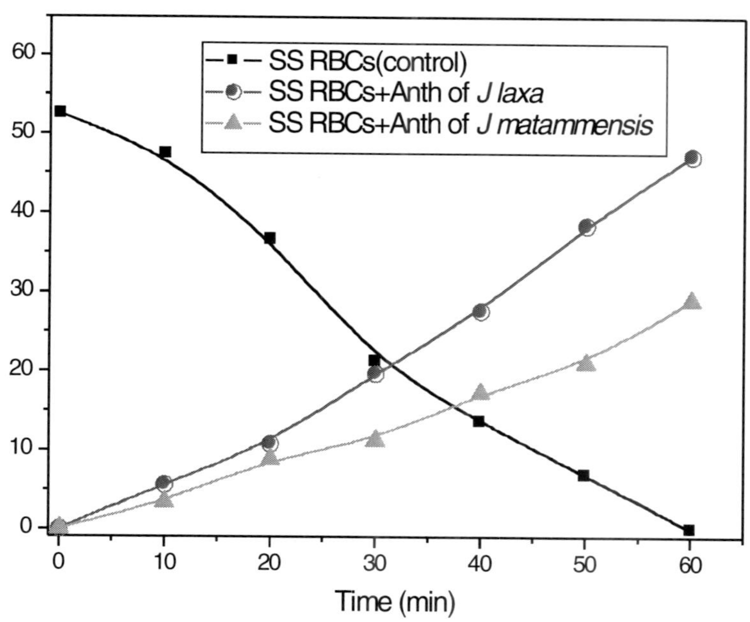

Figure 6. Effect of anthocyanin extracts of *J.laxa* and *J. matammensis* on the haemolysis of Sickle cells (NaCl 0, 9%; 2% sodium metabisulfite, wave length: 540nm).

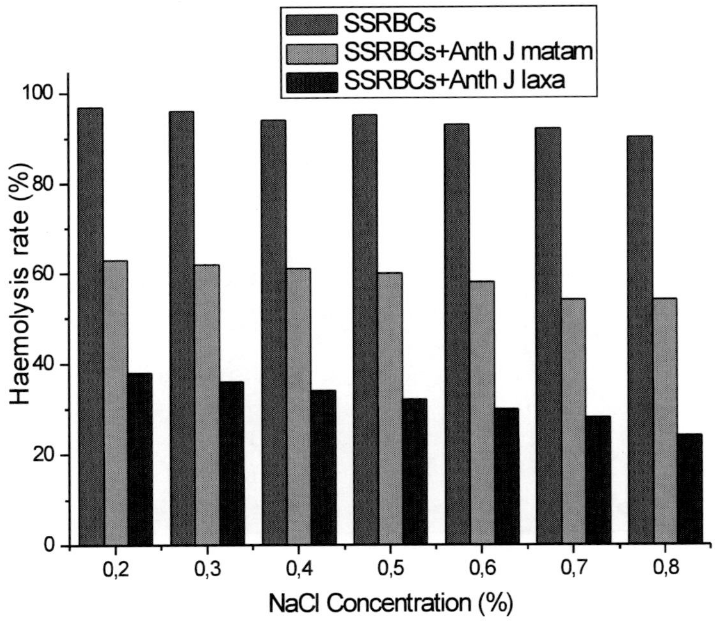

Figure 7. Haemolysis of untreated and treated sickle erythrocytes at different saline concentrations.

Figure 6 shows that the rate of inhibition of haemolysis decreases with the time in the suspension of ¦untreated SS red blood cells (control). But, after addition of anthocyanin extracts one could ¦ observe an increase of the inhibition rate for both *J.laxa* and *J. matammensis* anthocyanin extract. This increase was a little higher for *J. laxa* anthocyanins extract than for *J. matammensis* one. These results show that anthocyanin extracts possess anti-hemolytic effect. Indeed, red blood cells are unable to renew their enzymatic stock which would enable them to fight against oxidation injuries. So, if it is admitted that the anthocyanin extracts are not active, the haemolysis inhibition rate of sickle erythrocytes would decrease in the course of time [41, 42].

Such a significant activity of anthocyanin extracts can be explained by the fact that they might cross the erythrocytes' membrane for either interfering with intracellular polymerization, while be adsorbed on hemoglobin S or to scavenge the free radicals. It is thus probable that the anthocyanin extracts could exert this protective effect according to their reducing properties for preventing that the lipids membrane, hemoglobin and the enzymatic equipment are destroyed or inactivated by oxidation [43-45].

The ability of anthocyanin extracts in this study to reduce the sickle erythrocytes haemolysis could be deduced as antioxidative free radical properties. This was observed for anthocyanin extracts from other Congolese plants used to treat sickle cell anemia in Congolese traditional medicine [14-16].

Effect of Anthocyanin Extracts on RBC Osmotic Fragility

The effect of anthocyanin extracts on the sickle cells membrane stability could be evaluated by comparing the percentage of haemolysis of untreated and treated SS RBC using the osmotic fragility test [14,46]. Figure 7 shows the percentage of lysis of untreated and treated SS RBC at different saline concentrations.

This figure shows that haemolysis of sickle erythrocytes decreased with exposure to increasing concentrations of hypotonic saline solution. The percentage of lysis was higher for the sickle erythrocytes alone (control) than in the presence of anthocyanin extracts. This indicates that the anthocyanin extracts might improve the ability of sickle erythrocytes to take up water without lysis occurring. This stabilization effect could be explained by noting that anthocyanins rendered the SS RBC capable of withstanding higher concentrations of NaCl by increasing the volume of the RBC, reverting the sickling to produce biconcave cells, and, thereby, maintaining membrane integrity. Such effects have been reported for anthocyanin extracts from *Justicia secunda* [15]. These results might confirm the fact that anthocyanin extracts from *J. laxa* are more active than that from *J. matammensis.*

Conclusion

These results showed that anthocyanin extracts of *J. laxa* and *J. matammensis* would act not only on hemoglobin S polymerization but also on cell membranes. The anthocyanins effect was studied using Emmel, haemolysis and osmotic fragility tests. This confirms earlier reports on the antisickling properties of anthocyanins of Congolese plants justifying their use in the management of SCD by Congolese traditional healers.

Acknowledgments

The authors are indebted to the International Foundation for Science (IFS) and the Organization for the Prohibition of Chemical Weapons (OPCW) for the Research Grant given to one of them (Dr. NGBOLUA KOTO-TE-NYIWA, IFS Research Grant F/4921-1)

References

[1] Buchanan GR, De Baun MR, Quinn CT, Steinberg MH. Sickle Cell Disease, *Hematology* 35-47, 2004.

[2] Girot R, Bégué P. *La Drépanocytose*. John Libbey-EUROTEXT , Paris, 2003.

[3] Mehanna AS. Sickle cell anaemia and antisickling agents then and now. *Cur. Med. Chem.* 8, 79-88, 2001.

[4] Voet D and Voet JG *Biochimie*. De Boeck Université. 2nd édition Paris, 2005.

[5] Eaton WA, Hofrichter J. Hemogobin gelation and sickle cell disease. *Blood*; 70, 1245-66, 1987.

[6] Elekwa I, Monanu MO and Anosike EO. Effects of aqueous extracts of *Garcinia kola* seeds on membrane stability of HBAA, HBAS and HBSS human erythrocytes. *Gl. J. Med. Sci.* 2(2) 97 -101, 2003.

[7] Egunyomi A, Moody JO and Eletu OM. Antisickling activities of two ethnomedicinal plant recipes used for management of sickle cell anemiain Ibadan, Nigeria *Af. J. Biotechn.* 8, 20-25, 2009.

[8] Ibrahim H, Sani FS, Danladi BH & Ahmadu AA. Phytochemical and antisickling studies of the leaves of *Hymenocardia acida* Tul (Ehphorbiaceae) *Pak. J. Biol. Sci.* 10(5), 788-791, 2007.

[9] Iyamu EW, Turner EA and Asakura T. *In vitro* effects of NIPRISAN (Mix-0690), a naturally occurring, potent antisickling agent. *British Journal of Haematology,* 118, 337-343, 2002.

[10] Joppa KM, Vovor A, Eklu-Gadegbeku K, Agbonon A, Aklikokou K and Gbeasor M. Effet de *Morinda lucida* Benth(Rubiaceae) et de *Newbouldia leavis* P.Beauv.(Bignoniaceae) sur la falciformation , *Médecine Tropicale* 68(3),251-256, 2008.

[11] Moody JO, Ojo OO, Omotade OO, Adeyemo AA, Olumese PE and Ogundipe OO. Anti-sickling potential of a Nigerian herbal formula (Ajawaron HF) and the Major plant component (*Cissus populnea* L. CPK). *Phytother. Res.* 17 , 1173-1176, 2003.

[12] Ogunyemi CM, Elujoba AA, Durosinmi MA. Antisickling properties of *Carica papaya* Linn *J. Nat. Prod.* 1, 56-66, 2008.

[13] Neuwinger HD. *African traditional medicine*, Mepharm Scientifique Publisher, Suttgart, 2000.

[14] Mpiana PT, Bokota MT, Ndjele MBL, Mudogo V, Tshibangu DST, Ngbolua KN, Atibu EK, Kwembe JTK and Makelele LK. Antisickling activity of three species of justicia from Kisangani (DR Congo): *Justicia tenella, J. gendarussa and J.insularis. Int. J. Biol. Chem. Sci.* 4(6), 1953- 1961, 2010.

[15] Mpiana PT, Ngbolua KN, Bokota MT, Kasonga TK, Atibu EK and Mudogo V. In *vitro* Effetcts of Anthocyanins Extracts From *Justicia secunda* VAHL on the Solubility of Hemoglobin *S* and Membrane Stability of Sickle Erythrocytes. *Blood Transfusion* 8, 248-254, 2010.

[16] Mpiana PT, Makelele LK, Oleko RW, Bokota MT, Tshibangu DST, Ngbolua KN, Mbala MB, Atibu EK and Nsimba SM. Antisickling activity of medicinal plants used in the management of Sickle cell Disease in Tshopo district, D.R.Congo, *Australian Journal of Medical Herbalism*, 22(4), 132-137, 2010

[17] Mpiana PT, Mudogo V, Tshibangu DST, Ngbolua KN, Mangwala PK, Atibu EK, Kakule MK, Makelele LK and Bokota MT. Antisickling Activity and Thermodegradation of an Anthocyanin fraction from *Ocimum basilicum L.* (LAMIACEAE).In: Comp.Bio.Nat. Prod. Vol.3: Effects, safety & clinical Evaluation (part II) VK Gupta (Ed): Studium Press LLC, USA, 2010. pp. 279-287.

[18] Mpiana PT, Mudogo V, Ngbolua KN, Tshibangu DST & Atibu EK. In vitro antisickling activity of anthocyanins extracts from *Morinda lucida* Benth (RUBIACEAE) *in* Medicinal Plants: Phytochemistry, Pharmacology and Therapeutics, VK Gupta, GD

Singh, Surjeet Singh & A Kaul (Eds): Daya Publishing House, New Delhi, 2010. pp.330- 337

[19] Mpiana PT, Balangayi EK, Kanangila AB, Kalonda EM, Ngbolua KN, Tshibangu DST, Atibu EK & Lumbu JBS, Activité antidrépanocytaire et thermodégradation des anthocyanes extraits de *Sterculia quinqueloba* et *Ficus capensis. Int. J. Biol Chem. Sci.* 3(3), 551-560, 2009.

[20] Mpiana PT, Kitwa EK, Kanangila AB, Tshibangu DST, Ngbolua KN, Atibu EK & Kalonda EM. Influence de la photodégradation sur l'activité antifalcémiante des

anthocyanes extraits de quelques plantes de Lubumbashi (RD Congo). *Ann. Fac. Sci.* 1, 45-53, 2009.

[21] Mpiana PT, Kitwa EK, Kanangila AB, Tshibangu DST, Ngbolua KN, Atibu EK & Kalonda EM. Activités antifalcémiantes de quelques plantes médicinales de Lubumbashi (RD Congo). *Ann. Fac. Sci.* 1, 103-112, 2009.

[22] Mpiana PT, Mudogo V, Kabangu YF, Tshibangu DST, Ngbolua KN, Atibu EK, Mangwala KP, Mbala MB, Makelele LK, Bokota MT, Antisickling Activity and Thermostability of Anthocyanins Extract from a Congolese plant, *Hymenocardia acida Tul.* (Hymenocardiaceae), *Int. J. Pharmacol.* 5 (1), 65-70, 2009.

[23] Mpiana PT, Mudogo V, Ngbolua KN, Tshibangu DST, Atibu EK, Kitwa EK & Kanangila AB. *In vitro* antisickling activity of anthocyanins extracts of *Vigna unguiculata* (L.) walp., *In* Recent progress *in* Medicinal Plants: Chemistry and Medicinal Value, JN Govil & VK Singh (Eds): Daya Publishing House, New Delhi , 2009. pp. 91- 98.

[24] Mpiana PT, Mudogo V, Ngbolua KN, Tshibangu DST, Shetonde OM., & Mbala BM. *In vitro* Antisickling Activity of Anthocyanins from *Ocimum basilicum L.* (Lamiaceae). *Int. J. Pharmacol.* 3 (4), 371-374, 2007.

[25] Mpiana PT, Mudogo V, Nyamangombe L, Kakule MK, Ngbolua KN, Atibu EK, Mbala MB, Mbongo AK & Ntumba JN. Antisickling Activity and Photodegradation Effect of Anthocyanins Extracts from *Alchornea cordifolia (*SCHUMACH & Thonn.*)* and *Crotalaria retusa L. Ann.Afr.Med.* 2(4), 240-245, 2009.

[26] Mpiana PT, Mudogo V, Tshibangu DST, Kitwa EK, Kanangila AB, Lumbu JBS, Ngbolua KN, Atibu EK & Kakule MK , Antisickling Activity of Anthocyanins from *Bombax pentadrum, Ficus capensis, Ziziphus mucronata*: Photodegradation effect. *J.Ethnopharmacol.*120, 413-418, 2008.

[27] Mpiana PT, Mudogo V, Tshibangu DST, Ngbolua K.N, Shetonde OM, Mangwala PK & Mavakala BK. *In vitro* Antisickling Activity of Anthocyanins Extracts of a Congolese Plant: *Alchornea cordifolia M.Arg. J.Med.Sci.*7 (7), 1182-1186, 2007.

[28] Mpiana PT, Mudogo V, Tshibangu DST, Ngbolua KN, Atibu EK, Kitwa EK, Kanangila AB & Makelele L.K. Activité antifalcémiante et thermodégradation d'une fraction d'anthocyanes extraits de *Zizyphus mucronata*. *Ann.Afr.Med.*2(2) , 91-97, 2009.

[29] Mpiana PT, Mudogo V, Tshibangu DST, Ngbolua KN, Tshilanda DD & Atibu EK. Antisickling activity of anthocyanins extract of *Jatropha curcas* L. *In* Recent progress *in* Medicinal Plants: Chemistry and Medicinal Value, JN Govil, VK Singh (Eds): Publishing House, New Delhi, 2009. pp. 104-108.

[30] Mpiana PT, Tshibangu DST, Shetonde OM, Ngbolua KN. *In vitro* antidrepanocytary activity (antisickle cell anemia) of some Congolese plants. *Phytomedicine* 14 (2-3), 192- 195, 2007.

[31] Mpiana PT, Ngbolua KN, Mudogo V, Tshibangu DST & Atibu EK. The potential effectiveness of medicinal plants used for the treatment of Sickle cell Disease in the Democratic Republic of Congo folk medicine: A review. In: Progress *in* Traditional and Folk herbal medicine: Publishing House, New Delhi, (*in press*).

[32] Mpiana PT, Ngbolua KN, Mudogo V, Tshibangu DST, Atibu EK , Tshilanda DD, & Misengabu NM. Anti sickle erythrocytes haemolysis properties and inhibitory effect of anthocyanins extracts of *Trema orientalis* (ULMACEAE) on the aggregation of human deoxyhemoglobin S *in vitro. J. Med.Sci.* 11(3), 129-137, 2011.

[33] Bitsch R, Netzel M, Sonntag S, Strass G, Frank T and Bitsch I. Bioavailability and biokinetics of anthocyanins from red grape juice and red wine. *J. Biomed. Biotechn.* 5, 343-345, 2004.

[34] Francis FJ. Food colorants: anthocyanins. *Crit.Rev.Food.Sci.Nutr.* 28, 273-314, 1989.

[35] Konczak I and Zhang W. Anthocyanins-More than nature's colours. *J. Biomed. Biotechn.* 5, 239-240, 2004.

[36] Clifon PM, Effect of grape seed extract and quercin on cardiovascular and endothelial parameters in high-risk subjects *J. Biomed. Biotechn.* 5, 272-273, 2004.

[37] Lila MK. Anthocyanins and human health an in vitro investigative approach. *J. Biomed. Biotechn.* 5, 306-313, 2004.

[38] Hou DX, Fuju M, Terahara N, Yoshimoto M. Molecular mechanisms behind the chemopreventive effects of anthocyanins. *J. Biomed. Biotechn.* 5, 321-325, 2004.

[39] Bruneton J. Pharmacognosie: Phytochimie-plantes médicinales, Tec & Doc, Paris, 1999.

[40] Azoulay E and Cohen O. *Cours et Exercices de statistique*, Société d'Edition d'enseignement supérieur, Paris, 1969.

[41] Mills G CJ. Hemoglobin catabolism. I. Glutathione peroxidase, an erythrocyte enzyme which protects hemoglobin from oxidative breakdown. *J Biol Chem.* 229(1), 189-197, 1957.

[42] Giulivi C, Hochstein P, Davies KJA. Hydrogen peroxide production by red blood cells. *Free Radic.Biol.Med.* 16, 123–129, 1994.

[43] Galvez J, De la Cruz JP, Zarzuelo A and de la Cuesta FS, Flavonoid inhibition of enzymic and nonenzymic lipid peroxidation in rat liver differs from its influence on the glutathione-related enzymes. *Pharmacology,* 51(2), 127-133, 1995.

[44] Erben-Russ M; Bors W; Saran M. Reactions of linoleic acid peroxyl Radicals with phenolic antioxidants: a pulse Radiolysis study. *Int. J. Radiat. Biol.* 52(3), 393-412, 1987.

[45] Yoshino K; Tomita I; Sano M . Effects of long-term dietary supplement of tea polyphenols on lipid peroxide levels in rats. *Media, Pennsylvania.* 17(3), 79-85, 1994.

[46] Papart AK, Lorenzo PB, Papart ER, Gregg JR and Chase AM. The osmotic resistance (fragility) of human red cells *J.Clin.Invest.* 26, 636-638, 1947.

In: Anthocyanins: Structure, Biosynthesis and Health Benefits ISBN: 978-1-62257-329-5
Editor: Noboru Motohashi

Chapter 6

STABILTY OF ANTHOCYANIN IN STRAWBERRY JAM AND THE APPLICATION OF SACCHARIDES

Yoshiyuki Watanabe**, *Yoshiharu Okada and Masato Nomura
Department of Biotechnology and Chemistry, Faculty of Engineering,
Kinki University, Higashi-Hiroshima, Japan

ABSTRACT

Anthocyanins are susceptible to heat and light, which cause color deterioration of processed food containing it, such as strawberry jam, during manufacturing, transportation, and preservation. In this study, the effects of temperature and the relative humidity during the preservation on the stability of anthocyanin in strawberry jam were evaluated, and the first-ordered rate constant for the degradation of anthocyanins was estimated. The rate constant increased as temperature and the relative humidity increased. α-, β-, or γ-Cyclodextrin (CDs) exhibited the suppressive effect on the anthocyanin degradation in the jam, but the addition of glucose, which composes the cyclodextrins, to the jam did not improve the stability of anthocyanins. This might indicate that the degradation was suppressed by the formation of the inclusion complex of anthocyanins with CD due to its cyclic molecular structure or lowering the water activity in the jam. The impregnation of strawberry with sucrose in advance of the jam preparation stabilized anthocyanins in the jam. The rate constants of the Weibull model for the degradation of anthocyanins and the chromatic deterioration of the jam were estimated, and the relationship between the two rate constants was found to be almost linear. It was indicated that the impregnation of strawberry with sucrose was effective for improving the stabilities of both anthocyanins and the colors in the jam.

Keywords: Anthocyanin; Color stability; Impregnation; Saccharide; Strawberry jam

---

* Author to whom correspondence should be addressed (Telephone to: +81-82-434-7000; Fax: +81-82-434-7890; E-mail: wysyk@hiro.kindai.ac.jp).

INTRODUCTION

Anthocyanins are the most abundant flavonoid constituents of fruits and vegetables, and are bioactive compounds. However, anthocyanins are susceptible to heat and light during manufacturing, transportation, and preservation. Their anthocyanins cause color deterioration of processed food containing anthocyanins, such as strawberry jam. Therefore, the improvement of the stability of anthocyanin in food is important in both jam production and the food industry, and many studies on the color stability of strawberry jam have been reported [1-6]. The main anthocyanin involved in strawberry is known to be pelargonidin 3-glucoside (Figure 1) [7, 8], though the type of cultivar, ripening degree, storage conditions, and other components in strawberry affect the amounts of the anthocyanin [9]. The effects of high-pressure processing in jam manufacture [10], as well as various saccharides, not only monosaccharides but oligo- and polysaccharides [11] on the stabilization of anthocyanin have been examined. Saccharides, such as cyclodextrins and maltodextrin, are effective for the stabilization of anthocyanin in food [12]. Cyclodextrins (CDs) (Figure 2) are cyclic oligosaccharides containing a minimum of six D-(+)-glycopyranose units attached by α-1,4-linkages produced by the action of the cyclodextrin-*trans*-glycosidase on a medium containing starch [13]. The outside of the CD is hydrophilic while the inside of the cavity is hydrophobic in nature. If a molecule fits entirely or at least partially into the cavity, an inclusion complex may be formed. In general, hydrophobic molecules, rather than hydrophilic ones, have a higher affinity to the cyclodextrin cavity in aqueous solutions. The cyclodextrin complexes thus formed are stabilized by various intermolecular forces, such as a hydrophobic interaction, van der Waals forces, hydrogen bonding, release of high energy water molecules during complex formation and the release of strain energy in the macromolecular cyclodextrin ring. Many reports about the solubilization and stabilization of lipids and lipophilic vitamin by inclusion in cyclodextrins have been published [14-15]. The improvement of stability of pelargonidin 3-glucoside using cyclodextrin has also been reported [16]. Impregnation is a technique for replacing gas and liquid through the internal pores of a material by an external liquid [17, 18]. Impregnation can be used to accelerate mass transfer and promote a more homogeneous distribution in fruits, because solutes in the solution can be introduced more efficiently into the tissue. Immersing of the material in the solution and vacuum processing are needed, and this process can enrich food products with nutrients and vitamins, and aid the incorporation of useful additives. In addition, it may improve the sensorial taste, smell, and appearance properties of food.

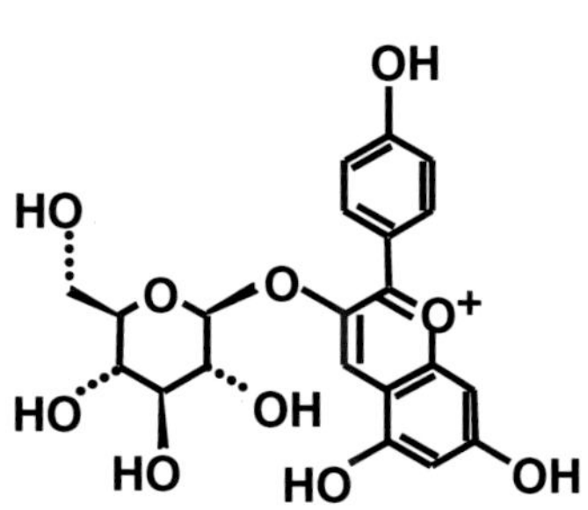

Figure 1. Molecular structure of pelargonidin 3-glucoside.

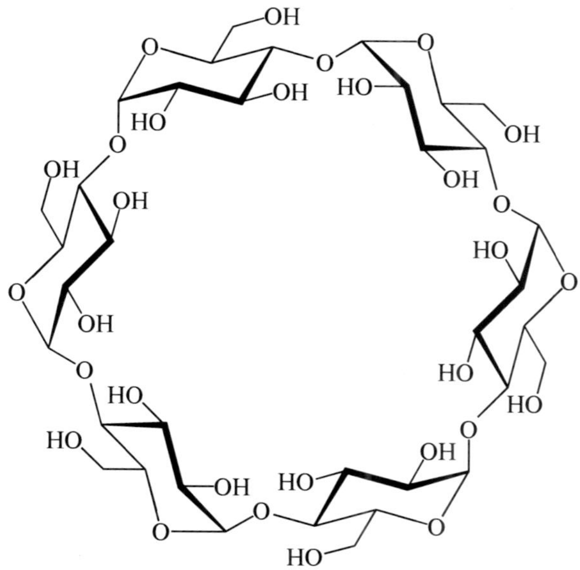

Figure 2. Molecular structure of α-cyclodextrin (α-CD).

In Section 1, the effects of humidity and temperature on the stability of anthocyanins in strawberry jam were examined. The jam was preserved at the range from 37 to 70°C. This temperature was selected because the examination was more easily to be conducted in an accelerated experiment in which bacterial deterioration of the jam hardly occurred. In terms of the state of current knowledge, only limited information is available on the temperature stability of anthocyanins-derived from food, and it is not possible to predict the effect of thermal treatment on anthocyanin retention [19]. Therefore, kinetic study on the thermal stability of anthocyanins in jam should contribute to the assessment of the thermal processing of foods. In Section 2, the stabilization of anthocyanins in the jam by the addition of CDs was evaluated. In Section 3, the stability of anthocyanins in the jam prepared through impregnation of strawberry using a sucrose solution and its color stability are described. The degradation of anthocyanins and the color deterioration processes were expressed by the Weibull equation, and the kinetic parameters were determined.

1. DEPENDENCE OF THE STABILITY OF ANTHOCYANIN IN STRAWBERRY JAM ON THE PRESERVATIVE CONDITIONS

The strawberry cultivar Chandler, harvested in China or Mexico, was used for the preparation of jam [5]. Frozen strawberry was defrosted using cold water, and was milled and dried. Forty grams of powdery strawberry, 36 mL of distilled water, 16.2 g of sucrose, and 1 g of pectin were mixed and heated at 102°C for 3 min. One gram of the jam was weighed in flat-bottomed glass cups (1.5 cm internal diameter (i.d.) and 3.0 cm height). The cups were placed in a plastic container with a Petri dish containing each saturated aqueous solution of inorganic compound to regulate the relative humidity. The container was stored in the dark at a specific temperature for 50 to 100 hr. The cup was removed at appropriate intervals for observation of the transient changes in the total amount of anthocyanins, and 1 mL of the hydrochloric acid/methanol mixture (5:95 by vol.) was added to the cup to fully dissolve anthocyanins in the strawberry jam. Then, the mixture was centrifuged using a centrifugal separator. The extraction and centrifugation were repeated 5 times. The total concentration of anthocyanins in the extract was measured at 510 nm using a UV-visible spectrophotometer [20]. The total amount of anthocyanins in the jam was measured 2 to 3 times.

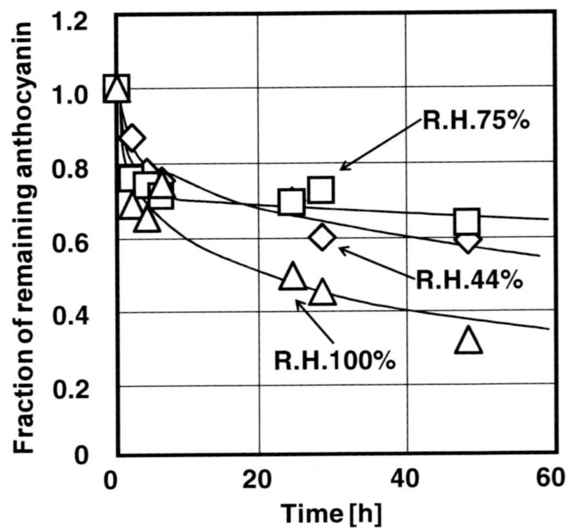

Figure 3. Degradation processes of anthocyanins in strawberry jam at 60°C and (◇) 44 %, (□) 75 %, and (Δ) 100 % relative humidity. The solid curves were calculated using the estimated kinetic parameters of the first-order kinetics.

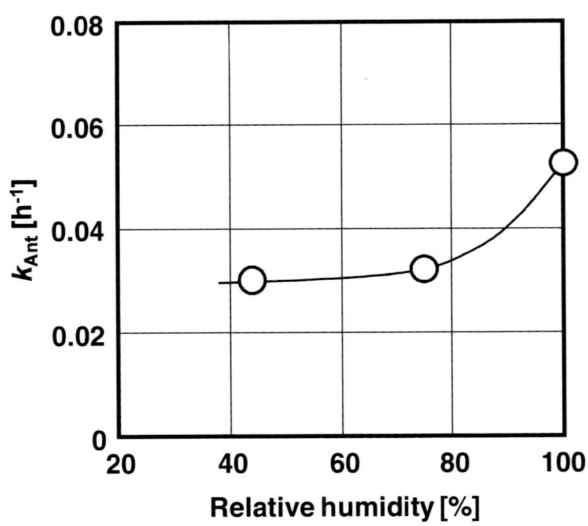

Figure 4. The relationship between the rate constant, k_{Ant}, for the degradation of anthocyanins in strawberry jam at 60°C and the relative humidity. The curves were empirically drawn.

Figure 3 shows the degradation processes of anthocyanins in strawberry jam at 60°C and 44, 75, and 100 % relative humidity. The faction of remaining anthocyanins was about 0.7 for

6 hr at all the humidity tested, and subsequently was decreased slowly. The degradation kinetics of the anthocyanins were empirically expressed by the Weibull equation (Eq. 1), which is flexible and can be used to describe many deterioration kinetics [21, 22]:

$$\frac{A}{A_0} = \exp[-(k_{\mathrm{Ant}}t)^n] \tag{1}$$

where A_0 and A are the initial amount of anthocyanins and that at time t, respectively, k_{Ant} is the rate constant, and n is the shape constant. The kinetic parameters, k and n, were estimated by fitting the experimental results by nonlinear regression using Solver of Microsoft® Excel 2007. The curves in Figure 3 were drawn on the basis of the first-order equation, which is expressed at n = 1 in Eq. 1, using the estimated parameter. Figure 4 shows the relationship between the k_{Ant} value for the degradation of anthocyanins and the relative humidity. There was no deference between the k_{Ant} values at 44 and 75 % relative humidity, but the k_{Ant} values at 100 % relative humidity was higher than those. The high humidity would cause the relative decrement of the viscosity of the jam, following that the mass transfer of the dissolved oxygen and the trace metals, which affect the oxidative decomposition of anthocyanins, in the jam is facilitated.

Figure 5 shows the effect of temperature on the fraction of remaining anthocyanins in the jam at 100 % relative humidity. The stability of anthocyanins was lower at higher temperature. At 37°C anthocyanins more than 80% remained for 2 days, though less than 40% at 70°C. From these results in Figure 5, the k_{Ant} values (n = 1) were calculated (Figure 6). The temperature-dependency of the rate conctant for the degradation of anthocyanins was expressed by Arrhenius equation (Eq. 2)

$$\ln k = \ln k_0 - \frac{E_a}{RT} \tag{2}$$

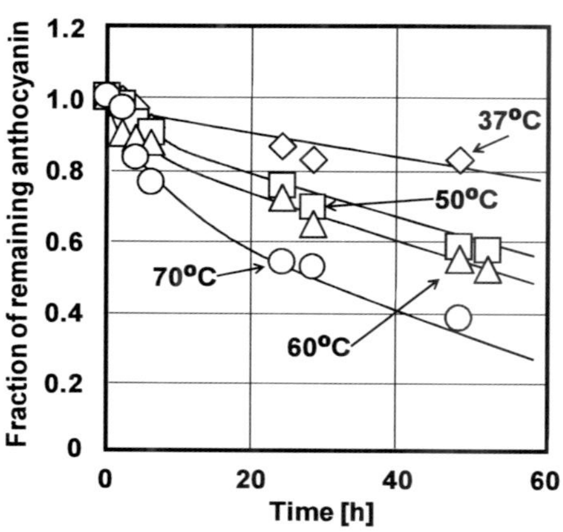

Figure 5. Effect of temperature on the fraction of remaining anthocyanin in strawberry jam at 100 % relative humidity. Preservation temperature were (◇) 37°C, (□) 50°C, (△) 60°C, and (○) 70°C. The solid curves were calculated using the estimated kinetic parameters of the first-order kinetics.

where k_0 is the frequency or pre-exponential factor, E_a is the apparent activation energy, R is the gas constant, T is temperature. The apparent activation energy was estimated to be 43.3 kJ/mol from the slope of the Arrhenius plots in Figure 6.

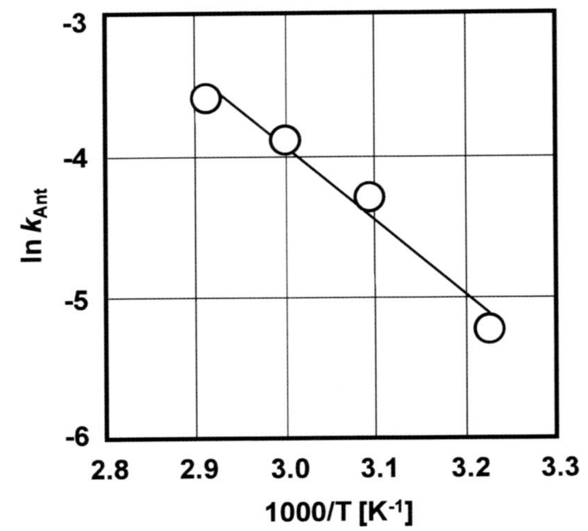

Figure 6. Arrhenius plots for the degradation rate constant of anthocyanins in strawberry jam at 100% relative humidity.

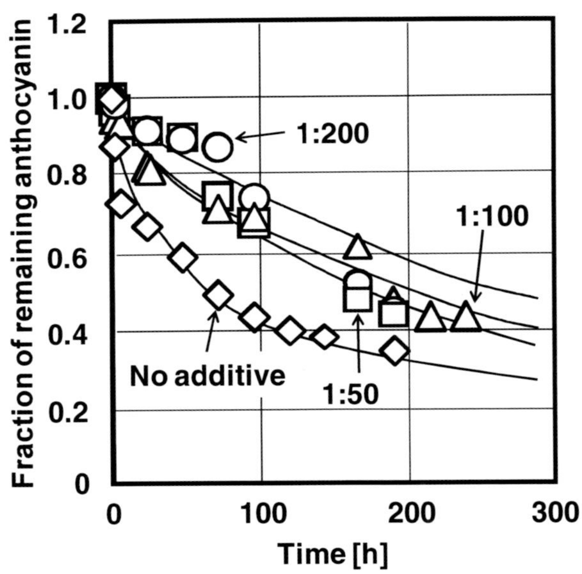

Figure 7. Transient changes in remaining anthocyanin in strawberry jam with β-cyclodextrin (β-CD) at 60°C and 100 % relative humidity. β-CD was added at the molecular ratio of β-CD to anthocyanins of (◇) 0, (□) 50, (△) 100, and (○) 200. The solid curves were calculated using the estimated kinetic parameters of the first-order kinetics.

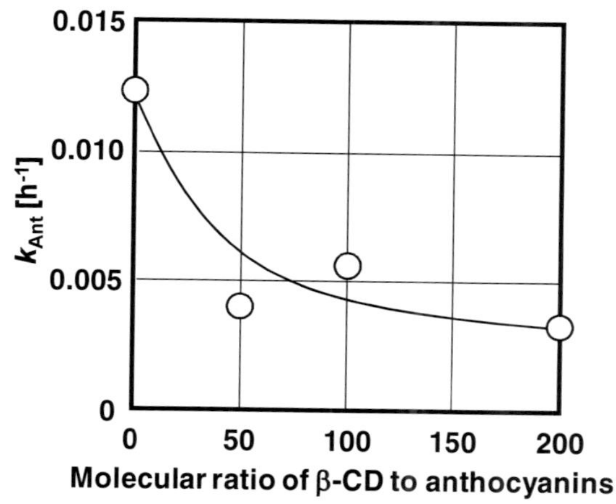

Figure 8. Dependence of the rate constant for the degradation of anthocyanins in strawberry jam containing β-CD at 60°C and 100 % relative humidity on the molecular ratio of β-CD to anthocyanins. The curves were empirically drawn.

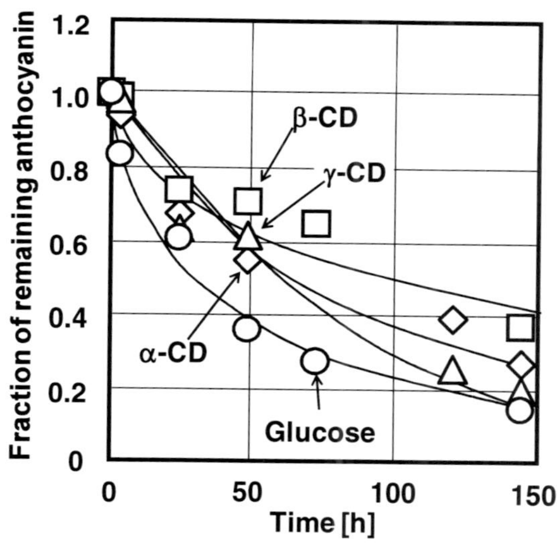

Figure 9. Time course for the degradation of anthocyanins in strawberry jam with (◇) α-CD, (□) β-CD, (Δ) γ-CD, and (○) glucose at the molecular ratio of 50 at 60°C and 100 % relative humidity. The solid curves were calculated using the estimated kinetic parameters of the Weibull model.

2. EFFECT OF THE ADDITION OF CYCLODEXTRIN ON THE STABILITY OF ANTHOCYANIN IN STRAWBERRY JAM

Figure 7 shows the transient changes in remaining anthocyanins in strawberry jam with β-cyclodextrin (β-CD), at 60°C and 100 % relative humidity. β-CD was added at the molecular ratio of β-CD to anthocyanins of 50, 100 and 200 before heating. Kinetic analysis was introduced to the results in Figure 7 as Section 1, the k_{Ant} values ($n = 1$) were determined. Figure 8 shows the dependence of the rate constant for the degradation of anthocyanins in the jam containing β-CD at 60°C and 100 % relative humidity on the added amount of β-CD. The addition of β-CD contributed to the stabilization of anthocyanins, but there was no deference among the stabilities at the molecular ratios of β-CD to anthocyanins more than 50. In this experimental system, the molecular ratio of 50 was sufficient amount for the stabilization of anthocyanins in the jam.

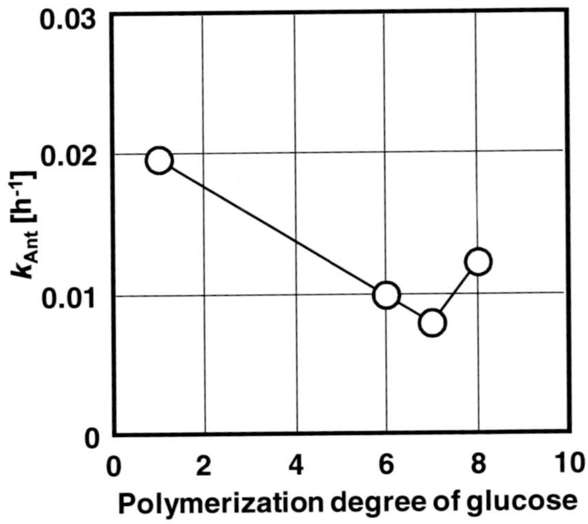

Figure 10. The relationship between the k_{Ant} value of the Weibull equation for the degradation of anthocyanins in strawberry jam at 60°C and 100 % relative humidity and the polymerization degree of glucose. The curves were empirically drawn.

Figure 9 shows the time course for the degradation of anthocyanins in the jam with α-CD, β-CD, γ-CD, and glucose at the molecular ratio of 50 at 60°C and 100 % relative humidity. The k_{Ant} and n values were estimated by the Weibull model (Eq. 1), and the relationship between the k_{Ant} value and the polymerization degree of glucose for the added saccharides is shown in Figure 10. The addition of glucose to the jam could not stabilize anthocyanins. On the other hand, CDs enhanced the anthocyanins stability, and the anthocyanins were most stabilized by β-CD, which is known to be most stable cyclic-structure in three CDs [23]. Therefore, the anthocyanin degradation might be suppressed by the formation of the inclusion complex of anthocyanins with CDs due to its cyclic molecular structure or lowering the water activity in the jam. The n values for the degradation of anthocyanins in the jam with α-CD, β-CD, γ-CD, and glucose were 0.749, 1.07, 1.14, and 0.618, respectively.

3. Impregnation of Strawberry Using Sucrose and Anthocyanin Stability in the Jam

A piece of strawberry (*ca.* 10 g), the moisture on the surface of which was removed using a wiper, was added to 30 mL of sucrose solution at concentrations from 0.29 to 2.34 mol/L [6]. Impregnation from 5 to 60 min was carried out under reduced pressure (4 kPa). The impregnated strawberry was ground down in a mortar with 0.1 g of sea sand. After filtration, the distilled water was added to the filtrate to obtain 20 mL of the extract. The concentration of sucrose in the extract was measured using HPLC with an YMC-Pack NH_2 column (4.6 mmϕ x 150 mm) and an RI detector. After dilution 2 to 10 times with an eluent, acetonitrile/water mixture (75:25 by voL for HPLC analysis, the diluted extract was passed through a disposable membrane filter cartridge, which held the hydrophilic polytetrafluoroethylene membrane with a pore diameter of 0.20 μm. 20 μL of the filtrate were applied to the column and eluted at 1.0 mL/min. Each measurement of the concentration was carried out in triplicate and the mean value was calculated. 15 mL of distilled water, 25 g of sucrose, and 0.5 g of pectin were added to 50 g of the impregnated strawberry. The mixture was heated at 102°C for 2 min. 1 g of the jam was weighed in flat-bottomed glass cups (1.5 cm i.d. and 3.0 cm height). The cups were placed in a plastic container with a Petri dish containing the distilled water to regulate the relative humidity at 100%. The container was stored in the dark at 65°C for 50 to 100 hr. Color analysis of the jam was conducted with a UV-visible colorimeter. The instrument was calibrated with a standard white and black reflective plate. The instrumental values, L^*, a^*, and b^*, were recorded and from these the color difference, ΔE ($\sqrt{\Delta L^{*2} + \Delta a^{*2} + \Delta b^{*2}}$), the chroma, C^*($\sqrt{a^{*2} + b^{*2}}$), and the hue angle, θ (arctan b^*/a^*), were determined. All color measurements were carried out in triplicate, and the mean value was reported.

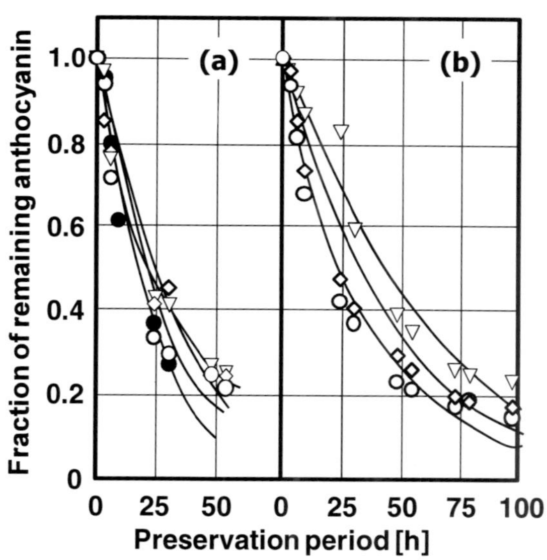

Figure 11. Transient changes in remaining anthocyanin in strawberry jam at 65°C and 100% relative humidity. The sucrose solutions were (a) mixed with the jam or (b) impregnated into strawberry for 30 min before the jam preparation at the concentrations of (○) 0.29, (◇) 0.88, and (▽) 1.46 mol/L. The control (●) represents no additive sucrose to the jam. The solid curves were calculated using the estimated kinetic parameters of the Weibull model.

First, the relationship between the concentration of sucrose in impregnating solution and the amount of sucrose in strawberry extract were examined with impregnation carried out for 30 min. For the range from 0 to 1.75 mol/L, the amount of sucrose extracted from the strawberry was proportional to the sucrose concentration in the impregnating solution. At concentrations over 1.75 mol/L, the amount of sucrose in the used strawberry reached the maximum under these processing conditions. It may be impossible to transfer the sucrose from the solution to the strawberry owing to the high viscosity of the impregnating solution. Second, the dependence of the amount of sucrose impregnated into the strawberry on the operation time was measured at the concentration of 1.17 mol/L. The impregnation rapidly proceeded for 10 min, whereas the amount of sucrose in the strawberry extract was almost constant from 10 to 30 min. In processing for more than 40 min, the amount markedly increased again. During the initial term of the impregnation, the sucrose in the solution was promptly transferred from the outer pericarp to unoccupied positions, to which it could easily move, in the strawberry. Afterwards, the transfer gradually slowed as the empty spaces were filled with the sucrose from the solution. Further impregnation under reduced pressure for a long time might break the internal structure of the strawberry, resulting in the large increment of sucrose in it. Thereafter, the processing at 30 min was adopted for evaluation of the effects of the impregnation on the stability of anthocyanins in the strawberry jam and the color stability of the jam.

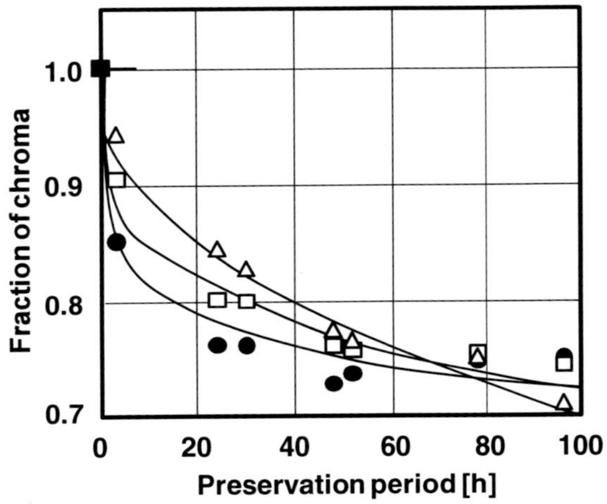

Figure 12. The chromatic deterioration of strawberry jam impregnated using (□) 0.88 and (Δ) 1.46 mol/L sucrose solution at 65°C and 100% relative humidity. The control (●) represents no impregnation into strawberry. The solid curves were calculated using the estimated kinetic parameters of the Weibull model for the fraction of chroma.

Strawberry jam mixed with or produced from strawberry impregnated with sucrose from 0.29 to 1.46 mol/L was prepared. Figure 11 shows the transient changes in the fractions of remaining anthocyanins in the strawberry jam mixed with or produced from strawberry impregnated with 0.29, 0.88, and 1.46 mol/L sucrose at 65°C and 100% relative humidity. The relationship between the rate constant, k_{Ant}, or the shape constant, n, of the Weibull model for the degradation of anthocyanin and the concentration of sucrose in the solution was examined. The k values both with and without impregnation decreased with the increasing concentration of sucrose, but the decrement with the impregnation was larger than that without sucrose. This suggested that the impregnation of sucrose in advance of the jam preparation stabilized the anthocyanin in the jam more strongly than the mere addition of sucrose during the preparation. Furthermore, a high concentration of sucrose in the impregnation was effective for the stabilization of anthocyanins in jam preservation. The intensity of color of anthocyanin is known to increase by the addition of saccharides such as glucose, maltose and sucrose [11]. The removal of water by saccharides caused this increase through the displacement of the hydration/dehydration equilibrium towards the colored species. The cause of the protective effect of sucrose was also suggested: inhibition of pigment degradative enzymes, namely, polyphenol oxidase and β-glucosidase, steric interference with condensation, namely, anthocyanin-phenolic polymerization; anthocyanin-ascorbate, or provision of a partial oxygen barrier [24]. These would support on the effectiveness of the impregnation for the stability of anthocyanin in the jam. Regardless of the impregnation processing, the *n* value for each concentration of the sucrose was about 1, indicating that the degradation of anthocyanins could be approximately expressed by the first-order kinetics as described in Sections 1 and 2.

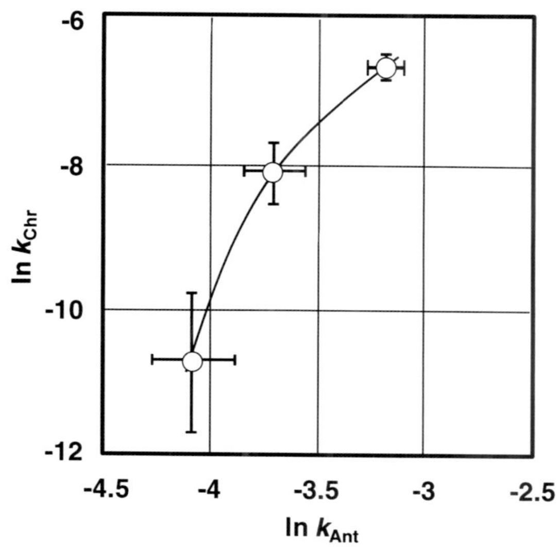

Figure 13. The relationship between the rate constants, k_{Ant}, and k_{Chr} [h^{-1}], of the model for anthocyanins degradation and the chromatic deterioration at 65°C and 100% relative humidity, respectively. Statistical significance was determined at $p < 0.05$. The curve was empirically drawn.

The transient changes in color parameters for strawberry jam impregnated using the sucrose solution at several concentrations were measured. The preservative conditions of the jam were 65°C and 100% relative humidity. The a^* value for each jam rapidly decreased during the early stage of the preservation and was almost constant after 50 hr. The b^* value for the impregnated jam slowly decreased, while that for the non-impregnated one did not change markedly. For the chroma, C^*, derived from the a^* and b^* values, a similar tendency was shown to that of the a^* value, but the difference between the C^* values for the jams with the different concentrations of sucrose solution was larger than that for the a^* values. For each color parameter, a higher value was observed at a higher concentration of the sucrose solution in the impregnation at almost all time points. The L^* values for the jam without impregnation also decreased from 33.1 to 15.8. However, there was no difference among the values with and without impregnation, and the value did not depend on the sucrose concentration in the impregnating solution. Furthermore, their differences for the color difference, ΔE, and the hue angle, θ, for the jam were not found. The ΔE value did not differ by the larger contribution of L^* value, compared with those of the a^* and b^* values, to the ΔE value. Thus, the C^* value was used as an index for the color stability of the impregnated jam. Figure 12 shows the chromatic deterioration of strawberry jam impregnated using 0.88 and 1.46 mol/L sucrose solution at 65°C and 100% relative humidity. The above-mentioned Weibull model was applied for the fraction of chroma, and the rate constant, k_{Chr}, for the chromatic deterioration of the jam was estimated. As shown in Figure 13, the relationship between the two rate constants, k_{Ant} and k_{Chr}, was almost linear. It has been reported that the decrease in the pigments in strawberry jam at temperatures from 20 to 37°C is not always correlated with a decrease in the a^* value [3]. Since the jam was preserved at higher temperature in this study, the thermal degradation of anthocyanins rather than enzymatic reaction in the strawberry might have resulted in the different influence on the color property of the jam. It was shown that the stabilization of anthocyanin by the impregnation of sucrose into strawberry prior to the production of the jam also contributed to the color stability, especially the chroma.

CONCLUSION

The stability of anthocyanin in strawberry jam affects not only the color of the jam but also the functions, such as the antioxidative activity [6]. The addition of CDs stabilized the anthocyanins in the jam, but it was found by kinetic analysis that the effect was weak. On the other hand, impregnation of sucrose solution into strawberry is efficient for the improvement of these properties. Sucrose is an essential component of the jam at all times. The Brix of the jam impregnated at various concentrations of sucrose were from 30 to 33%, which is not too high as a jam generally produced. In future, more detailed information about the behavior of sucrose in the jam, such as an inversion from sucrose to glucose and fructose, will be needed for the efficient application of the impregnated jam. Impregnation should be available for the preservation of fruits and the various processed foods that contain anthocyanins.

ACKNOWLEDGMENTS

The authors would like to thank all the laboratory members of Department of Biotechnology and Chemistry, Faculty of Engineering, Kinki University.

REFERENCES

[1] Abers, J. E., & Wrolstad, R. E. (1979). Causative factors of color deterioration in strawberry preserves during processing and storage. *Journal of Food Science, 44(1)*, 75-78, 81.

[2] O'Beirne, D., & Kelly, G. (1985). Some physical and organoleptic effects of adding skim milk powder and cream to fruit preserves, *Lebensmittel-Wissenschaft und -Technologie*, *18*, 47-51.

[3] García-Viguera, C., Zafrilla, P., Romero, F., Abellán, P., Artés, F., & Tomás-Barberán, F. A. (1999). Color stability of strawberry jam as affected by cultivar and storage temperature. *Journal of Food Science*, *64(2)*, 243-247.

[4] Wicklund, T., Rosenfeld, H. J., Martinsen, B. K., Sundfør, M. W., Lea, P., Bruun, T., Blomhoff, R., & Haffner, K. (2005). Antioxidant capacity and colour of strawberry jam as influenced by cultivar and storage conditions. *Lebensmittel-Wissenschaft und - Technologie*, *38*, 387-391.

[5] Watanabe, Y., Kurata, A., Yamanaka, K., & Suginaka, T. (2007). Effects of preservation and preparation conditions on the color stability of strawberry jam (in jamanese). *New Food Industry*, *49(4)*, 31-35.

[6] Watanabe, Y., Yoshimoto, K., Okada, Y., & Nomura M. (2011). Effect of impregnation using sucrose solution on stability of anthocyanin in strawberry jam. *LWT - Food Science and Technology*, *44(4)*, 891-895.

[7] Sondheimer, E., & Kertesz, Z. I. (1948). The anthocyanin of strawberries, *Journal of the American Chemical Society*, *70*, 3476-3479.

[8] Viguera, C. G., Zafrilla, P., & Barberan, F. T. (1997). Determination of authenticity of fruit jams by HPLC analysis of anthocyanins, *Journal of Science and Food Agriculture*, *73*, 207-213.

[9] Olsson, M. E., Ekvall, J., Gustavsson, K. E., Nilsson, J., Pillai, D., Sjöholm, I., Svensson, U., Åkesson, B., & Nyman, M. G. L. (2004). Antioxidants, low molecular weight carbohydrates, and total antioxidant capacity in strawberries (*Fragaria* X *ananassa*): Effects of cultivar, ripening, and storage. *Journal of Agricultural and Food Chemistry*, *52*, 2490-2498.

[10] Dervisi, P., Lamb, J., & Zabetakis, I. (2001). High pressure processing in jam manufacture: effects on textural and colour properties. *Food Chemistry*, *73*, 85-91.

[11] Lewis, C. E., Walker, J. R. L., & Lancaster, J. E. (1995). Effect of polysaccharides on the colour of anthocyanins. *Food Chemistry*, *54*, 315-319.

[12] Chandra, A., Nair, M. G., & Iezzoni, A. F. (1993). Isolation and stabilization of anthocyanins from tart cherries (*Prunus cerasus* L.), *Journal of Agricultural and Food Chemistry*, *41*, 1062-1065.

[13] Bekers, O., Uijtendaal, E. V., Beijnen, J. H., Bult, A., & Underberg, W. J. M. (1991). Cyclodextrins in the pharmaceutical field. *Drug Development and Industrial Pharmacy*, *17*, 1503-1549.

[14] Okada, Y., Tachibana, M., & Koizumi, K. (1990). Solubilization of lipid-soluble vitamins by complexation with glucosyl-β-cyclodextrin. *Chemical & pharmaceutical bulletin*, *38*, 2047-2049.

[15] Kim, S. J., Park, G. B., Kang, C. B., Park, S. D., Jung, M. Y., Kim, J. O., & Ha, Y. L. (2000). Improvement of oxidative stability of conjugated linoleic acid (CLA) by microencapsulation in cyclodextrins. *Journal of Agricultural and Food Chemistry*, *48*, 3922-3929.

[16] Dangles, O., Stoeckel, C., Wigand, M. C., & Brouillard, R. (1992). Two very distinct types of anthocyanin complexation: copigmentation and inclusion. *Tetrahedron Letters*, *33(36)*, 5227-5230.

[17] Hújica-Paz, H., Vaidez-Fragoso, A., López-Malo, A., Palou, E., & Welti-Chanes, J. (2003). Impregnation and osmotic dehydration of some fruits: effect of the vacuum pressure and syrup concentration. *Journal of Food Engineering*, *57*, 305-314.

[18] Castelló, M. L., Fito, P. J., & Chiralt, A. (2010). Changes in respiration rate and physical properties of strawberries due to osmotic dehydration and storage. *Journal of Food Engineering*, *97*, 64-71.

[19] Patras, A., Brunton, N. P., O'Donnell, C., & Tiwari, B. K. (2010). Effect of thermal processing on anthocyanin stability in foods; mechanisms and kinetics of degradation. *Trends in Food Science & Technology*, *21*, 3-11.

[20] Dangles, O., Wigand, M. C., & Brouillard, R. (1992). Anthocyanin anti-copigment effect, *Phytochemistry*, *31*, 3811-3812.

[21] Cunha, L. M., Oliveira, F. A. R. & Oliveira, J. C. (1998). Optimal experimental design for estimating the kinetic parameters of processes described by Weibull probability distribution function. *Journal of Food Engineering*, *37*, 175-191.

[22] Watanabe, Y., Nakanishi, H., Goto, N., Otsuka, K., Kimura, T., & Adachi, S. (2010). Antioxidative properties of ascorbic acid and acyl ascorbates in ML/W emulsion. *Journal of American Oil Chemists' Society*, *87(12)*, 1475-1480.

[23] Terao, K., Konishi, M., Nakata, D., & Uekaji, Y. (2005). Long known but brand new useable tool for food science -New use of natural cyclodextrins- (in jamanese). *Foods and Food Ingredients Journal of Japan*, *210(3)*, 222-243.

[24] Wrolstad, R. E., Skrede, G., Lea, P., & Enersen, G. (1990). Influence of Sugar on Anthocyanin Pigment Stability in Frozen Strawberries. *Journal of Food Science*, *55(4)*, 1064-1065.

In: Anthocyanins: Structure, Biosynthesis and Health Benefits ISBN: 978-1-62257-329-5
Editor: Noboru Motohashi

Chapter 7

ENVIRONMENTALLY FRIENDLY TECHNOLOGIES FOR OBTAINING ANTHOCYANINS FROM AN UNUSUAL SOURCE

S. P. Jesus, D. T. Santos and M. A. A. Meireles*

LASEFI (LAboratory of Supercritical technology: Extraction, Fractionation, and Identification of vegetable extracts) / DEA (Department of Food Engineering) / FEA (School of Food Engineering) / UNICAMP (University of Campinas), Campinas, São Paulo, Brazil

ABSTRACT

The consumers' constantly growing interest in healthier products has been motivating the search for new sources of bioactive compounds such as anthocyanins. Academic and industrial researches have been intensively investigating the anthocyanins due to their valuable functional properties. Although these natural pigments are widely spread in nature, their extraction procedure demands special attention to avoid degradation. For this reason, the extraction method plays a fundamental role in the recovery of anthocyanins from natural sources. Conventional extraction techniques are both time- and solvent-consuming, besides that they usually have a negative impact on the environment since they generate large amounts of organic solvent waste. On the other hand, there is an increasing demand for developing alternative methods in order to achieve faster, more efficient, cost-effective and environmentally friendly extraction processes. This chapter presents the recent work done by our research group (LASEFI/DEA/FEA/UNICAMP) for obtaining anthocyanins from skins of a Brazilian fruit named jabuticaba (*Myrciaria cauliflora*). Several results from conventional and innovative extraction technologies are summarized and discussed with the aim of comparing extraction performances in terms of anthocyanins recovery. This chapter provides a more complete understanding about the influence of different extraction methods on the recovery of anthocyanins.

---

* E-mail address: meireles@fea.unicamp.br.

INTRODUCTION

The consumers' desire for a healthier lifestyle has motivated the search for new sources of natural ingredients [1]. Anthocyanins are water-soluble pigments found in many fruits, vegetables and flowers [2]. Besides being colorants, anthocyanins have other bioactive properties, of which the strong antioxidant capacity is one of the most significant ones [3]. It is known that anthocyanins may promote important health benefits because they present antioxidant, anti-inflammatory and antimutagenic activities [4]. Therefore, the academic and industrial interest in anthocyanins has been increasing in recent decades since these substances have a high potential for application in food, cosmetic and pharmaceutical products.

Jabuticaba (*Myrciaria cauliflora*) is a Brazilian native fruit that seems to be a promising source for anthocyanins extraction. This fruit has a white gelatinous flesh covered by a dark purple to almost black skin, of which the intense color is attributed to the high content of anthocyanins [1]. The extraction method is a key factor for obtaining anthocyanins from natural sources. Taking into account that conventional extraction methods are commonly time- and solvent-consuming [5], there is an increasing demand for developing faster, more efficient and environmentally friendly extraction procedures [6]. An efficient extraction method should maximize anthocyanins recovery with concomitant minimal degradation of these unstable pigments [1].

This chapter presents some fundamental aspects about anthocyanins extraction and degradation, considering the influence of different extraction techniques on anthocyanins recovery. The focus is on discussing some environmentally friendly extraction methods that have been recently used by our research group (LASEFI/DEA/FEA/UNICAMP) in order to recover anthocyanins from jabuticaba skins.

ANTHOCYANINS DEGRADATION

Anthocyanins are highly unstable compounds, being very susceptible to degradation. Their stability is affected by many factors, such as temperature, pH, light, oxygen, anthocyanins concentration, enzymes, and the presence of accompanying substances (metallic ions, sugars, copigments, among others) [7, 8]. It is known that anthocyanins are particularly sensitive to heat, alkaline pH, and light exposure. Thus, in order to minimize thermal degradation, conventional extraction techniques are generally conducted at low temperatures combined with long processing times. Besides time-consuming, these methods have a negative impact on the environment because they generate large amounts of solvent wastes [9].

It is well established that anthocyanins degradation is particularly dependent of both temperature and time [9]. Low temperature and short extraction time tend to result in poor efficiency in extracting anthocyanins. On the other hand, high temperature and long extraction time lead to significant anthocyanins degradation. Therefore, an optimal combination between temperature and time must be found in order to achieve maximum anthocyanins extraction together with minimum anthocyanins degradation [10].

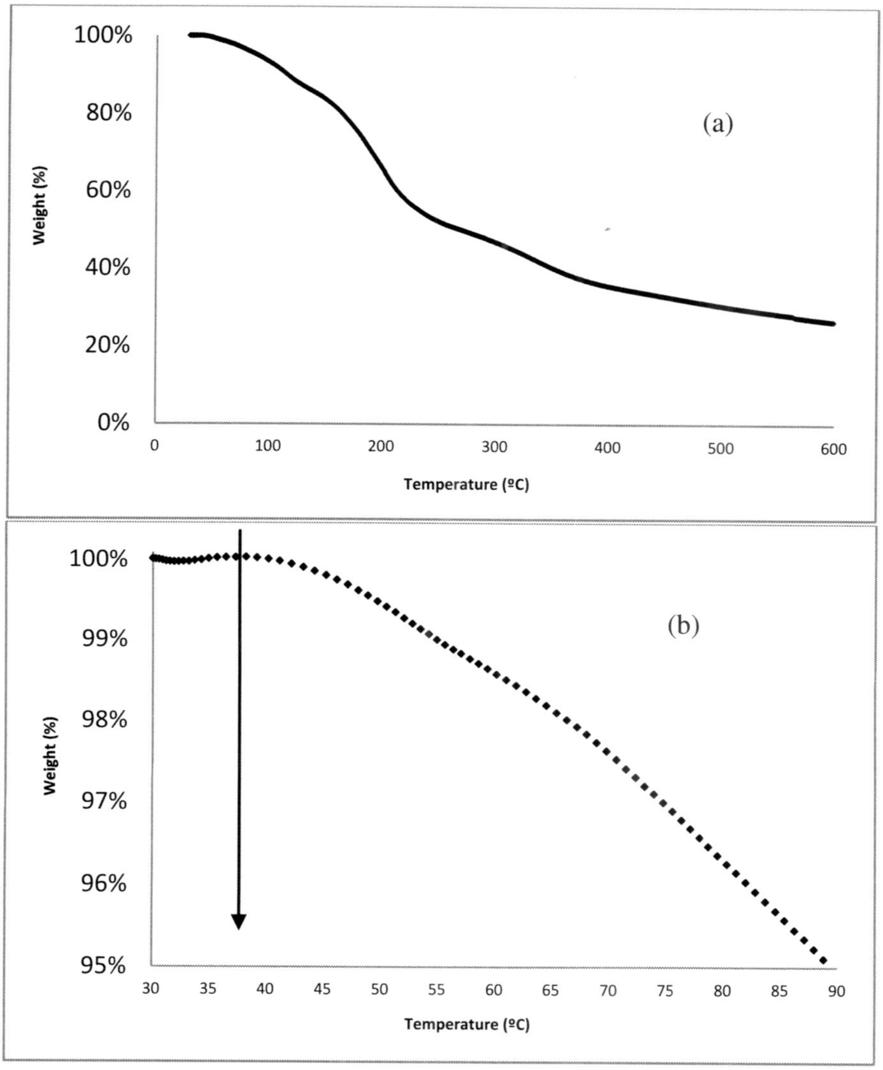

Figure 1. (a) Thermogravimetric analysis (TGA) of an ethanolic extract obtained from jabuticaba skins (analysis was conducted in nitrogen atmosphere). (b) Image zoom.

Santos and co-authors [1] used thin layer chromatography (TLC) to fractionate some ethanolic extracts obtained from jabuticaba skins. The results from qualitative TLC analysis indicated that cyanidin-3-glucoside and delphinidin-3-glucoside were the anthocyanins possibly present in the extracts [1]. The thermal stability of an anthocyanin-rich extract obtained from jabuticaba skins was determined by thermogravimetric analysis (TGA), which is presented in Figure 1. The TGA curve showed that the thermal degradation started at approximately 45 °C and increased continually as the temperatures increased.

The total yield of a target compound will remain constant (if no degradation occurs) after some time of extraction, when the available compound has been entirely exhausted from the sample. Actually, when working with unstable substances, the yield of the extraction process will be a function of both extraction and degradation phenomena. Since anthocyanins are particularly unstable, a maximum level of anthocyanins will be reached after a certain time of extraction. After this time, degradation effects will probably overcome the extraction effects, decreasing the total anthocyanins content [9].

Petersson and co-authors [9] investigated the extraction/degradation curves of anthocyanins from extracts of red onion. They used an elevated temperature to perform a batch extraction in a pressurized vessel at 2–4 bar. These authors successfully distinguished extraction and degradation effects and proposed a model to predict the real extraction curve, which might be similar to the theoretical one if there was no degradation phenomenon competing with extraction. The obtained curves are schematically represented in Figure 2. The experimental data demonstrated that a very short extraction time (13–25 min, including pre-heating period) must be used in order to avoid significant degradation effects when extracting anthocyanins at high temperature (110ºC). A dynamic extraction process may probably result in less degradation at the same temperature, because residence time of the solvent is shorter in a semi-continuous solid-liquid extraction [9]. Thus, to avoid significant thermal degradation, anthocyanins extraction techniques performed at elevated temperature should always be associated with short process time.

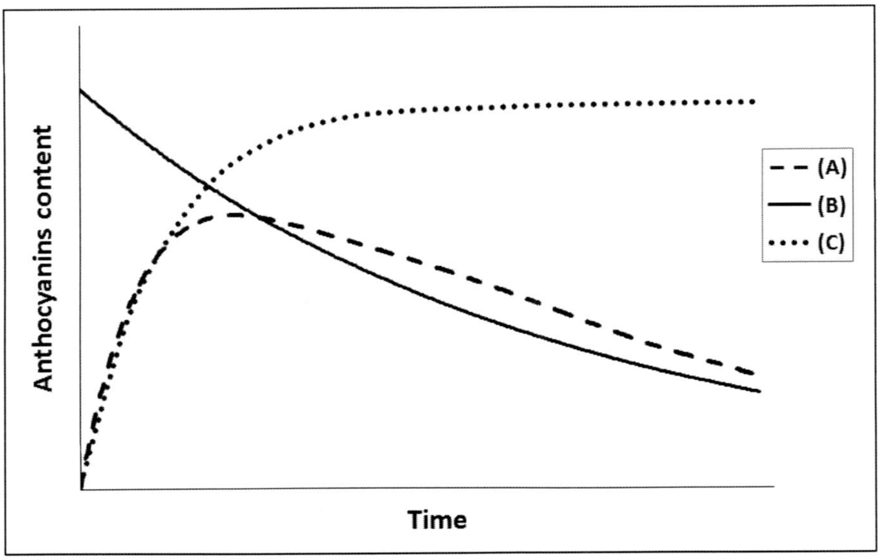

Figure 2. Extraction and degradation phenomena competing during high temperature extraction of anthocyanins: (A) combined extraction/degradation curve; (B) degradation curve; (C) total extraction curve (predicted). (Schematic figure based on the hypothesis proposed by Petersson and co-authors [9].

Anthocyanins Extraction Methods

The recovery of anthocyanins from plant materials has been traditionally performed by conventional solid-liquid extraction, which usually requires a long processing time that demands large amounts of organic solvents. Because anthocyanins are polar molecules, the most common solvents are ethanol, methanol, and acetone, or aqueous solutions of these solvents [5,7]. The solvents are commonly acidified in order to increase both extraction efficiency and anthocyanins stability [2,5]. The extraction efficiency is improved because acidified solvents denature cellular membranes and facilitate solubilization of anthocyanins

[2]. The pH decrease also enhances stability since in acidic conditions (pH < 2), anthocyanins are in their most stable form (flavylium cation) [5,8].

There are several extraction techniques that have been used for obtaining target compounds from plant matrices. These methods may be divided into two main groups: Low Pressure Solvent Extraction (LPSE) and High Pressure Solvent Extraction (HPSE). As the names indicate, the LPSE is conducted at atmospheric pressure, while HPSE requires a pressurized system (pressure ranges vary according to the applied method). The LPSE group includes a wide variety of extraction methods, from very traditional (Soxhlet, agitation, maceration, percolation, centrifugation, among others) to more recent ones (ultrasound, microwave, and pulsed electric field assisted extractions). The HPSE group involves modern developed technologies, of which Supercritical Fluid Extraction (SFE) is the most studied and industrially consolidated. The HPSE includes also emerging technologies such as Pressurized Liquid Extraction (PLE), High Hydrostatic Pressure Extraction (HHPE), and High Pressure Carbon Dioxide Assisted Extraction (HPCDAE).

Although various procedures for anthocyanins extraction have been investigated by different research groups [11–18], in this chapter our focus is to discuss some methods that have been recently used by our research group (LASEFI/DEA/FEA/UNICAMP) for recovering anthocyanins from jabuticaba skins. The evaluated LPSE methods were agitated bed extraction (ABE), ultrasound-assisted extraction (UAE), Soxhlet, and percolation. Beyond that, a novel extraction process was developed applying an ultrasound treatment followed by agitated bed extraction (UAE + ABE). In the HPSE group, the applied methods were pressurized liquid extraction (PLE) and high pressure carbon dioxide assisted extraction (HPCDAE). A schematic summary showing important parameters of the investigated LPSE and HPSE methods is presented in Table 1.

INFLUENCE OF EXTRACTION METHODS ON ANTHOCYANINS RECOVERY

The composition of a natural plant extract is strongly dependent on three fundamental factors: plant material characteristics, solvent properties and extraction process conditions. The characteristics (yield and chemical profile) of extracts obtained by different techniques should only be effectively compared when using both the same solvent and plant material. According to literature information, special attention should be dedicated to plant material, since it is essential to minimize the possible significant influence of origin, year of production, storage conditions and pretreatment of the plant material [6].

This section presents recently published data [1,6,10] concerning the anthocyanins extraction from jabuticaba skins. Table 2 and Table 3 contain a data compilation including experimental LPSE and HPSE results expressed in terms of anthocyanins recovery. It is important to highlight that only "green" solvents were used in all extraction procedures. Since the focus was the extraction technique, and not particularly the solvent composition optimization, different solvent mixtures were not tested. It is well established that the quality of a natural extract is directly associated to its functional properties, which are a function of chemical composition [6]. For this reason, we considered that the most efficient extraction process would be the one which leaded to the highest anthocyanins content in the extract.

Table 1. General process parameters according to the extraction method

Extraction Group	Extraction method	Combination of temperature and time	Operation mode	Solid (plant material) configuration	Solvent residence time	Specific characteristics
LPSE	UAE	low T + long t	batch	dispersed particles	long t_R	ultrasonic frequency (cavitation bubbles)
LPSE	ABE	low T + long t	batch	dispersed particles	long t_R	continuous agitation
LPSE	UAE + ABE	low T + long t	batch	dispersed particles	long t_R	ultrasonic frequency + agitation
LPSE	Soxhlet	high T + long t	batch	compact fixed bed	long t_R	solvent recirculation; exhaustive extraction
LPSE	Percolation	low T + long t	semi-continuous	compact fixed bed	short t_R	solvent recirculation
HPSE	PLE	high T + short t	semi-continuous	compact fixed bed	short t_R	dynamic extraction; combination of T + P + t
HPSE	HPCDAE	high T + short t	batch	dispersed particles	long t_R	explosive effect (CO_2 expansion); pH decrease

LPSE, Low Pressure Solvent Extraction; HPSE, High Pressure Solvent Extraction; UAE, Ultrasound Assisted Extraction; ABE, Agitated Bed Extraction; PLE, Pressurized Liquid Extraction; HPCDAE, High Pressure Carbon Dioxide-Assisted Extraction; CO_2, carbon dioxide; T, temperature; P, pressure; t, time; t_R, residence time of solvent.

Table 2. LPSE of anthocyanins from jabuticaba skins using ethanol as solvent and solvent to feed ratio equal to 10: compilation of experimental results from LASEFI/DEA/FEA/UNICAMP

IC	Extraction method	Plant material moisture (%, w/w)	Process parameters	TMA (mg cy-3-glu eq/g dry material)[(a)]	Ref.
E1	UAE	80.8	T = 25 °C; t_E = 2 h; f = 40 kHz	4.7 ± 0.2	[1]
E2	ABE	80.8	T = 30 °C; t_E = 2 h; AGR = 150 rpm	6.2 ± 0.3	[1]
E3	Combined UAE + ABE	80.8	UAE (10 min) + ABE (2 h)	5.3 ± 0.2	[1]
E4	Soxhlet	80.8	t_E = 8 h	5.0 ± 0.2	[1]
E5	Percolation	80.8	T = 23 °C; t_E = 2 h; SFR = 28 mL/min	1.90 ± 0.09	[26]
E6	Percolation	66.5	T = 23 °C; t_E = 2 h; SFR = 28 mL/min	1.2 ± 0.2	[10]

LPSE, Low Pressure Solvent Extraction; IC, identification code; UAE, Ultrasound-Assisted Extraction; ABE, Agitated Bed Extraction; PM, Plant Material; S/F, solvent to feed ratio; T, temperature; tE, extraction time; f, frequency; AGR, agitation rate; SFR, solvent flow rate; (mg cy-3-glu eq/g dry material), mg of cyanidin-3-glucoside equivalent/g of jabuticaba skins (dry basis).

(a) The content of anthocyanins is expressed in terms of Total Monomeric Anthocyanins (TMA), which was determined using the pH differential method as described by Santos and co-authors [1]. A detailed discussion about methods of anthocyanins analysis can be found in the reference [27].

The LPSE experiments E1 to E5 (Table 2) were conducted using both the same plant material (fresh jabuticaba skins, 80.8 % moisture) and solvent. Therefore, this data may be compared in order to evaluate the extraction methods performances. Taking into account the anthocyanins content, the most efficient LPSE method was ABE. The high-temperature and long-time characteristics of Soxhlet process probably induced significant anthocyanins degradation, resulting in lower anthocyanins recovery than ABE. When comparing percolation and ABE, although both were conducted in low temperature and had the same extraction time, the two important differences (the solid material configuration and the residence time of solvent) may have affected the anthocyanins extraction. The compact fixed bed and the short residence time are factors that imply disadvantages for mass transfer in the percolation method. These process differences may explain why percolation provided much lower anthocyanins recovery than ABE.

Ultrasound has been shown to improve the extraction of target compounds from various plant materials by reducing extraction times and increasing yields [19]. The enhancement of extraction efficiency of organic compounds by UAE is attributed to the phenomenon of cavitation, which is produced when an ultrasonic wave passes through the solvent. Cavitation bubbles are produced and compressed during the application of ultrasound treatment. When the bubbles and the gas within them are compressed, there is an important increase in temperature and pressure causing the collapse of the bubble. This collapse results in a significant disruption of cells and produces a "shock wave" that passes through the solvent enhancing the mixing. Ultrasound also exerts a mechanical effect, allowing a more intense penetration of solvent into the plant matrix. These effects provide enhanced mass transfer and improve the release of intracellular products into the bulk medium [19–21].

Although UAE enhances mass transfer and may improve the extraction of organic compounds [19–21], the experimental data (Table 2) showed that ultrasound did not have a positive influence on anthocyanins recovery from jabuticaba skins. Considering results from E1 and E3 (Table 2), the applied ultrasonic frequency may have favored the degradation of highly sensitive anthocyanins. This effect justifies why the short time ultrasonic treatment (UAE + ABE) was less aggressive to anthocyanins recovery than UAE.

Table 3. HPSE of anthocyanins from jabuticaba skins: compilation of experimental results from LASEFI/DEA/FEA/UNICAMP

IC	Extraction method	PM moisture (%, w/w)	Solvent	S/F	Process parameters	TMA (mg cy-3-glu eq/g dry material)[(a)]	Ref.
E7	PLE	66.5	Ethanol	4.4	T = 80 °C; P = 50 bar; t_S = 9 min; t_D = 12 min; SFR = ~ 1.7 mL/min	2.4 ± 0.5	[10]
E8	PLE	59.2	Ac. water[(b)] (pH = 2.5)	4.4	T = 80 °C; P = 117 bar; t_S = 5 min; t_D = 20 min; SFR = ~ 1.0 mL/min	0.58 ± 0.02	[6]
E9	HPCDAE	59.2	Ac. water[(b)] (pH = 2.5)	10	T = 80 °C; P = 117 bar; $R_{S\text{-}L/CO2}$ = 20 %; t_E = 20 min	2.2 ± 0.3	[6]
E10	HPCDAE Control[(c)]	59.2	Ac. water[(b)] (pH = 2.5)	10	T = 80 °C; P = 1 bar; t_E = 20 min	1.6 ± 0.1	[6]

HPSE, High Pressure Solvent Extraction; IC, identification code; PLE, Pressurized Liquid Extraction; HPCDAE, High Pressure Carbon Dioxide Assisted-Extraction; PM, Plant Material; S/F, solvent to feed ratio; T, temperature; P, pressure; SFR, solvent flow rate; t_S, static extraction time; t_D, dynamic extraction time; t_E, extraction time; $R_{S\text{-}L/CO2}$, volume ratio of solid-liquid mixture/pressurized CO_2; (mg cy-3-glu eq/g dry material), mg of cyanidin-3-glucoside equivalent/g of jabuticaba skins (dry basis).

[(a)] The content of anthocyanins is expressed in terms of Total Monomeric Anthocyanins (TMA), which was determined using the pH differential method as described by Santos and co-authors [1]. A detailed discussion about methods of anthocyanins analysis can be found in the reference [27].

[(b)] Distilled water acidified by hydrochloric acid addition.

[(c)] HPCDAE Control: an extraction under atmospheric pressure using the HPCDAE system without CO_2 addition and pressurization.

The PLE is an emerging technology that has been intensively studied in recent years [2,5,11,12,22–25]. The major advantage of PLE process is that the pressurized solvent remains in the liquid state well above its boiling point, allowing for high-temperature extraction [2]. It is known that increasing temperature improves extraction efficiency by enhancing solubility and diffusion rate of compounds into the solvent, reducing solute-matrix interactions, decreasing viscosity and surface tension of the solvent [9,12]. If water or aqueous solutions are used as solvents, a particularly important effect is that elevated temperatures modify the dielectric constant of the water resulting in the possibility of tuning its polarity [24]. However, elevation of temperature simultaneously results in increasing degradation rates of thermolabile compounds such as anthocyanins [1, 2, 9]. Therefore, a high-temperature process must be combined with short extraction time because anthocyanins degradation rates are temperature and time dependent [9].

In the case of anthocyanins recovery from jabuticaba skins by PLE, a process parameters optimization study has been recently published by Santos and co-authors [10]. The evaluated independent variables were temperature (T), pressure (P), and static extraction time (t_S). The data from PLE experiment (E7 – Table 3) represent the optimal extraction conditions considering the investigated range of each variable (T = 40–120°C, P = 50 –100 bar, t_S = 3–15 min). The optimization study demonstrated that extraction temperature strongly affected the anthocyanins recovery, being clearly the most significant extraction variable. In the lowest range (T=40–80°C), the temperature presented a positive influence on anthocyanins extraction, while in the highest range (T=80–120°C), the influence was negative since increasing temperature caused a decrease in anthocyanins recovery. In addition, when working in the highest temperature range studied (T=80–120°C), the elevation of static extraction time presented a negative influence on anthocyanins extraction [10].

In order to compare PLE (E7 – Table 3) to a LPSE technique, a percolation (E6– Table 2) extraction was conducted using the same plant material (dried jabuticaba skins, 66.5% moisture) and solvent. This LPSE method was chosen for comparison because PLE and percolation are semi-continuous processes, in which the solvent flows through a solid fixed bed. The experimental data demonstrated that the PLE procedure was more effective in extracting anthocyanins from jabuticaba skins, since the PLE extract presented significantly higher anthocyanins content than the percolation extract [10]. The better performance presented by PLE is probably due to the use of elevated temperature combined with system pressurization and short extraction time. Several published research articles have demonstrated that the above mentioned combination enhances the extraction of anthocyanins from various sources [2,5,11,12].

The HPCDAE is an innovative technique that exploits the explosive effect of high pressure carbon dioxide (CO_2) combined with its ability to acidify the extraction medium. The pH decrease is attributed to the generation of "in situ" carbonic acid and/or alkyl carbonic acid when CO_2 is inserted into the water or aqueous solution solvent [6]. The rapid release of gas pressure results in the explosive effect of high pressure CO_2, which may improve the extraction efficiency of target compounds by modifying cell membranes, decreasing the intracellular pH, disordering the intracellular electrolyte balance and removing vital constituents from cells and cells membranes [6, 18]. According to Xu and co-authors [18], in HPCDAE, there are probably five forms associated with CO_2, including supercritical CO_2, H_2CO_3 and its dissociated products such as H^+, HCO_3^-, and CO_3^{-2}. These different forms may play important roles in anthocyanins extraction, which were not fully understood until now [6].

Santos and Meireles [6] have published a work concerning the optimization of HPCDAE process parameters in order to achieve anthocyanin-rich extracts. The investigated independent variables were temperature (T=40–80°C), pressure (P=65–135 bar), and volume ratio of solid-liquid mixture/pressurized CO_2 ($R_{S\text{-}L/CO2}$ = 20–80%). The experimental results demonstrated that the recovery of anthocyanins was significantly affected by both temperature and pressure. However, the effect of the volume ratio of solid- liquid mixture/pressurized CO_2 was not significant in terms of anthocyanins extraction. The

HPCDAE data from Table 3 (E9) present the determined optimal conditions for anthocyanins recovery from jabuticaba skins [6].

A comparison between HPCDAE and PLE was conducted using the same plant material (dried jabuticaba skins, 59.2% moisture), solvent, temperature and pressure in both techniques [6]. Results (Table 3, E8 and E9) showed that HPCDAE provides much higher anthocyanins recovery than PLE in similar process conditions. The better performance of HPCDAE is probably due to the explosive effect caused by fast decompression of CO_2, which may have improved mass transfer from plant matrix to extracting solvent because of plant cell destruction during CO_2 expansion [6]. Besides that, the pH decrease was possibly favorable for anthocyanins extraction and stability. When considering the comparison between HPCDAE and HPCDAE-control (E9 and E10, respectively), the positive influence of pressure and CO_2 addition was confirmed since the extract obtained by HPCDAE presented significant higher anthocyanins content [6].

It is possible that the anthocyanins recovery by PLE with acidified water (Table 3, E8) might be improved if a study for temperature optimization had been performed. Some authors have demonstrated that the optimal extraction condition may be achieved in temperatures higher than 80ºC when using acidified water or acidified water solutions as PLE solvents [11,12].

The HPSE procedures are certainly promising techniques for obtaining anthocyanins from various sources. HPSE techniques usually require smaller amounts of solvents (S/F, solvent to feed ratio) and shorter extraction times than LPSE, which are important factors from an industrial point of view.

The plant material plays a noticeable influence on the extraction process as can be demonstrated by comparing data from E5 and E6 (Table 2). Although applying the same process parameters (temperature, time, and solvent) in both extractions, the percolation experiments E5 and E6 were conducted using jabuticaba skins with different moisture contents (80.8 and 66.5 %, respectively). The results showed that a better recovery of anthocyanins was achieved when using fresh jabuticaba skins, revealing the possible adverse effects of the drying pre-treatment [6, 10]. The heating applied in the drying process (conducted at T = 45 ºC) may have caused anthocyanins degradation [6, 10]. According to Monrad and co-authors [12], a certain water quantity is necessary to facilitate anthocyanins extraction. Therefore, the water from plant material possibly acted as a co-solvent improving anthocyanins recovery.

CONCLUSION

It is well established that the extraction method exerts a strongly influence over the recovery of anthocyanins from plant matrices. In general, from all discussed extraction process parameters, the temperature seems to be the most significant one. Therefore, despite working with LPSE or HPSE processes, the temperature always plays a fundamental role in anthocyanins extraction. Since degradation of anthocyanins is temperature- and time-dependent, the extraction time also requires special attention, particularly when working with high temperature processes.

Anthocyanins extraction has been traditionally conducted by LPSE at low temperatures, probably because in conventional techniques it is usual to utilize long extraction times. The HPSE procedures are certainly promising techniques for obtaining anthocyanins from various sources. One of the major advantages of HPSE methods is the possibility of using only "green" solvents (such as water, ethanol and CO_2) still providing a high recovery of anthocyanins. In addition, HPSE techniques usually require smaller amounts of solvents and shorter extraction times than LPSE, which are important factors from an industrial point of view.

ACKNOWLEDGMENT

The authors are grateful to CNPq and FAPESP for financial support. Susana P. Jesus thanks CNPq (141828/2010-2); Diego T. Santos thanks CNPq (141894/2009-1) for the doctorate fellowships and FAPESP (10/16485-5) for the Postdoc fellowship.

REFERENCES

[1] Santos, D.T., Veggi, P.C., Meireles, M.A.A. (2010). Extraction of antioxidant compounds from Jabuticaba (*Myrciaria cauliflora*) skins: Yield, composition and economical evaluation. *Journal of Food Engineering*, 101, 23-31.

[2] Ju, Z.Y., Howard, L.R. (2003). Effects of solvent and temperature on pressurized liquid extraction of anthocyanins and total phenolics from dried red grape skin. *Journal of Agricultural and Food Chemistry*, 51, 5207-5213.

[3] Nicoué, E.E, Savard, S., Belkacemi, K. (2007). Anthocyanins in wild blueberries of Quebec: extraction and identification. *Journal of Agricultural and Food Chemistry*, 55, 5626-5635.

[4] Santos, D.T., Meireles, M.A.A. (2009). Jabuticaba as a source of functional pigments. *Pharmacognosy Reviews*, 3, 127-132.

[5] Gizir, A.M., Turker, N., Artuvan, E. (2008). Pressurized acidified water extraction of black carrot [*Daucus carota* ssp *sativus* var. *atrorubens* Alef.] anthocyanins. *European Food Research and Technology*, 226, 363-370.

[6] Santos, D.T., Meireles, M.A.A. (2011). Optimization of bioactive compounds extraction from jabuticaba (*Myrciaria cauliflora*) skins assisted by high pressure CO2, *Innovative Food Science and Emerging Technologies*, 12, 398-406.

[7] Castañeda-Ovando, A., Pacheco-Hernández, M.L., Páez-Hernández, M.E., Rodríguez, J. A. & Galán-Vidal, C. A. (2009). Chemical studies of anthocyanins: A review. *Food Chemistry*, 113, 859-871.

[8] Cavalcanti, R.N., Santos, D.T., Meireles, M.A.A. (2011). Non-thermal stabilization mechanisms of anthocyanins in model and food systems-An overview. *Food Research International*, 44, 499-509.

[9] Petersson, E.V., Liu, J., Sjöberg, P.J.R., Danielsson, R., Turner, C. (2010). Pressurized Hot Water Extraction of anthocyanins from red onion: A study on extraction and degradation rates. *Analytica Chimica Acta*, 663, 27-32.

[10] Santos, D.T., Veggi, P.C., Meireles, M.A.A. (2011). Optimization and economic evaluation of pressurized liquid extraction of phenolic compounds from jabuticaba skins. *Journal of Food Engineering*, in press.

[11] Arapitsas, P., Turner, C. (2008). Pressurized solvent extraction and monolithic column-HPLC/DAD analysis of anthocyanins in red cabbage. *Talanta*, 74, 1218-1223.

[12] Monrad, J.K., Howard, L.R., King, J.W., Srinivas, K., Mauromoustakos, A. (2010). Subcritical solvent extraction of anthocyanins from dried red grape pomace. *Journal of Agricultural and Food Chemistry*, 58, 2862-2868.

[13] Corrales, M., García, A.F., Butz, P., Tauscher, B. (2009). Extraction of anthocyanins from grape skins assisted by high hydrostatic pressure. *Journal of Food Engineering*, 90, 415-421.

[14] Corrales, M., Toepfl , S., Butz, P., Knorr, D., Tauscher, B. (2008). Extraction of anthocyanins from grape by-products assisted by ultrasonics, high hydrostatic pressure or pulsed electric fields: A comparison. *Innovative Food Science & Emerging Technologies*, 9, 85-91.

[15] Seabra, I.J., Braga, M.E.M., Batista, M.T., Sousa, H.C. (2010). Effect of solvent (CO2/ethanol/H2O) on the fractionated enhanced solvent extraction of anthocyanins from elder berry pomace. *Journal of Supercritical Fluids*, 54, 145-152.

[16] Seabra, I.J., Braga, M.E.M., Batista, M.T.P., Sousa, H.C. (2011). Fractioned high pressure extraction of anthocyanins from elderberry (*Sambucus nigra* L.) pomace. *Food Bioprocess Technology*, 3, 674–683.

[17] Perez-Serradilla, J.A., Luque de Castro, M.D. (2011). Microwave-assisted extraction of phenolic compounds from wine lees and spray-drying of the extract. *Food Chemistry*, 124, 1652-1659.

[18] Xu, Z., Wu, J., Zhang, Y., Hu, X., Liao, X., Wang, Z. (2010). Extraction of anthocyanins from red cabbage using high pressure CO2. *Bioresource Technology*, 101, 7151-7157.

[19] Paniwnyk, L., Beaufoy, E., Lorimer, J.P., Mason, T.J. (2001). The extraction of rutin from flower buds of *Sophora japonica*. *Ultrasonics Sonochemistry*, 8, 299-301.

[20] Rostagno, M.A.; Palma, M., Barroso, C.G. (2003). Ultrasound-assisted extraction of soy isoflavones. *Journal of Chromatography A*, 1012, 119-128.

[21] Wu, J., Lin, L., Chau, F. (2001). Ultrasound-assisted extraction of ginseng saponins from ginseng roots and cultured ginseng cells. *Ultrasonics Sonochemistry*, 8, 347-352.

[22] Ibáñez, E., Kubátová, A., Señoráns, F.J., Cavero, S., Reglero, G., Hawthorn, S.B. (2003). Subcritical water extraction of antioxidant compounds from rosemary plants. *Journal of Agricultural and Food Chemistry*, 51, 375-382.

[23] García-Marino, M., Rivas-Gonzalo, J.C., Ibáñez, E., García-Moreno, C. (2006). Recovery of catechins and proanthocyanidins from winery by-products using subcritical water extraction. *Analytica Chimica Acta*, 563, 44-50.

[24] Rodríguez-Meizoso, I., Jaime, L., Santoyo, S., Señoráns, F.J., Cifuentes, A., Ibáñez, E. (2010). Subcritical water extraction and characterization of bioactive compounds from *Haematococcus pluvialis* microalga. *Journal of Pharmaceutical and Biomedical Analysis*, 51, 456-463.

[25] Teo, C.C., Tan, S.N., Yong, J.W.H., Hew, C.S., Ong, E.S. (2010). Review: Pressurized hot water extraction (PHWE). *Journal of Chromatography A*, 1217, 2484-2494.

[26] Santos, D.T., Veggi, P.C., Meireles, M.A.A. (2009). "Avaliação Técnico-Econômica da Extração de Antocianinas por Percolação em Leito Fixo". In *Proceedings of the XXXIV Brazilian Congress on Particulate Systems*, Oct. 2009 (Campinas, Brazil).

[27] Giusti, M.M., Jing, P. (2008). Analysis of Anthocyanins. In C. Socaciu (Ed.), *Food Colorants: Chemical and Functional Properties* (pp. 479-505). Boca Raton, FL: CRC Press.

In: Anthocyanins: Structure, Biosynthesis and Health Benefits
Editor: Noboru Motohashi
ISBN: 978-1-62257-329-5

Chapter 8

ADVANCED ANALYTICAL TECHNIQUES FOR THE CHARACTERIZATION OF ANTHOCYANINS IN FRUITS AND VEGETABLES

Ihsan Iswaldi, David Arráez-Román, Ana María Gómez-Caravaca, Antonio Segura-Carretero [*] ***and Alberto Fernández-Gutiérrez***

Department of Analytical Chemistry, Faculty of Sciences,
University of Granada, Granada, Spain
Research and Development Functional Food Center (CIDAF), Granada, Spain

ABSTRACT

Over 600 anthocyanins have been identified to date. These structures have been determined by high-performance liquid chromatography (HPLC) coupled to UV-Vis (UV) detector, diode-array detector (DAD), and different mass spectrometry (MS) analyzers, such as ion trap (IT), quadrupole (Q), time-of-flight (TOF) and hybrid analyzers Q-TOF. MS has generally been coupled to different ion sources such as electrospray ionization (ESI) and matrix-assisted laser desorption/ionization (MALDI). Moreover, capillary electrophoresis (CE) coupled to DAD and MS detectors has also been used to characterize anthocyanins.

Among these analytical methods, the UV-Vis spectra represents a valuable analytical tool for identifying anthocyanins because they have a specific and intense absorbance band in the range of 520 to 560 nm; meanwhile, MS is a powerful tool which can provide information on molecular mass and also structural features of the different constituents.

This chapter provides an overview of the uses of HPLC and CE coupled to different detection systems, discussing the fundamentals of sample extraction and separation.

* E-mail address: ansegura@ugr.es.

INTRODUCTION

Anthocyanins, water-soluble pigments found in flowers, fruits, leaves, grains, and other plant organs, are responsible for bright colors such as orange, red, purple, and blue. These pigments are found in the vacuoles and at the surface of the fruit and vegetable skin. Meanwhile, in flowers, they are located in epidermal cells and rarely in the mesophyll. Because anthocyanins act as pigments in a variety of plant organs, the color intensity, hue, and stability are key properties which are highly influenced by structure, pH, oxygen, and a large number of other factors [1,2].

The basic chemical structure of an anthocyanin is an aglycone (anthocyanidin), with one or more sugars attached at the 3, 5, or 7 positions and possibly esterification on the sugars. The six most common anthocyanidins occurring in nature include delphinidin, cyanidin, pelargonidin, petunidin, peonidin, and malvidin (Figure 1) while the most common sugars in natural anthocyanins are glucose, galactose (hexoses), rhamnose, xylose, and arabinose (pentoses). Many anthocyanins have been found to be acylated by aliphatic or aromatic acids. The most frequent acylations are those with acetic, malonic, *p*-coumaric, and caffeic acids, whereas acylations with ferulic, *p*-hydroxybenzoic, succinic, oxalic, gallic, vanillic, syringic, protocatechuic, and malic acids are more restricted [1,3]. There are over 600 anthocyanins which have been identified from natural sources considering all these factors.

In aqueous media, anthocyanins undergo structural transformations with a change in pH in four basic structures (Figure 2), where the colored oxonium form predominates at a pH of approximately 2.0 or lower, which is orange to purple, and the colorless carbinol pseudo-base form at pH 4.5. When the pH rises, most of flavylium cation changes to other anthocyanin forms. The first reaction gives a colorless carbinol pseudo-base, which can undergo ring opening to a yellow chalcone, and later reactions give rise to a quinonoidal base with the formation of purplish-blue resonance-stabilized quinonoid anions [4].

Anthocyanins are highly reactive and easily degraded. Other factors affecting the anthocyanin stability, besides chemical structure and pH, include storage temperature, concentration, light, oxygen, solvents, enzymes, flavonoids, proteins, and metal ions [5]. Turker *et al.*, studying temperature in relation to the stability of shalgam anthocyanins, found that the monomeric anthocyanin content and color density decreased with time as a function of storage temperature whereas the percentage of polymeric color and browning increased [6]. Furthermore, the effect of light, temperature, pH, and species on the stability of anthocyanin pigments in four *Berberis* species has also been investigated, revealing that increased pH, temperature or light exposure can degrade the anthocyanin molecule [7].

The great interest in anthocyanins emerged because of their possible human-health benefits such as antiartherogenic, anticancer, antidiabetic, anti-inflammatory and antioxidant activities [8,9,10]. Recently, studying the red petals of *Rosa hybrida*, Lee *et al.* have reported that cyanidin 3,5-di-*O*-glucoside, the predominant constituent, showed high antioxidant activity, while pelargonidin 3,5-di-*O*-glucoside exhibited potent anticancer effects against cell cultures of LNCap (a human prostate cell line), ACHN (a human renal cell line), and MOLT-4F (a human leukemia cell line) [11].

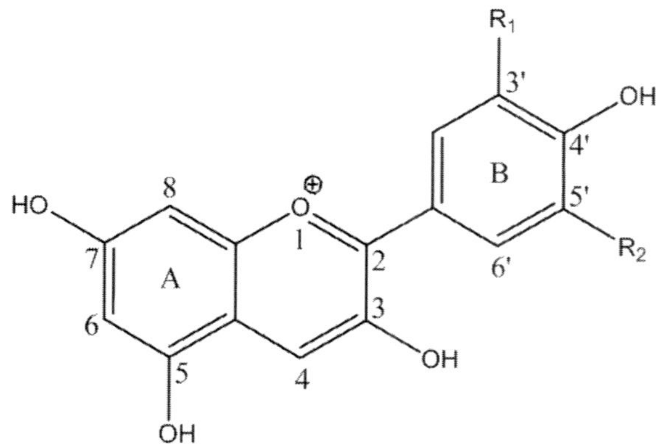

Anthocyanidins	R1	R2	Substitution pattern	MW
Pelargonidin (Pg)	H	H	3,5,7,4'-OH	271
Cyanidin (Cy)	OH	H	3,5,7,3',4'-OH	287
Delphinidin (Dp)	OH	OH	3,5,7,3',4',5'-OH	303
Peonidin (Pn)	OMe	H	3,5,7,4'-OH; 3'-OMe	301
Petunidin (Pt)	OMe	OH	3,5,7,4',5'-OH; 3'-OMe	317
Malvidin (Mv)	OMe	OMe	3,5,7,4'-OH; 3',5'-OMe	331

Figure 1. Chemical structures of six common anthocyanidins.

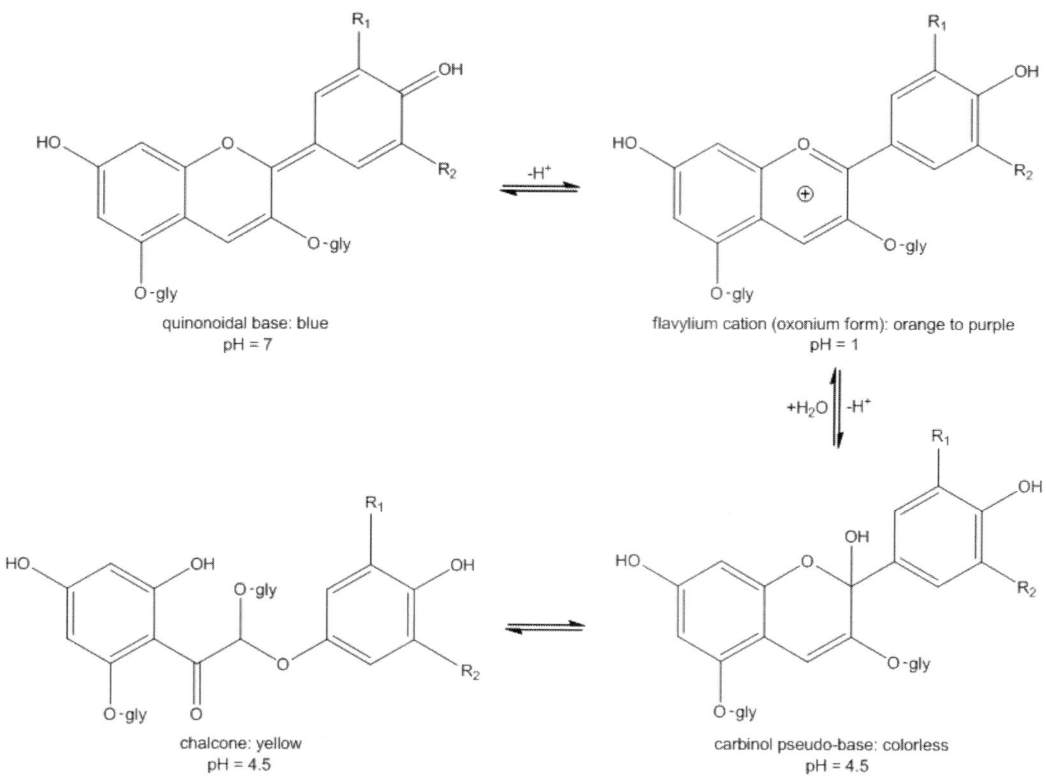

Figure 2. Structural transformations of cyanidins with change in pH.

Sample Extraction Methods

Plant matrices are usually so complex that direct analysis of the matrix is not possible. This makes it necessary to extract the components of interest —in this case, phenolic compounds. This first step is especially critical in the case of quantitative analysis. Many procedures for the polyphenol extraction from plant matrices have been described, most using liquid-liquid extraction (LLE) or solid-phase extraction (SPE) [12].

Whichever method is preferred, SPE or LLE, it must allow the recovery of the anthocyanins without any chemical modification. Anthocyanins are commonly extracted by organic solvents such as ethanol, methanol or acetone due to their property as polar molecules. The solvent destroys the cell membranes and simultaneously dissolves the anthocyanins and stabilizes them [13]. The extracting solution should be slightly acidic to maintain the flavylium cation form, which is red and stable in a highly acidic medium. The most common acids used for the extraction are hydrochloric acid or formic acid. However, the use of strong acid media should be avoided because the acyl moieties in acylated anthocyanins might be degraded by hydrolysis [14]. Acetone is normally used for LLE, offering more efficient and reproducible results than using methanol or ethanol. This method works well because it can be performed at lower temperatures and, furthermore, it can avoid trouble with the presence of pectins found in the walls of some anthocyanins such as in berries. In acidic extraction environments, they form a turbid extract that slows the process. However, acetone contains pectin-clotting properties and rapidly produces a clean anthocyanin extract. For SPE, popular methods utilize C_{18} cartridges or Sephadex LH-20 for the initial separation of the anthocyanin extracts from the crude plant sample. The polar anthocyanins bind to these adsorbents through unsubstituted hydroxyl groups and are separated from other plant compounds using solvents of increasing polarity. Studies have shown that SPE is fast and highly efficient for separating crude anthocyanin solutions from plant material. With the collection of the anthocyanin extract, the next step is to remove the solvent to a known amount and continue with the analysis. Samples which need to be stored are generally stored as dried samples, evaporated and lyophilized, and storage in cold and dark environments to avoid decomposition of the anthocyanins.

Preliminary Separation

After an anthocyanin extract is separated from plant tissue, it is necessary to evaluate whether purification is required. Generally, additional separation methods are followed to purify the sample before analysis because the extraction methods are not selective for anthocyanins and give crude anthocyanin extracts that might be co-extracted with sugars or organic acid. The appropriate techniques for a preliminary separation include large-scale chromatographic techniques, such as normal-phase column chromatography, counter-current chromatography (CCC), and preparative-HPLC.

The classic column chromatography, originally grew in popularity because of the need to obtain pure anthocyanin compounds in sufficient quantities for comprehensive characterization and to be used as reference standards. Normal-phase chromatography is normally used to separate non-ionic compounds with moderate molecular weight. Separation by CCC is based on the partition of an analyte between two immiscible liquids by vigorous mixing in a separatory funnel, where one liquid works as the stationary phase and the other as the mobile phase. This chromatographic technique allows non-destructive isolation of the anthocyanin compounds. High-speed counter-current chromatography (HSCCC), a relatively new technique and the most advanced CCC, requires the use of two immiscible solvents, generally an organic and an aqueous phase, one being used as the stationary phase and the other as the mobile phase. HSCCC does not use a solid support as the stationary phase and therefore has several advantages over conventional chromatography: (1) the elimination of sample losses caused by irrecoverable adsorption to the solid support matrix; (2) easy scale-up to a larger fractionation system by simply changing the Teflon tubing coil (column) to larger sizes; (3) low-cost for not using expensive absorbents and columns; and (4) reflecting the real distribution profile of phytochemicals in a sample [15].

A common multisolvent system composed of *tert*-butyl methyl ether (TBME)/*n*-butanol/acetonitrile/water acidified with 0.1% (v/v) trifluoroacetic acid (TFA) has been proven to be efficient regarding anthocyanin fractionation from a number of different plant matrices by using CCC [16]. A similar solvent system containing less TFA (0.01% v/v) can be used to avoid degradation of the labile malonylated anthocyanins [17]. Hillebrand *et al.* have used HSCCC and low-speed rotary CCC for the preparative isolation of anthovyanins in blue-fleshed potatoes. The common multisolvent system above was used. By means of HSCCC, it was possible to isolate the major pigment in large yields and good purity. Further isolation by using preparative-HPLC (described below) made it possible to isolate minor pigments from different CCC fractions [18].

Preparative HPLC is used to isolate and/or purify compounds. For identification and structure elucidation of unknown compounds in crude plant extracts it is necessary to obtain pure compounds in amounts ranging from one to a few milligrams. This approach makes it possible to apply a large amount of sample to the stationary phase. The columns that are used can be constituted by the same packing material, but with a larger dimension, as the material used for the analytical HPLC methods and the same mobile phases with simple gradient or isocratic system. The collected fractions from preparative HPLC can be monitored by TLC and the similar TLC plates will correspond to the fractions that can be combined, concentrated, and used to proceed with isolation [19].

SEPARATION METHODS

The analysis of anthocyanins starts with the extraction of anthocyanins from plant tissues, followed by their isolation and purification from crude plant extract and, ultimately, the identification of each anthocyanin. This analysis is quite complex because of possible structural transformations and complexation reactions. These compounds are difficult to measure separately from other flavonoids because they have similar structures and reactivity characteristics. Finally, pure anthocyanin standards, which are integral for accurate quantification purposes, are not readily available for purchase and are not easily purified from plant sources for use in research, partly due to the aforementioned instability issues.

Table 1. Recent reports on the analysis of anthocyanins by HPLC

Sample	Sample preparation and extraction conditions	Separation techniques and detector	Stationary phase	Mobile phase	Reference
Vitis vinifera L.	Extraction with MeOH:HCl (1000:1 v/v) centrifuged, filtered	HPLC–DAD–ESI-Q-MS	Waters Novapack C_{18} (150 × 3.9 mm, 4 μm)	A: water/formic acid, 90:10 B: water/MeOH/formic acid, 45/45/10, v/v	[25]
Blackberry, black raspberry, blueberry, cranberry, red raspberry and strawberry	Extraction with MeOH containing 0.1% HCl	HPLC–DAD–ESI-IT-MS	Symmetry C_{18} (250 × 4.6 mm, 5 μm)	A: water, 1% formic acid B: ACN	[26]
Grape pomace	Extraction with MeOH with 1% 1N HCl and followed by hexane	HPLC–UV-DAD HPLC–ESI-MS	Phenomenex Luna C_{18} (250 × 4.6 mm, 5 μm)	A: water/formic acid, 9:1 (v/v) B: ACN/formic acid, 9:1 (v/v)	[27]
Strawberry fruit (*Fragaria ananassa*)	Liquid-liquid extraction with MeOH:0.1% HCl (99:1) followed by removal of lipids by *n*-hexane. Lipid-free extract was subjected to silica-gel-column chromatography to separate sugars and acid; whereas carotenoids were isolated from anthocyanins using TLC	HPLC UV-Vis	Particil C_{18} (250 x 4.6 mm, 5 μm)	A: 0.1% TFA in water B: ACN	[28]
Commercial fruit juices and fruit drinks	Samples were centrifuged, filtered and injected directly to HPLC system	HPLC-PDA-MS^2	Synergi Max-RP column (250 x 4.6 mm i.d., 4 μm)	A: 1% formic acid B: ACN	[29]
Bilberry (*Vaccinium myrtillus* L.);	Extraction with solvent A-B (10:90 v/v); ACN-MeOH (85:15 v/v) (solvent A) and 8.5% aqueous formic acid (solvent B)	HPLC-DAD	Phenomenex Gemini C_{18} (150 x 4.6 mm i.d., 5 μm) equipped with 4 x 3 mm C_{18} precolumn	A: ACN-MeOH (85:15 v/v) B: 8.5% aqueous formic acid	[30]
Delonix regia Flowers	Extraction with deionized tap water acidified with citric acid (0.05 N), filtration	HPLC-DAD-ESI-MS^2	HPLC-DAD: RP-18 Satisfaction column, (250 x 4.6 mm, 0.45 μm) HPLC-MS: Merck LiChrospher 100-RP-18 (column 250 x 2 mm, 5 μm)	HPLC-DAD A: 10:90 formic acid/water B: 10:90 formic acid/ACN HPLC-MS A: water and formic acid (98:2,v/v) B: water:ACN:formic acid (18:80:2, v/v)	[31]

Sample	Sample preparation and extraction conditions	Separation techniques and detector	Stationary phase	Mobile phase	Reference
Vegetables, Nuts, and Grains	Extraction with MeOH:water:acetic acid (85:15:0.5; v/v, MeOH:H_2O:AcAc), dilution with acidic MeOH	HPLC-DAD-ESI-MS^2	Zorbax Stablebond Analytical SB-C_{18} column (4.6 x 250 mm, 5 μm)	A: 5% formic acid aqueous solution) B: MeOH	[32]
Blue-fleshed potatoes (*Solanum tuberosum* L.)	Extraction with water; water/HCl (19/1, v/v)	Preliminary separation: 1. Amberlite XAD-7 column. The column was washed with water and the anthocyanins were eluted with a mixture of MeOH/acetic acid (19:1, v/v). 2. Countercurrent Chromatography (CCC): High-Speed CCC and Low Speed Rotary CCC. 3. Preparative HPLC HPLC-DAD, ESI-MS^n and NMR	Luna RP-18 column (250 x 4.6 mm, 5 μm)	A: water/ACN/formic acid: 87/3/10 v/v B: water/ACN/formic acid: 40/50/10 v/v	[18]
Muscadine grapes	Extracted with 80:20 (v/v) MeOH–water solution containing 0.1 mL/L HCl. The suspension was loaded to Waters C_{18} Sep-Pak SPE and eluted with MeOH	HPLC–DAD–ESI-Q-MS	Symmetry C_{18} (150 × 3.0 mm, 5 μm)	A: water/formic acid, 95:5 B: MeOH/formic acid, 95:5	[33]
Purple Rice Bran	Extraction of lipophilic sample with hexane. Extraction of hydrophilic sample with MeOH	HPLC-PDA	Discovery C_{18} column (3 mm i.d. × 25 cm)	A: 0.4% TFA in water B: ACN	[34]
Organic commercial juices	Samples were filtered, diluted, and injected directly into the HPLC system	LC-DAD-ESI-Qq-TOF-MS	Phenomenex Synergy Fusion RP100A (50 × 2 mm, 2.5 μm)	A: water B:ACN, both with 0.1% (v/v) formic acid.	[35]
Blueberry	Extraction with a mixture of MeOH:water:TFA (70:30:1, v/v/v)	HPLC-UV HPLC-ESI-IT-TOF-MS	Method 1: ODS Hypersil column (200 × 4.6 mm, 5 μm) Method 2: Prodigy 5 mm ODS3 reversed phase silica (250 x 4.6 mm i.d.) and followed by Zorbax SB-C_{18} column (150 ×4.6 mm, 3.5 μm)	Method 1 and 2: A: water:THF:TFA (98:2:0.1, v/v/v) B: ACN A: 10% formic acid in water B: ACN	[36]

Table 1. (Continued)

Sample	Sample preparation and extraction conditions	Separation techniques and detector	Stationary phase	Mobile phase	Reference
Caucasian blueberries (*Vaccinium arctostaphylos* L.)	Extraction with solvent A-B (10:90 v/v); ACN-MeOH (85:15 v/v) (solvent A) and 8.5% aqueous formic acid (solvent B).	HPLC-DAD-ESI-MS^3	Gemini C_{18} (150 x 4.6 mm i.d., 5 μm) fitted with 4 x 3 mm i.d. C_{18} guard column	A: MeCN/MeOH (85:15 v/v) B: aqueous 8.5% formic acid	[37]
Hibiscus sabdariffa	The powder was mixed with water and injected directly to HPLC system	HPLC-DAD-ESI-TOF/IT-MS	Zorbax Eclipse Plus C_{18} (4.6 x 150 mm, 1.8 μm)	A: 10% formic acid B: ACN	[38]
Chinese sugarcane (*Saccharum sinensis* Roxb.) tips, stems, roots and leaves	Sugarcane tips, stems, roots: Ultrasonic maceration with MeOH/water (6:4, v/v) Sugarcane leaves: washing with petroleum ether to remove chlorophyll and ultrasonic maceration with MeOH/water (6:4, v/v). TLC analysis using n-BuOH/EtOH/water/ formic acid (16:4:8:1, v/v)	HPLC-UV diode-array detector HPLC-IT-MS/MS	Waters C_{18} Symmetry column (150 x 4.6 mm i.d., 4.6 μm)	A: 0.1% formic acid in water B: MeOH	[39]
Fruit peel of *Vitis coignetiae* Pulliat (meoru)	Extraction with MeOH containing 0.1% HCl. The solid residue was washed with MeOH. The extract was loaded onto an Amberlite XAD-7 column and washed with water and ethyl acetate. The anthocyanin pigments were eluted with MeOH containing 1% aqueous HCl.	HPLC-PAD and ESI-IT-MS	Zorbax StableBond Analytical SB-C_{18} column (4.6 x 250 mm, 5 μm)	A: aqueous 0.5% v/v formic acid B: MeOH	[40]
Mulberry Fruit (*Morus alba* L.)	Ectraction with 95% alcohol/0.1% HC l (1:1, ratio) and isolation using C_{18} Sep-Pak cartridge	HPLC-PAD-ESI-IT-MS and ^1H NMR	Waters RP C_{18} column (250 x 4.6 mm i.d., 5 μm)	A: water containing 0.1% TFA B: ACN containing 0.1% TFA	[41]

Sample	Sample preparation and extraction conditions	Separation techniques and detector	Stationary phase	Mobile phase	Reference
Purple-Fleshed Sweet-potatoes	Extraction with 7% acetic acid in 80% MeOH	HPLC-DAD/ESI-MS/MS	Synergi Polar-RP C_{18} (250 x 2 mm i.d., 4 μm) equipped with a guard column of 7.5 x 4.6 mm i.d., Allsphere Phenyl 5 μm.	A: 1% v/v formic acid in water B: ACN containing 1% formic acid)	[42]
Berberis boliviana Lechler	Extraction with aqueous acetone/water (30:70 v/v), filtration.	Anthocyanin purification: The aqueous extract was passed through a C_{18} mini column and eluted with 0.01% aqueous HCl. HPLC-PDA, HPLC- triple quadrupole - MSMS	Sample was semi-purified with a C_{18} cartridge and the phenolic fraction (containing anthocyanins) was eluted with 1% HCl-acidified MeOH. Waters Symmetry C_{18} column (4.6 x 150 mm, 3.5 μm)	A: 10% formic acid in water B: ACN	[43]
Waxberry (*Myrica rubra* Sieb. et Zucc.)	Extraction with 0.1% HCl in EtOH	(HPLC) with photodiode-array detection (PAD) and (MALDI-TOF-MS) ^{1}H NMR	Preparative HPLC (separation and isolation): (250 × 21.2 mm i.d., 7 μm) Zorbax Stable Bond-C_{18} preparative column, coupled to a (50 × 21.2 mm i.d., 5 μm). Zorbax Stable Bond-C_{18} guard column. The crude anthocyanin extract was purified by C_{18} Sep-Pak cartridge open-column chromatography. Analytical HPLC-PAD: Kromasil C_{18} (250 × 4.6 mm i.d., 5 μm) with (4 × 3 mm i.d.) Phenomenex C_{18} guard cartridge.	Preparative HPLC:formicacid/water/MeOH (10:75:15 v/v). Analytical HPLC-PAD: A: 0.1% TFA in water B: 0.1% TFA in ACN	[44]
Skin extracts of red grape (*Vitis vinifera* and *Vitis labrusca*) pomace	Extraction with 50 mL of a solvent mixture comprising MeOH/ water/acetic acid (80:20:5). Extract was added to polyamide SC6 column and eluted with MeOH	HPLC-DAD	Prodigy ODS3 reversed-phase silica column (250 × 4.6 mm i.d., 5 μm)	A: water:THF:TFA (98:2:0.1) B: MeOH: THF:TFA (98:2:0.1).	[45]

Table 1. (Continued)

Sample	Sample preparation and extraction conditions	Separation techniques and detector	Stationary phase	Mobile phase	Refer-ence
Purple Bordeaux Radish (*Raphanus sativus* L.)	Extraction with MeOH/water/HCl (60:40-0,03, v/v). The residue was mixed with 2 N NaOH and HCl (37%) was added to the reaction mixture and passed through a Waters OASIS HLB cartridge and washed with water and then with MeOH-1% HCl	LC-PDA-ESI/MS^n For accurate mass measurements: UPLC-HRMS	Symmetry 100 RP-18 (150 x 2.1 mm i.d., 3.5 μm) with (10 mm x 2.1 mm i.d.) guard column of the same material. For accurate mass measurements: column (200 x 2.1 mm i.d., 1.9 mm) Thermo Hypersil Gold AQ RP- C_{18} with an HPLC/UHPLC precolumn filter	A: 0.1% formic acid in water B: 0.1% formic acid in ACN	[46]
Strawberry and elderberry extracts	Extraction with MeOH containing 0.1% HCl	RP-HPLC at pH 1.8, and compared with separations achieved by capillary zone electrophoresis using a standard silica capillary and pH 8.0 running buffer	ODS-Hypersil reversed phase column (200 x 2.1 mm, 5 pm)	*Strawberry* A: aqueous 0.6% $HC1O_4$ B: MeOH, *Elderberry* A: aqueous 0.6% $HC1O_4$ B: THF CZE: The fused silica capillary was 57 cm total length (50 cm to detector) x 75 pm i.d., with 150 mM sodium borate buffer (pH 8.0)	[47]

* This table contains a representative selection of references. Additional references are available in the references cited.

High-Performance Liquid Chromatography (HPLC)

HPLC methods tend to utilize gradient solvent systems of acetonitrile water or methanol water with a small amount of acid to lower the solution pH and bolster the stability of the anthocyanins. These solvents are commonly used due to their compatibility with gradient methods for isolation and the various detection systems used for identification that are able to be coupled to the HPLC. For reproducible results using this instrumentation, the pH of the mobile phase and temperature of the column must be controlled due to the instability of anthocyanin compounds in changing pH and temperature environments.

In reverse-phase HPLC, the most common mode for separating anthocyanins, the order of elution is delphinidin, cyanidin, petunidin, pelargonidin, peonidin, and malvidin derivatives, consecutively. This can be explained by the fact that the retention time decreases with increasing polarity signed by raising the number of hydroxyl moieties in the flavylium ion.

On the contrary, acylation and the presence of sugars lengthens the retention time where diglycosides elute before monoglycosylated anthocyanins [20]. Normal-phase systems, or those using unmodified silica gel, are not effective for separating anthocyanins due to the polarity of these compounds.

The analytical separation and identification of anthocyanins have been summarized in recent review articles [4,21,22,23,24], reflecting that liquid chromatography occupies a primary position among the other analytical techniques. In general, HPLC separations are performed by using C_{18} reverse-phase columns and a binary solvent gradient. The mobile phases used consist of an acidified aqueous solution and an organic solvent such as acetonitrile and methanol. HPLC is most frequently coupled to simple ultraviolet (UV), photodiode-array detection (DAD) and MS detector. This technique has proven to be a powerful analytical tool for identifying anthocyanins, particularly. Table 1 lists the most representative references on anthocyanin analysis by HPLC.

Capillary Electrophoresis

Another analytical technique such as capillary electrophoresis (CE), has proven to be an advantageous alternative to HPLC in anthocyanin analysis for a very small sample size, low operation cost, high efficiency, shorter analytical time, and reagent use. CE separates compounds based on differences in their electrophoretic mobility and offers excellent mass sensitivity, high resolution, low sample consumption, and minimal generation of solvent waste. In fact, for anthocyanins, CE can be highly efficient separation technique and ideal for charged compounds such as anthocyanins (Table 2).

There are different modes in CE separations. Capillary-zone electrophoresis (CZE) is the most common mode used for separating anthocyanins, the migration of a particular compound depending on its charge-to-size ratio. That is, the total migration time for positively charged smaller molecules is longer than that for molecules of lesser charge and/or larger size. The separation is usually performed in the positive polarity mode followed by a positive electroosmotic flow with migration of the compounds from anode to cathode. The applicability of CZE in basic media is limited by the instability of anthocyanins in basic environments; therefore, the separation of anthocyanins from plant extracts by CZE use acidic media and configure the system to run from cathode to anode. The acidic media will ensure that the anthocyanins remain in the protonated flavylium form and the migration of anthocyanins will be towards the anode, reversing the electroosmotic flow. The speed by which they are drawn to the anode depends on their charge-to-size ratio, as seen with normal electroosmotic flow. With acidic media and the acidic form of anthocyanins, another problem arises with the formation of ionic interactions between the flavylium cation and the anionic silanol groups covering the capillary tube.

DETECTION AND IDENTIFICATION

UV-Vis Spectroscopy

Spectral characteristics of anthocyanins can be detected in two distinctive absorption bands, *i.e.* in the UV-region (260-280 nm) and in the visible region (490-550 nm). It should be taken into consideration when comparing the same anthocyanins that they must be in the same solvent because the solvent will affect the position of absorption bands.

Table 2. Recent reports on the analysis of anthocyanins by capillary electrophoresis

Sample matrix	Sample preparation and extraction condition	Separation conditions	Reference
Wine and wine must	Extraction with SP (polymer phase, Strata SDB-L) and elute with 0.1% (v/v) formic acid in MeOH	CE/MS^n Fused-silica separation capillary without internal coating (60 cm x 50 μm i.d.) Buffer: 1) acidic one (chloroacetate ammonium, pH 2.0) and 2) basic one with high selectivity towards derivatives containing vicinal hydroxyl groups (borate ammonium, pH 9.0).	[48]
Red wines	Fractionation of sample by Gel Permiation Chromatography (GPS) TSK Toyopearl gel HW-50F, column (120 mm ×12 mm i.d.). First fraction (F1) was eluted with EtOH/water/TFA (55:45:0.05, v/v/v) and second fraction (F2) with acetone/water (60:40, v/v). Sample fractions were dissolved in synthetic wine (12% (v/v) EtOH in aqueous solution containing 6 g L^{-1} tartaric acid, pH 3.5).	CZE-DAD Uncoated fused-silica capillary Buffer: Sodium tetraborate buffer solutions (50 mM) of pH 9.4 with 10% MeOH (v/v) content, and 56 cm (effective length) capillary.	[49]
Hibiscus sabdariffa L.	Extraction A: with MeOH/HCl (99:1 v/v) for 4 h. Extraction B: with MeOH/HCl (99:1 v/v) by sonication for 30 min Extraction C: with acetic acid (15% v/v) for 48 h. Filtrate was mixed with Amberlite XAD-2, packed into a glass column and eluted with EtOH (70% v/v)–acetic acid (1% v/v)	CE-TOF/IT-MS Fused-silica capillaries with 50 μm i.d. x 80 cm Buffer: 200 mM boric acid (pH 9.0)	[50]

Sample matrix	Sample preparation and extraction condition	Separation conditions	Reference
Strawberry	Extraction with MeOH, followed with 95% aqueous MeOH. The extract was dissolved in acidified water (3% formic acid) and passed through an SPE Strata C_{18}-E 55 µm 70 A cartridge (500 mg/3 mL) (Phenomenex) and recovered with MeOH containing 3% formic acid.	CZE UV–vis (DAD) Capillary cartridge: polyimide-coated fused-silica tube (375 µm o.d., 50 µm i.d.) Total capillary Length: 57 cm with effective length: 50 cm. Buffer: 250 mM monobasic sodium phosphate containing 30% v/v ACN (pH 1.4).	[51]
Red onions (*Allium cepa* L.)	Extraction with (EtOH and formic acid 99:1 v/v)	CZE -ESI-TOF-MS Bare and poly-LA 313-coated fused-silica capillaries with 50 µm i.d. x 365 mm o.d. and 57 cm length. Buffers: ammonium acetate, ammonium formate, borate, acetic acid, formic acid and mixtures of formic acid and hydrochloric acid, with ionic strength (IS) in the range 1–75 mM and with pH in the range 1.6–11.0	[52]
Grape (*Vitis vinifera* L. cv. Malbec)	Extraction with synthetic solution (EtOH, 12%; tartaric acid, 6 mg/mL; SO_2, 100 µg/mL; pH 3.2) and diluted 1:50 (v/v) with acidified distilled water (1% v/v HCl)	CZE-UV fused-silica capillary, 67 cm full length, 50 cm effective length, 75 µm i.d., 375 µm o.d. Buffer: 20 mM sodium tetraborate containing 30% MeOH, pH 9.00	[53]

Mass Spectrometry

Mass spectrometry is a technique that can be used for the structural elucidation or confirmation of organic, bioorganic, and organometallic compounds, as well as for quantitative analysis. The first step in MS measurement is the conversion of neutral molecules to charged species (*i.e.* ions), which are then separated according to their mass-to-charge (*m/z*) ratio in a mass analyzer. The relative abundances of individual *m/z* values are recorded by a suitable detector to produce what is known as a mass spectrum. MS can be coupled to both gas-phase and liquid-phase separation techniques, enabling the structural analysis of complex mixtures after their chromatographic separation without time-consuming off-line isolation.

Different MS analyzers have been used in the determination of anthocyanins such as triple quadrupole (QqQ), ion trap (IT) and time-of-flight (TOF). The QqQ was generally employed to search for anthocyanins with experiments such as neutral loss scan and product ion scanning, while IT allows structure fragmentation of sample by MS^n. Both analyzers, however, provide nominal mass accuracy alone and may not be well applied in untargeted phenolic compounds. Orthogonal acceleration TOF-MS provides much better accuracy and precision of mass information generated. These accurately measured mass values with a mass error less than 5 ppm can be used to produce candidate empirical formulae and identify the potential substance with elemental composition analysis [54].

The main ion sources used to separate and determine phenolic compounds are electrospray ionization (ESI), atmospheric-pressure chemical ionization (APCI), continuous-flow fast-atom bombardment (CF-FAB) and matrix-assisted laser desorption ionization (MALDI). These interphases supply energy to the solid sample or liquid in different ways, so that the direct formation of gaseous ions occurs while avoiding the volatilization, thereby allowing the subsequent ionization of the compounds. APCI is commonly used for detecting low-molecular-weight polar and nonpolar compounds. MALDI, as a special soft ionization technique is applied mostly to the analysis of large organic molecules such as proteins, peptides, and polysaccharides. ESI is one of the most versatile ionization sources that provides a simple, real-time means of analyzing a wide range of polar molecules (100-200,000 Dalton range). It becomes the preferred choice for detecting polar compounds separated by liquid chromatography and capillary electrophoresis.

Nuclear Magnetic Resonance

Nuclear magnetic resonance (NMR) is highly useful for identifying compounds when standards are not available. To facilitate the interpretation of NMR spectra, acidified media such as DCl or CF_3COOD need to be used in order to displace the equilibrium, and therefore only the flavylium cation exists in the solution.

The hyphenated technique of HPLC-NMR is now well established as a powerful technique for the structure elucidation of unknown compounds, including the identification of various forms of anthocyanins in crude plant matrices. The physical interfacing of these two instruments is not difficult. The development of solvent-suppression technique can solve the problems in observing of analyte response in the presence of non-deuterated solvents.

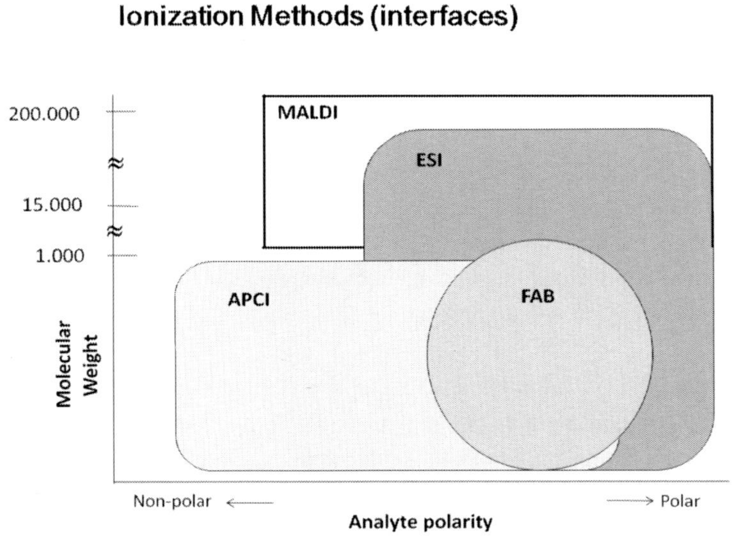

Figure 3. The application spectrum of different ion-source techniques.

APPLICATIONS

HPLC

The most frequently used detection system for HPLC is UV/Vis spectrophotometry. This is due to the prevalence of spectrophotometric instrumentation in laboratories studying anthocyanin chemistry, as well as in the industrial and academic sectors. Anthocyanins have a unique absorption maximum at ~520 nm, which sets these compounds apart from other flavonoids in the plant extract and simplifies the resulting chromatograms for isolation and purification.

HPLC hyphenated to mass spectrometry (MS) is one of the most important analytical techniques used for detecting phenolic compounds. Recently, chromatographic performance has been improved by using columns packed with small particles. The newest column packed with a particle size of less than 2 μm and operated at a pressure of up to 600 bar enables high-resolution analysis in shorter times. The on-line coupling of HPLC with MS using ESI as an interface constitutes a powerful method because of its highly efficient resolution and accurate characterization of a wide range of polar compounds. ESI, one of the most versatile ionization techniques, is preferred for detecting polar compounds separated by liquid chromatography. The advantages of MS detection includes the ability to determine molecular weight and to gain structural information. TOF-MS offers excellent mass accuracy over a wide dynamic range while offering measurements of the isotopic pattern and providing important additional information for determining the elemental composition. Furthermore, TOF-MS permits the rapid and efficient confirmation of the elemental composition of ions in fragmentation studies and also provides high selectivity in the determination of compounds in complex matrices using the extracted ion chromatogram (EIC) mode when there are overlapping peaks. Otherwise, IT-MS can be used to fragment ions of structural relevance for identifying target compounds in a highly complex matrix.

Figure 4 shows an example of the chromatograms resulting from the coupling HPLC-DAD and MS (TOF/IT) to analyze the aqueous extract of *Hibiscus sabdariffa* (roselle). Dry leaves of the plant were manually milled to a fine powder. An amount of *H. sabdariffa* was weighed and mixed with ultrapure water at room temperature. The aqueous extract of *H. sabdariffa* was freshly prepared for each analysis. The aqueous extract was stirred in a vortex, filtered, and directly injected into the HPLC system. In the positive ionization mode, the two most representative anthocyanins were identified, namely delphinidin 3-sambubioside (peak 16) and cyanidin 3-sambubinoside (peak 17) using the base peak chromatogram (BPC). This was consistent with the UV spectra at 520 nm.

The application of tandem MS (MS-MS or MS^n) can be used to determine individual compounds in a mixture or to identify the structure of compounds by separate ionization and fragmentation steps. An important mass spectrometer for structure elucidation is the triple quadrupole ion trap because it can serve as an exceptionally high specificity detector due to a mass filter capable of transmitting only the ion of choice. There are three quadrupoles in which the first (Q1) and third (Q3) quadrupoles act as mass filters, and the middle (Q2) quadrupole is employed as a collision cell, where the ionization takes place. This filtering process works by mass to charge ratios (*m/z*) and allows the study of fragments which are crucial in structural elucidation. For example, the Q1 can be set to filter out an ion of a known mass, which is fragmented in Q2, and Q3 can then be set to scan the entire *m/z* range, giving information on the sizes of the fragments made as well as information on one specific ion fragment. Thus, the structure of the original ion can be deduced. However, as mentioned above, ESI has been an efficient ionization technique for HPLC-MS because it is a soft technique and does not tend to produce many fragments, which are essential for MS/MS. Because of this, HPLC-MS/MS uses two ionization methods: ESI to introduce a parent ion into the mass spectrometer, and then collision-induced dissociation (CID) to enhance the fragmentation of the parent ion.

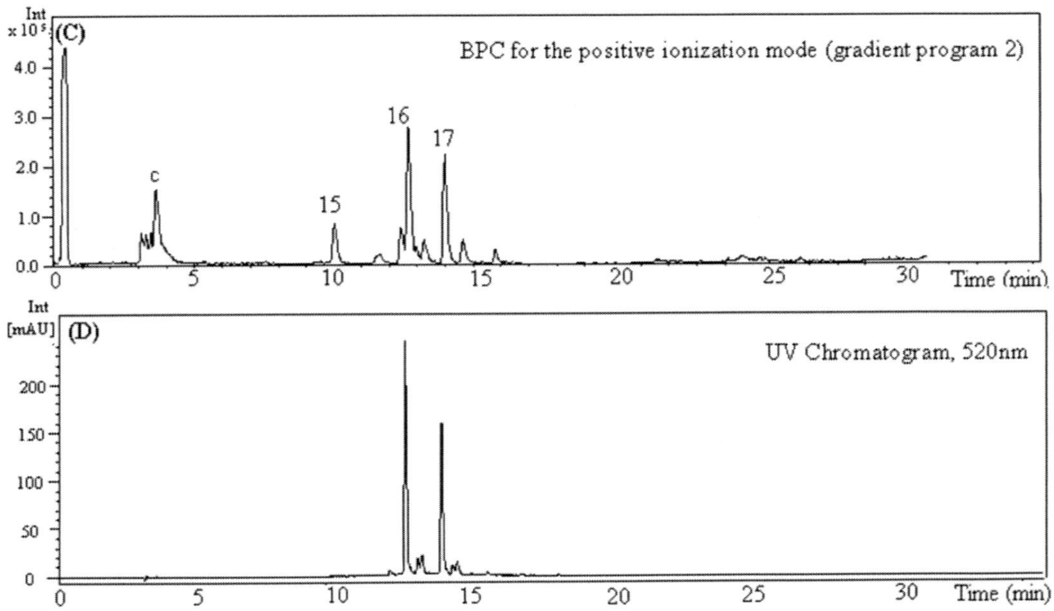

Figure 4. Chromatographic profiles an aqueous extract of *Hibiscus sabdariffa* (C) Base peak chromatogram (BPC) in positive mode. (D) UV chromatogram at 520 nm. Reproduced from Rodríguez-Medina *et al.* (2009) [35] with permission from John Wiley and Sons.

CE

An example of a complete anthocyanin analysis would be selective extraction, separation, and identification of anthocyanins in roselle using SPE-CE-MS(TOF-IT) [50]. Three extraction procedures (extraction procedures A-C) were tested in this study in order to compare the one that showed the highest capacity to extract mainly anthocyanins. With extraction procedure A, the sample was mixed with MeOH/HCl (99:1v/v) for 4 h at room temperature. During this time the solution was magnetically stirred. In extraction procedure B, the sample was extracted with MeOH/HCl (99:1 v/v) by sonication for 30 min and filtered. Finally, with extraction procedure C, the sample was mixed with acetic acid (15% v/v) for 48 h with magnetic stirring and then filtered. The filtrate was mixed with Amberlite XAD-2 for 40 min. The amberlite particles were then packed into a glass column. The anthocyanins remained absorbed on the column they were eluted with 1 L of EtOH (70% v/v) and acetic acid (1% v/v).

As reflected in Table 2, a major number of compounds were extracted by employing procedure C. The analysis by CE-ESI-MS using IT and TOF analyzers showed that anthocyanins glycoside derivatives were present in the methanolic extract of dried roselle calyces. Figure 5 shows the base peak electropherogram (BPE) and the extracted ion electropherogram (EIE) of the identified compounds of the extract C under CE-ESI-MS conditions, and revealed two main anthocyanins with retention times at 11.8 min (Cy-3-sambubioside) and 11.9 min (Dp-3-sambubioside).

Table 2. The *m/z* ratio found in different roselle extracts. Reproduced from Segura-Carretero *et al.* (2008) [50Error! Bookmark not defined.] with permission from John Wiley and Sons

Extract A	Extract B	Extract C	Analytes
-	-	465.1	delphinidin-3-*O*-glucoside
581.2	581.2	581.2	cyanidin-3-sambubioside
-	-	595.1	cyanidin-3-rutinoside
597.2	597.2	597.2	delphinidin-3-sambubioside
-	-	611.2	cy-3,5-*O*-diglicoside

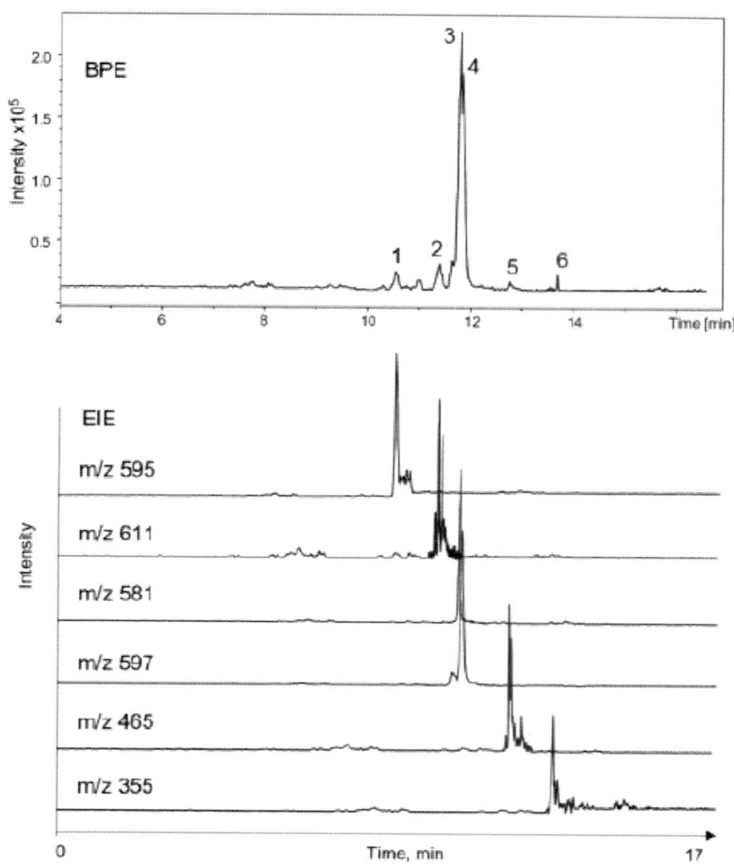

Figure 5. Base-peak electropherogram and extracted ion electropherogram of roselle. Reproduced from Segura-Carretero *et al.* (2008) [50] with permission from John Wiley and Sons.

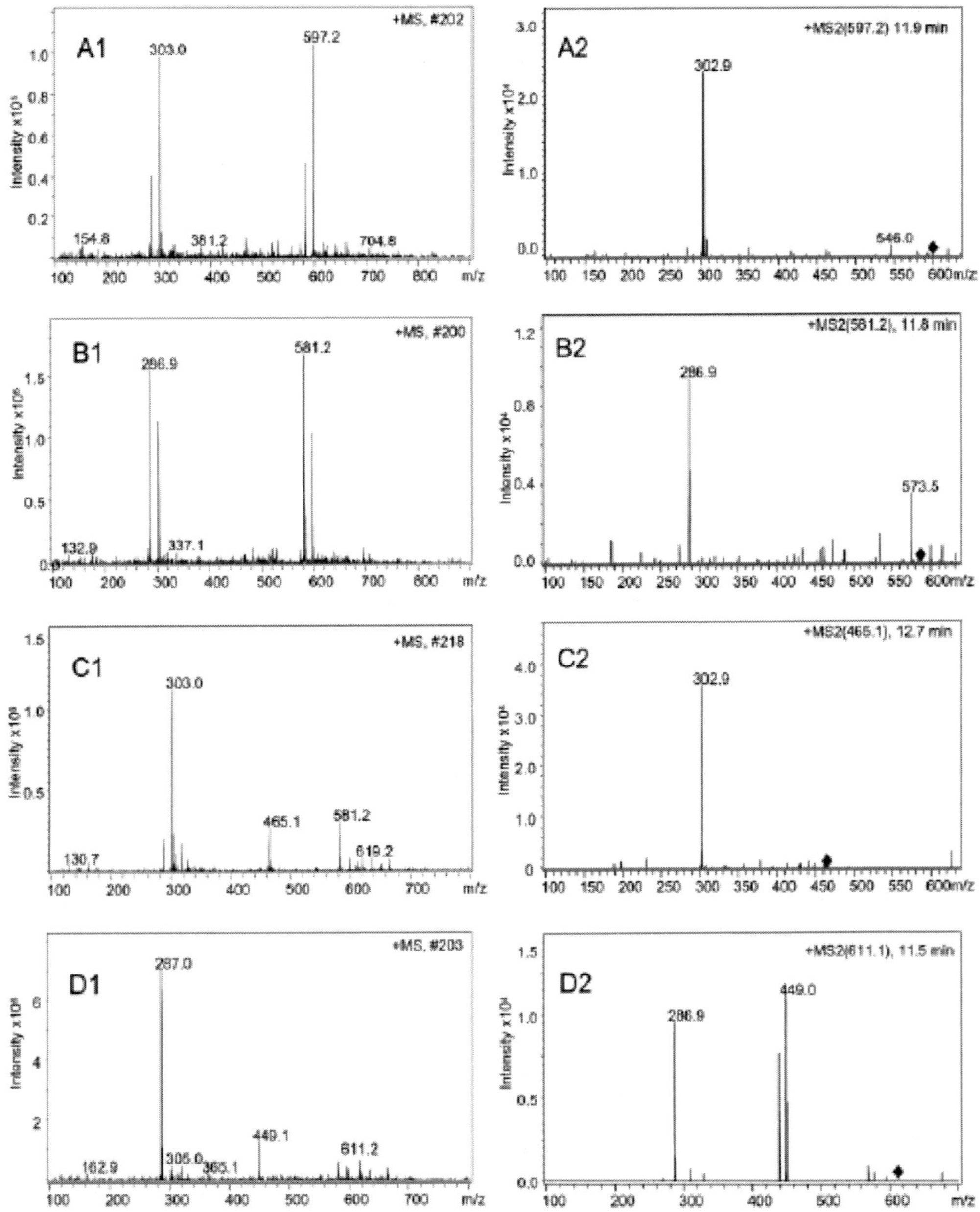

Figure 6. MS (A1, B1, C1, D1) and MS^2 (A2, B2, C2, D2) spectra of the most important dried calyces of roselle compounds. MS/MS of compounds in extract C. (A) Dp-3-sambubioside, (B) Cy-3-ambubioside, (C) Dp-3-glucoside, and (D) Cy-3,5-diglucoside. Reproduced from Segura-Carretero *et al.* (2008) [50] with permission from John Wiley and Sons.

In Figure 6, the MS and MS/MS spectra of the most important compounds in dried roselle flower are shown and resulted in clear and characteristic fragmentation patterns.

Conclusion

In this chapter, different powerful analytical tools, such as HPLC and CE coupled to DAD, MS and NMR have been described to separate and determine anthocyanins from complex matrices. These techniques have proved very useful for determining most unknown compounds without reference in complex matrices. The difficulty to obtain reference compounds and the spectral similarities of the anthocyanins represents a major drawback. Therefore, MS and NMR have become the preferred techniques for anthocyanin identification.

Given the beneficial effect for the health of anthocyanins such as anti-artherogenic, anticancer, antidiabetic, anti-inflammatory and antioxidant activities, and their use as a natural colorant of processed foods, the interest in anthocyanins continues to grow. The implementation of better extraction, purification, and identification methodologies will have an impact in the establishment of new tools for the food-product authentication.

The analytical methodologies described here represent the current and foremost methodologies for the complete analyses of anthocyanins. As we look forward to the development of HPLC/CE-MS, the coupling with other efficient structural analysis techniques, especially NMR, can offer advantages, thanks to several improvements that have been made to increase the sensitivity of NMR, bringing it more into line with MS sensitivity and enabling the analysis of anthocyanins at low concentration.

Acknowledgments

The authors are grateful to the Spanish Ministry of Education and Science (AGL2008-05108-C03-03 and AGL2011-29857-C03-02), the Andalusian Regional Government Council of Innovation and Science (P09-CTS-4564, P10-FQM-6563) and the University of Andalas for the Directorate General of Higher Education (DGHE) scholarship (Ref.: 2609/D4.4/2008).

References

[1]

[1] F. Delgado-Vargas, A. R. Jiménez, O. Paredes-López and F. J. Francis. Natural pigments: carotenoids, anthocyanins, and betalains - Characteristics, biosynthesis, processing, and stability. *Critic. Rev. Food Sci. Nutr.* 40, 173 (2000).

[2] K. R. Markham, K. S. Gould, C. S. Winefield, K. A. Mitchell, S. J. Bloor and M. R. Boase. Anthocyanic vacuolar inclusions - Their nature and significance in flower colouration. *Phytochemistry* 55, 327 (2000).

[3] R. E. Wrolstad. Anthocyanin pigments - Bioactivity and coloring properties. *J. Food Sci.* 69, C419 (2004).

[4] C. R. Welcha, Q. Wub and J. E. Simonb. Recent advances in anthocyanin analysis and characterization. *Curr. Anal. Chem.* 4, 75 (2008).

[5] M. Rein, Copigmentation reactions and color stability of berry anthocyanins, University of Helsinki, Helsinki (2005).

[6] N. Turker, S. Aksay, H. I. Ekiz. Effect of storage temperature on the stability of anthocyanins of a fermented black carrot (*Daucus carota* var. L.) beverage: shalgam. *J. Agric. Food Chem.* 52, 3807 (2004).

[7] G. H. Laleh, H. Frydoonfar, R. Heidary, R. Jameei and S. Zare. The effect of light, temperature, pH and species on stability of anthocyanin pigments in four *Berberis* species. *Pakistan J. Nutr.* 5, 90 (2006).

[8] A. Castañeda-Ovando, M. L. Pacheco-Hernandez, M. E. Paez-Hernandez, J. A. Rodriguez and C. A. Galan-Vidal. Chemical studies of anthocyanins: A review. *Food Chem.* 113, 859 (2009).

[9] R. L. Prior and X. Wu. Anthocyanins: Structural characteristics that result in unique metabolic patterns and biological activities. *Free Radical Res.* 40, 1014 (2006).

[10] H. Wang, G. Cao and R. L. Prior. Oxygen radical absorbing capacity of anthocyanins. *J. Agric. Food Chem. 45*, 304 (1997).

[11] J. H. Lee, H. -J. Lee and M. -G. Choung. Anthocyanin compositions and biological activities from the red petals of Korean edible rose (*Rosa hybrida* cv. Noblered). *Food Chem.* 129, 272 (2011).

[12] J. M. Harnly, S. Bhagwat and L. Z. Lin. Profiling methods for the determination of phenolic compounds in foods and dietary supplements. *Anal. Bioanal. Chem.* 389, 47 (2007).

[13] M. Naczk and F. Shahidi. Extraction and analysis of phenolics in food. *J. Chromatogr. A* 1054, 95 (2004).

[14] *P. G. Kapasakalidis, R. A. Rastall and M. H. Gordon. Extraction of polyphenols from processed black currant (*Ribes nigrum *L.) residues.* J. Agric. Food Chem. *54, 4016 (2006).*

[15] R. Tsao and Z. Deng. Separation procedures for naturally occurring antioxidant phytochemicals. *J. Chromatogr. B* 812, 85 (2004).

[16] Q. Du, G. Jerz and P. Winterhalter. Isolation of two anthocyanin sambubiosides from bilberry (*Vaccinium myrtillus*) by high-speed counter-current chromatography. *J. Chromatogr. A* 1045, 59 (2004).

[17] M. Schwarz, S. Hillebrand, S. Habben, A. Degenhardt and P. Winter-halter. Application of high-speed countercurrent chromatography to the large-scale isolation of anthocyanins. *Biochem. Eng. J.* 14, 179 (2003).

[18] S. Hillebrand, H. Naumann, N. Kitzinski, N. Köhler and P. Winterhalter. Isolation and characterization of anthocyanins from blue-fleshed potatoes (*Solanum tuberosum* L.). In*Potato III. Food 3 (Special Issue 1)*; N. Yee, W. Bussel, Eds. Global Science Books, Ltd: Miki cho, Japan 2009 pp 96−101.

[19] O. M. Andersen and K. R. Markham, Flavonoids: Chemistry, biochemistry and applications, CRC Press, Boca Raton (2006).

[20] C. T. da Costa, D. Horton and S. A. Margolis. Analysis of anthocyanins in foods by liquid chromatography, liquid chromatography–mass spectrometry and capillary electrophoresis. *J. Chromatogr. A* 881, 403 (2000).

[21] X. Sun, X. Yang and E. Wang. Chromatographic and electrophoretic procedures for analyzing plant pigments of pharmacologically interests. *Anal. Chim. Acta* 547, 153 (2005).

[22] R. H. F. Cheung, P. J. Marriott and D. M. Small. CE methods applied to the analysis of micronutrients in foods. *Electrophoresis* 28, 3390 (2007).

[23] J. Valls, S. Millán, M. P. Marti, E. Borrás and L. Arola. Advanced separation methods of food anthocyanins, isoflavones and flavanols. *J. Chromatogr. A* 1216, 7143 (2009).

[24] Z. Zhang and J. -J. Wang, "Study on the biological characteristic and antioxidant activity of anthocyanin in plant: A review", *The 4th International Conference on Bioinformatics and Biomedical Engineering,* 18-20, Jun. 2010 (Chengdu, China) art. no. 5516411.

[25] M. Monagas, I. Garrido, B. Bartolomé and C. Gómez-Cordovés. Chemical characterization of commercial dietary ingredients from *Vitis vinifera* L. *Anal. Chim. Acta* 563, 401 (2006).

[26] N. P. Seeram, L. S. Adams, Y. Zhang, R. Lee, D. Sand, H. S. Scheuller and D. Heber. Blackberry, black Raspberry, blueberry, cranberry, red raspberry, and strawberry extracts inhibit growth and stimulate apoptosis of human cancer cells in vitro. *J. Agric. Food Chem.* 54, 9329 (2006).

[27] G. Ruberto, A. Renda, C. Daquino, V. Amico, C. Spatafora, C. Tringali and N. D. Tommasi. Polyphenol constituents and antioxidant activity of grape pomace extracts from five Sicilian red grape cultivars. *Food Chem.* 100, 203 (2007).

[28] T. Fazeelat, W. Afzal, M. Asif, M. Zamir and H. Saleem. HPLC analysis of strawberry anthocyanins at partially ripe and ripe levels. *J. Chem. Soc. Pak.* 29, 243 (2007).

[29] W. Mullen, S. C. Marks and A. Crozier. Evaluation of phenolic compounds in commercial fruit juices and fruit drinks. J. Agric. Food Chem. 55, 3148 (2007).

[30] A. K. Lätti, K. R. Riihinen and P. S. Kainulainen. Analysis of anthocyanin variation in wild populations of Bilberry (*Vaccinium myrtillus* L.) in Finland. *J. Agric. Food Chem.* 56, 190 (2008).

[31] F. Adje, Y. F. Lozano, E. Meudec, P. Lozano, A. Adima, G. A. N'zi and E. M. Gaydou. Anthocyanin characterization of pilot plant water extracts of *Delonix regia* flowers. *Molecules* 13, 1238 (2008).

[32] X. Wu and R. L. Prior. Identification and characterization of anthocyanins by high-performance liquid chromatography−electrospray ionization− tandem mass spectrometry in common foods in the United States: vegetables, nuts, and grains. *J. Agric. Food Chem.* 53, 3101 (2005).

[33] Z. Huang, B. Wang, P. Williams and R. D. Pace. Identification of anthocyanins in muscadine grapes with HPLC-ESI-MS. LWT - *Food Sci. Tech.* 42, 819 (2009).

[34] S. Jang and Z. Xu. Lipophilic and hydrophilic antioxidants and their antioxidant activities in purple rice bran. *J. Agric. Food Chem.* 57, 858 (2009).

[35] I. C. Rodríguez-Medina, A. Segura-Carretero and A. Fernández-Gutiérrez. Use of high-performance liquid chromatography with diode array detection coupled to electrospray-Qq-time-of-flight mass spectrometry for the direct characterization of the phenolic fraction in organic commercial juices. *J. Chromatogr. A* 1216, 4736 (2009).

[36] J. S. Barnes, H. P. Nguyen, S. Shen and K. A. Schug. General method for extraction of blueberry anthocyanins and identification using high performance liquid chromatography–electrospray ionization-ion trap-time of flight-mass spectrometry. *J. Chromatogr. A* 1216, 4728 (2009).

[37] A. K. Lätti, P. S. Kainulainen, S. Hayirlioglu-Ayaz, F. A. Ayaz and K. R. Riihinen. Characterization of anthocyanins in caucasian blueberries (Vaccinium arctostaphylos L.) native to Turkey. *J. Agric. Food Chem.* 57, 5244 (2009).

[38] I. C. Rodríguez-Medina, R. Beltrán-Debón, V. M. Molina, C. Alonso-Villaverde, J. Joven, J. A. Menéndez, A. Segura-Carretero and A. Fernández-Gutiérrez. Direct characterization of aqueous extract of *Hibiscus sabdariffa* using HPLC with diode array detection coupled to ESI and ion trap MS. *J. Sep. Sci.* 32, 3441 (2009).

[39] X. Li, S. Yao, B. Tu, X. Li, C. Jia and H. Song. Determination and comparison of flavonoids and anthocyanins in Chinese sugarcane tips, stems, roots and leaves. *J. Sep. Sci.* 33, 1216 (2010).

[40] J. Y. Choi, S. J. Lee, S. J. Lee, S. Park, J. H. Lee, J. H. Shim, A. M. Abd El-Aty, J. S. Jin, E. D. Jeong, W. S. Lee and S. C. Shin. Analysis and tentative structure elucidation of new anthocyanins in fruit peel of *Vitis coignetiae* Pulliat (meoru) using LC-MS/MS: Contribution to the overall antioxidant activity. *J. Sep. Sci.* 33, 1192 (2010).

[41] C. Qin, Y. Li, W. Niu, Y. Ding, R. Zhang and X. Shang. Analysis and characterisation of anthocyanins in mulberry fruit. *Czech J. Food Sci.* 28, 117 (2010).

[42] V. D. Truong, N. Deighton, R. T. Thompson, R. F. McFeeters, L. O. Dean, K. V. Pecota and G. C. Yencho. Characterization of anthocyanins and anthocyanidins in purple-fleshed sweetpotatoes by HPLC-DAD/ESI-MS/MS. *J. Agric. Food Chem.* 58, 404 (2010).

[43] C. Del Carpio Jiménez, C. Serrano Flores, J. He, Q. Tian, S. J. Schwartz and M. M. Giusti. Characterisation and preliminary bioactivity determination of *Berberis boliviana* Lechler fruit anthocyanins. *Food Chem.* 128, 717 (2011).

[44] C. -G. Qin, Y. Li, W. Niu, Y. Ding, X. Shang and C. Xu. Composition analysis and structural identification of anthocyanins in fruit of waxberry. *Czech J. Food Sci.* 29, 171 (2011).

[45] I. Rockenbach, L. V. Gonzaga, V. M. Rizelio, A. E. D. S. S. Gonçalves, M. I. Genovese and R. Fett. Phenolic compounds and antioxidant activity of seed and skin extracts of red grape (*Vitis vinifera* and *Vitis labrusca*) pomace from Brazilian winemaking. *Food Res. Intern.* 44, 897 (2011).

[46] L. -Z Lin, J.Sun, P. Chen and J. A. Harnly. LC-PDA-ESI/MS identification of new anthocyanins in purple bordeaux radish (Raphanus sativus L. variety). *J. Agric. Food Chem.* 59, 6616 (2011).

[47] P. Bridle and C. García-Viguera. Analysis of anthocyanins in strawberries and elderberries. A comparison of capillary zone electrophoresis and HPLC. *Food Chem.* 59, 299 (1997).

[48] P. Bednář, B. Papoušková, L. Müller, P. Barták, J. Stávek, P. Pavloušek and K. Lemr. Utilization of capillary electrophoresis/mass spectrometry (CE/MS) for the study of anthocyanin dyes. *J. Sep. Sci.* 28, 1291 (2005).

[49] Z. Guadalupe, A. Soldevilla, M. -P, Sáenz-Navajas and B. Ayestarán. Analysis of polymeric phenolics in red wines using different techniques combined with gel permeation chromatography fractionation. *J. Chromatogr. A* 1112, 112 (2006).

[50] A. Segura-Carretero, , M. A. Puertas-Mejía, , S. Cortacero-Ramírez, R. Beltrán, C. Alonso-Villaverde, J. Joven, G. Dinelli and A. Fernández-Gutiérrez. Selective extraction, separation, and identification of anthocyanins from *Hibiscus sabdariffa* L. using solid phase extraction-capillary electrophoresis-mass spectrometry (time-of-flight /ion trap). *Electrophoresis* 29, 2852 (2008).

[51] P. Comandini, G. Blanda, A. Cardinali, L. Cerretani, A. Bendini and M. F. Caboni. CZE separation of strawberry anthocyanins with acidic buffer and comparison with HPLC. *J. Sep. Sci.* 31, 3257 (2008).

[52] E.V. Petersson, A. Puerta, J. Bergquist and C. Turner. Analysis of anthocyanins in red onion using capillary electrophoresis-time of flight-mass spectrometry. *Electrophoresis* 29, 2723 (2008).

[53] F. Berli, J. D'Angelo, B. Cavagnaro, R. Bottini, R. Wuilloud and M. F. Silva. Phenolic composition in grape (Vitis vinifera L. cv. Malbec) ripened with different solar UV-B radiation levels by capillary zone electrophoresis. *J. Agric. Food Chem.* 56, 2892 (2008).

[54] G. Xie, R. Plumb, M. Su, Z. Xu, A. Zhao, M. Qiu, X. Long Z. Liu and W. Jia. Ultra-performance LC/TOF MS analysis of medicinal *Panax* herbs for metabolomic research. *J. Sep. Sci.* 31, 1015 (2008).

In: Anthocyanins: Structure, Biosynthesis and Health Benefits ISBN: 978-1-62257-329-5
Editor: Noboru Motohashi

Chapter 9

THE HEALTH EFFECTS OF JUJUBES (*ZIZIPHUS* SPECIES) BASED ON THEIR EVIDENCES

Noboru Motohashi*
Meiji Pharmaceutical University, Noshio, Kiyose-shi,Tokyo, Japan

ABSTRACT

Jujubes have been traditionally used in the preventions and treatments offor stress, antifungal, antibacterial, anti-ulcer, anti-inflammatory, sedative, antispastic, antifertility or contraception, hypotensive and antinephritic, cardiotonic, antioxidant, immunostimulant, wound healing properties and others in China, India, Korea and their surrounding countires. Jujubes have abundant abundantly their diverse phytochemicals such as dietary fibers, alkaloids and terpenoids as well as phenolic and anthocyanins. Then, the purpose of this review chapter is to describe the ir jujube phytochemicals and their biological activities based on their evidences.

Keywords: *Zizyphus jujuba* (*Zizyphus vulgaris*), phenolics, anthocyanins, alkaloids, sedative drug, protection of gut, antiaging

ABBREVIATIONS

vitamin C (**1**)
vitamin B_1 (thiamine. **2**)
vitamin B_2 (riboflavin. **3**)
zizyphus-pectin A (**4**)
protocatechuic acid (**5**)
gallic acid (**6**)
chlorogenic acid (**7**)
caffeic acid (**8**)

* noborumotohashi@jcom.home.ne.jp.

12-*O*-tetradecanoylphorbol 13-acetate (TPA. **9**)
sativanine-M (**10**)
nummularine-P (**11**)
sativanine-H (**12**)
amphibine-B (**13**)
amphibine-C (**14**)
amphibine-D (**15**)
amphibine-E (**16**)
sativanine-N (**17**)
sativanine-O (**18**)
coclaurine (**19**)
isoboldine (**20**)
norisoboldine (**21**)
asimilobine (**22**)
zizyphine A (**23**)
juzirine (**24**)
frangufoline (sanjoinine A. **25**)
sanjoinine B (**26**)
sanjoinine F (**27**)
sanjoinine D (**28**)
sanjoinine G2 (**29**)
sanjoinenine (**30**)
mauritine J (**31**)
paliurine G (**32**)
paliurine H (**33**)
paliurine I (**34**)
nummularine H (**35**)
daechuine-S3 (**36**)
paliurine A (**37**)
paliurine B (**38**)
paliurine C (**39**)
paliurine F (**40**)
methohexital (**41**)
serotonin (**42**)
chlorpromazine (**43**)
daechuine S27 (**44**)
pimozide (**45**)
pentobarbital (**46**)
muscimol (**47**)
alphitolic acid (**48**)
33-*O*-(*cis*-*p*-coumaroyl)-alphitolic acid (**49**)
3-*O*-(trans-*p*-coumaroyl)-alphitolic acid (**50**)
betulinic acid (**51**)
betulonic acid (**52**)
3-*O*-(*cis*-*p*-coumaroyl)-maslinic acid (**53**)
3-*O*-(*trans*-*p*-coumaroyl)-maslinic acid (**54**)

oleanolic acid (**55**)]
oleanonic acid (**56**)]
colubrinic acid (**57**)
zizyberenalic acid (**58**)
betulin (**59**)
betulinic acid (**60**)
oleic acid (ω-9. **61**)
linoleic acid (ω-6. **62**)
myristic acid (**63**)
palmitic acid (**64**)
stearic acid (**65**)
behenic acid (**66**)
β-sitosterol (**67**)
Maslinic acid (**68**)]
3-*O*-(*cis*-*p*-coumaroyl)-maslinic acid (**69**)
3-*O*-(*trans*-*p*-coumaroyl)-maslinic acid (**70**)
oleanolic acid (**71**)
jujuboside A (**72**)
jujuboside B (**73**)
p-hydroxybenzoic acid (**74**)
cinnamic acid (**75**)
naringenin (**76**)
naringenin 2(*S*)-6,8-di-C-glucoside (6,8-di-C-glucosyl-2(*S*)-naringenin. **77**)
naringenin 2(*R*)-6,8-di-C-glucoside (6,8-di-C-glucosyl-2(*R*)-naringenin. **78**)
methicillin (**79**)
procyanidin B3 (**80**)
procyanidin B4 (**81**)
penicilline G (**82**)
oxacillin (**83**)
spinosin (**84**)
di-*O*-methylspinosin (**85**)
spinosin nona-acetate (**86**)
permethylspinosin (**87**)
6'''-feruloylspinosin (**88**
6'''-sinapolyspinosin (**89**)
6'''-*p*-coumaloylspinosin (**90**)
6'''-*p*-hydroxybenzoylspinosin (apigenin 6-*C*-[(6-*O*-*p*-hydroxybenzoyl) β-D-glucopyranosyl(1→2)] β-D-glucopyranoside) (**91**)
swertisin (**92**)
2''-*O*-glucosylisoswertisin (**93**)
vecenin-2 (di-C-β-D-glucopyranosyl-6,8 apigenin. **94**)
procyanidin B2 (**95**)
(-)-epicatechin (**96**)
unidentified flavonoid I (UIS-I. **97**)
quercetin-3-*O*-rutinoside (rutin. **98**)

quercetin-3-*O*-galactoside (hyperin. **99**)
kaempferol-glucosyl-(1→2)rhamnoside (**100**)
unidentified flavonoid II (UIS-II. **101**)
saponarin (**102**)
vitexin (**103**)
swertish (**104**)
6'''-hydroxybenzoylspinosin (**105**)
unidentified substance (UIS. **106**)
butylhydroxyanisole (BHA. **107**)
(+)-catechin (**108**)
α-tocopherol (**109**)
β-carotene (**110**)
fatty acid methyl esters (**111**)
p-coumaric acid (**112**)
ferulic acid (**113**)
apigenin-7-*O*-glucoside (**114**)
eriodictyol (**115**)
quercetin (**116**)
syringic acid (**117**)
palmitoleic acid (**118**)
kaempferol (**119**)
phloretin (**120**)

Photo 1. Jujubes nearin almost harvest time. at Laoling jujube orchard with over 1000 hectares at Laoling (Leling) in Dezhou, Shangdong, China. Photographed by Noboru Motohashi. 9/14/2011 Wed <477.1 IMGP6043.JPG>.

1. Origin and Scientic Name

Jujubes (Chinese dates) (Photos 1, 2, 3, 4 and 5) have been thought to be native to tropical or subtropical areas from India and southern China to Malaysia. Especially, Laoling of Shangding in China was well-known as the city withof the highest jujube productions worldwide.

The taxonomy, a classification of organisms into groups based on similarities of structure or origin and others, of genus *Zizyphus* might belong to around over 40 species.

Since ancient times, jujube has been cultivated and used in vogue both in China and India. At least, around 400 varieties of Chinese jujube haves been known [1, 2].

Photo 2. A momument *"King of Jujube"* , jujube trees growing over 1000 years at Laoling jujube orchard. Photographed by Noboru Motohashi. 9/14/2011 Wed <477.1 IMGP6059.JPG>.

Photo 3. Flowers of *"Li"* variety jujube. at the home orchard of the author. Photographed by Noboru Motohashi. 6/9/2007 Sat <140. IMGP5280(jujube flower, Li,6/9/07).JPG >.

Photo 4. *"Lang"* jujube (left) and *"Li"* jujube (right) harvested in mid-dle September. At the home orchard of the author. Photographed by Noboru Motohashi. 9/12/2009 Sat < 392. IMGP6884(Lang and Li Jujube, 9/12/09).JPG>.

Photo 5. Their croorss- sections of three *"Li"*, *"Lang"* and *"Natume"* jujube varieties harvested in mid-dle September. at the home orchard of the author. Photographed by Noboru Motohashi. 9/12/2009 Sat <392. IMGP6892(Li, Lang & natume Jujube cross sections, 9/12/09).JPG > .

The Sscientific name of jujube is *Zizyphus jujuba* or *Zizyphus vulgaris*. Jujube is a deciduous small tree which is also called as red date, Chinese date (taisoh in Chinese), Indian date and natsume (in Japanse) in Japan, belonging to Rhamnales order, Rhamnaceae family and *Zizyphus* genus [2, 3].

2. Introduction

Man is equipping the antioxidant defense system. This antioxidant defense system could be called as an armamentarium with enzymes and antioxidative compounds for their removal against the diverses harmful free radicals [4, 5, 6]. Their imbalance between the production and elimination of harmful free radicals leads to oxidative stress which might damage the cells and eventually cause their diverse diseases including age-realted diseases such as hypertension, diabetic and cancers. Therefore, their maintenance of antioxidant activity might be important in the prevention against the free radical-associated diseases and aging [7].

It is thought that traditional medicines have been used by around 60% of the world's population. *Ziziphus jujuba* fruits (*Ziziphus jujuba*) are very rich in vitamin C (**1**) (188-544 mg/100 g edible pulp) and vitamin B_1 (thiamine. **2**) and vitamin B_2 (riboflavin. **3**) [8]. Jujube berry fruit also has athe high content of a bioflavonoid vitamin P (354-888 mg/100 g edible pulp). A bioflavonoid vitamin P could enhance their action such as antibacterial, anti-inflammatory, antioxidant, the stimulation of bile production, promotion of circulation and prevention of allergies by vitamin C (**1**).

Additionally, Zizyphus-Pectin A (**4**) in *Ziziphus jujuba* fruit was found. Zizyphus-Pectin A (**4**) contains chemically 1,4,5-*tri-O*-acetyl-2,3,6-*tri-O*-methyl-D-galactitol units [9]. These pectins could represent many pharmaceutical properties such as binding with bile acid, lowering of plasma cholesterol and anti-diarrhoeal properties [10]. Jujube hasd antioxidative activity [11].

The curative properties of herbs and plants might be due to their diverse phytochemicals which contain secondary metabolites such as alkaloids, flavonoids, terpenoids, saponin and phenolic compounds.

Especially, in India, China and Korea, the barks, stems, roots, leaves, seeds and fruits of *Ziziphus jujuba* Mill (*Ziziphus jujuba* Mill) haves been used traditionally as hypnotic-sedatives [12, 13], their inhibitory effects against both cyclooxygenase-1 (COX-1) and cyclooxygenase--2 (COX-2) [14], anxiolytic effect [10], anti-plasmodial and anti-mycobacterial activities [15], inhibiting of the hemolytic activity of human serum against erythrocytes [16], antioxidant and anti-listerial effect [17], anti-cancer activity [18, 19], sedative and hypnotic effects [20, 21], inhibition of adipogenesis [22], anti-inflammatory activity [23], hair growth promoting activity [24], anti-hepatotoxic and immunological activity [25], antioxidative [26] and anti-cognitive, anti-epileptic and anti-convulsant effects [27]. These wide treatments of jujube might derive from the phytochemicals such as alkaloids, terpenoids and flavonoids.

Some analyses on radical quenching phytochemicals such as phenolics and flavonoids in jujubes have been conductedtried by today.

In 2010, a group of College of Food Science and Nutritional Engineering, China Agricultural University, Beijing, China studied their systematic evaluation of antioxidant capacities of the ethanolic extract of different tissues of jujube (*Ziziphus jujuba* Mill.) from China.

Their ethanolic extracts on the tissues of three jujube varieties were characterized on their antioxidant capacities, and antioxidative flavonoids, ascorbic acid, anthocyanins and phenolics. Their antioxidant capacity was measured using oxygen radical absorption capacity (ORAC) or trolox equivalent antioxidant capacity (TEAC) such as the DPPH (2,2-diphenyl-1-picrylhydrazyl) assay (DPPH antioxidative activity analysis) and FRAP (ferric-reducing antioxidant power) assay.

Jujube phenolic acids such as protocatechuic acid (**5**), gallic acid (**6**), chlorogenic acid (**7**) and caffeic acid (**8**) were analyzed by high-performance liquid chromatography. Their jujube peel of all cultivars had the highest antioxidant capacities because of the highest content of total phenolics, flavonoids and anthocyanins, suggesting Chinese jujube might be a significant potential berry for a natural antioxidant agent (Figure 1) [28].

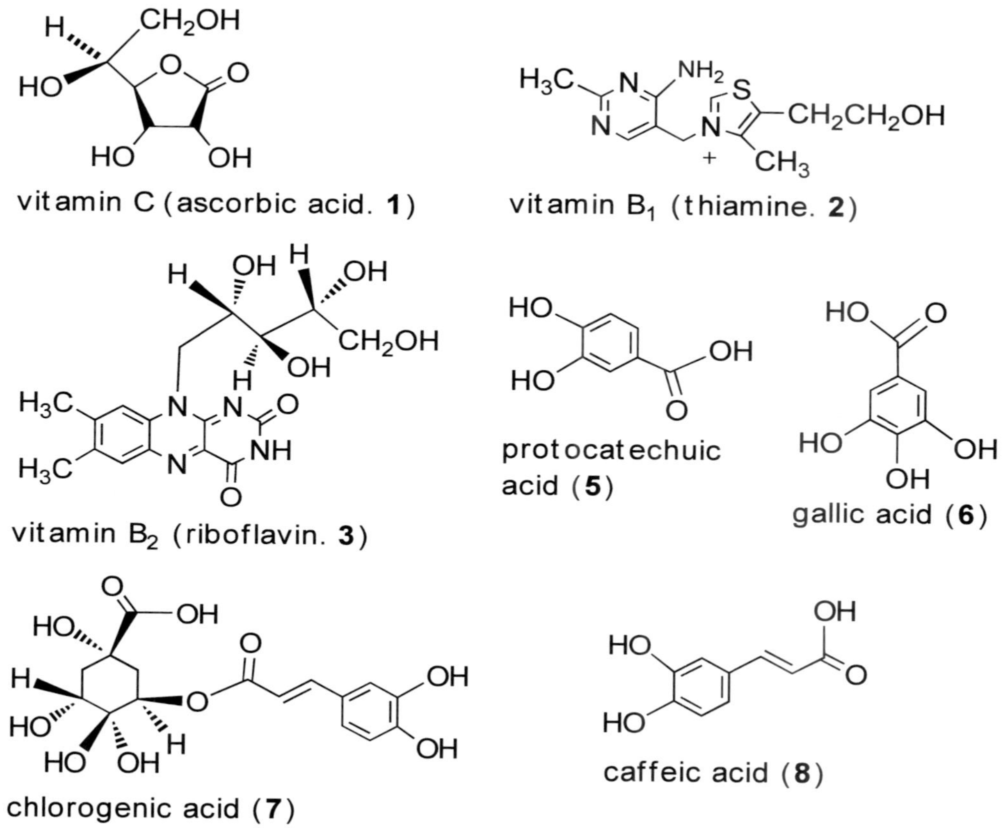

Figure 1 Vitamin C (ascorbic acid. **1**), vitamin B_1 (thiamine. **2**), vitam (riboflavin. **3**), protocatechuic acid (**5**), gallic acid (**6**), chlorogenic ac **and** caffeic acid (**8**)

In 2010, the anti-inflammatory activity of jujube seeds was measured using 12-*O*-tetradecanoylphorbol 13-acetate (TPA. tetradecanoylphorbol. **9**)-induced skin inflammation in experimental mice. As a result, it suggests that the essential oil of *Zizyphus jujuba* seeds might have a potential for expansion as new drugs for various inflammatory diseases (Figure 1) [23].

In 2008, a group from the Department of Gastroenterology and Hepatology, Meir Medical Center, Kfar Saba, Israel studied the safety and efficacy of an extract of *Ziziphus jujuba* for chronic constipation by their administrations of the liquid *Ziziphus jujuba* or placebo for 12 weeks to the constipated patients. Jujube extract groups were improved higher when compared to the control group. Then, it suggests that jujube extract might be an effective and safe treatment for chronic constipation [29, 30].

3. Phytochemicals and Their Functionality of Jujube

Diverese jujube components have been found as their functional phytochemicals. Since long ago, jujubes in China have been used in the improvement of vital energy, spleen improvement, alimentation to bloods and recovery, and a sedative drug which make to relux. Furthermore, jujubes are also carbohydrate (glucide), proteins, lipids, vitamins A, B_2, C, calcium (Ca), phosphorus (P), iron (Fe), malic acid and tartaric acid [31].

Additionally, jujubes have been involved in thebeen their improvement and recovery of the following symptoms: poor blood, blood deficiency, anaemia; nourishment; physical decline and weight loss; convalescing by hepatitis or tuberculosis; insomnia or interrupted sleep; allergic purpura; feeling of weakness with sweating; weak stomach and spleen leading to frequent stomach coldness and pain) and feces semi-liquid with diarrhoea; irritability during menopause and menopausal paranoid state [31].

By recent results, jujubes contain their diverse functionary components such as triterpene saponins, a sticky ingredient mucilage, tannins, flavonoids, isoquinoline alkaloids, peptide alkaloids, triperpenes, hydroxycoumarins, sugars and fruit acids [32].

3.1. Sugars: Convalescence and Prompt Nutritional Supplementation

Dry jujubes contain carbohydrates 71.4%/100g dry edible part. In carbohydrates, around 85% carbohydrates are soluble carbohydrates such as D-glucose and D-fructose. Besides, jJujubes might contain oligosaccharide which isare composed from D-glucose and D-fructose [3].

Therefore, it is thought that jujubes might be suitable as an alimentation berry for the early recovery of debilitated physical conditions.

3.2. Jujube Extract: Anti-Obesity Effect

3.2.1. Inhibition of Adipogenesis in 3T3-L1 Preadipocytes: Diet Effect

In 2009, the anti-obesity effect of *Zizyphus jujuba* extracts was examined on adipocyte differentiation of 3T3-L1 preadipocytes of a cell line derived from 3T3 cells. Then, it was found that the treatment with an extract of *Zizyphus jujuba* could suppress their lipid accumulation and glycerol-3-phosphate dehydrogenase (GPDH) activity without affecting the cell viability [22].

3.2.2. Oligochitosan: Diet Effect

Oligochitosan or chitosan (chitosans) is a carbohydrate oligosaccharide derived from chitin which is used sometimes used forto the absorption of fat. Chitin is a type of polysaccharide (sugar molecule) that is made by some plants and animals. The hard outer shelll of shrimp, lobsters, and many insects is made of chitin.

In 2011, a group from the of College of Food Science and Nutritional Engineering, China Agricultural University, Beijing 100083, People's Republic of China examined the effects of oligochitosan on post-harvest rot caused by alternaria alternata in Chinese jujube (*Zizyphus jujuba* Mill. cv. Dongzao) fruit. Chinese jujube (*Zizyphus jujuba* Mill. cv. Dongzao) fruit contained oligochitosan after post-harvest. Then, their biochemical evaluations showed that the activities of the main defense-related enzymes in the jujube fruits such as phenylalanine ammonia-lyase, which is an enzyme that catalyzes the chemical reaction, peroxidase, which are the enzymes occurring especially in plant cells that catalyze the oxidation of a peroxide, chitinase and β-1,3-glucanase were significantly enhanced by the oligochitosan treatment. Oligochitosan increased superoxide dismutase activity. However, the jujube fruit decreased the catalase activity and hydrogen peroxide levels increased in the fruit [33].

Oligochitosan has been known as a type of prebiotics [34] occurring before the advent of life, which favourably alters the intestinal microflora balance, inhibits the growth of harmful bacteria, promotes good digestion and boosts their immune function. Then, in 2007, based on these facts, a group of Key Laboratory of Food Science of Ministry of Education of China, Department of Science & Engineering, Nanchang University, Nanchang 330047, People's Republic of China confirmed that the dietary supplementation of oligochitosan could elevate their serum levels of immunoglobulins in birds by broiler chicks. From these results, it might be suggested that oligochitosan could improve the immune functions in birds. Through the positive modulation of the immune response, oligochitosan might have the great potential as an effective, antibiotic-like growth promoter for poultry [35]. These results also might also support Yam, J. *et al.*' experiments in 2011 [33].

3.3. Jujube Pectin: Protection of the Gut (Digestive Tract) and Enhancement of Digestion

In 1985, zizyphus-pectin A (**4**) was isolated from the fruits of *Zizyphus jujuba* MILLER var. inermis REHD.

It is generally known that the plant pectic polymers have the backbones consisting of rhamnogalacturonan and sides chains such as a branched arabinan and a linear 1→4 linked galactan of a polysaccharide consisting of many galactose residues [36].

However, pectins of the various water-soluble colloidal carbohydrates that occur in ripe fruit and vegetables with *O*-acetyl groups in pectic substances such as zizyphus-pectin A (**4**) might be a rare pectin. In fact, zizyphus-pectin A (**4**) has 2,3,6-tri-*O*-acetylated D-galactopyranosyl units in the side chains of pectic substances [9, 37].

In 2010, it is confirmed that soluble dietary fiberres such as β-glucan, pectin, guar gum of a heteropolysaccharide gum which extracted from the guar bean and guar gum is used as a food additive, and psyllium, which plantain of Mediterranean regions, whose seeds swell and become gelatinous when moist and are used as a mild laxative, could decrease the risk of cardiovascular disease. TBy the fact, the three main mechanisms on cholesterol-reducing effects of soluble dietary fibres were proposed as follows: The first is prevention of bile salt (The sodium salt of a bile acid; produced by the bile to neutralize the products leaving the stomach and to emulsify fatty material in it) reabsorption from the small intestine leading to an excess faecal bile salt excretion. Because, *in vivo, ex vivo* and *in vitro* experiments suggest that bile salt micelles (micelle is an electrically charged particle built up from polymeric molecules or ions and occurring in certain colloidal electrolytic solutions like soaps and detergents) might bind to soluble dietary fiberres which were preventing their reabsorption; The second is to reduce the glycemic response leading to lower the insulin-stimulation for the hepatic cholesterol synthesis. Therefore, their glycemic responses to soluble dietary fiberres have been extensively examined by using their nature of the interactions between bile salt micelles and soluble dietary fiberres that lead to the incomplete soluble dietary fibres reabsorption; The third is the physiological effects of mainly propionate, butyrate and acetate in fermentation products of soluble dietary fiberres [38].

On the effects of dietary fiberre on digestive tractgut function, propionate might be a major microbial fermentation metabolite in the human gut with the putative health effects which extended beyond the gut epithelium. Propionate has been thought to lower the lipogenesis, serum cholesterol levels, and carcinogenesis in other tissues [39].

3.3.1. Extracts of Jujube Berry Fruit: Improvement of Chronic Sudden Constipation

In 2008, a group from theof Department of Gastroenterology and Hepatology, Meir Medical Center, Kfar Saba, Israel examined on their treatment of chronic idiopathic constipation and the safety by *Ziziphus jujuba* extract. Their controlled clinical trial was examined on 18 subjects (male 3; female 15) of ages 34-77 with the chronic constipation during average 2.7 years and 19 control subjects (male 9; female 10) of ages 41-80 with chronic constipation during an average of 2.4 years. All 37 constipated patients received the liquid *Ziziphus jujuba* extract or placebo for 12 weeks. During the transit time of 12 weeks, two subjects (12%) of the study drug group and 16 (84%) of the control (placebo) group dropped out because of the severe constipation. By a SF-36 (MOS *36*-item *S*hort *F*orm Health Survey) QOL (quality of life) questionaire, the severe time of subjects with the severe constipation decreased from 12.2 scale at week 0 to 3 scale at week 11. Their symptom severity ratings on constipation could decrease from 6 and 6.2 to 2 and 5, and the quality of life score could improve from 1.9 and 2.3 to 1.3 and 1.4 in the study and control groups, respectively. Originally, since the extracts of *Ziziphus jujuba* are derived from a naturtally growing berry fruit, extracts of *Ziziphus jujuba* contains several ingredients. However, it is at least not known that the extracts of *Ziziphus jujuba* contains the any active ingregients in improving the constipation. The ingredients might be a form of anthraquinone which are contained in senna extract (senokot), cascara (casanthranol) extract and other plant-derived laxatives. Since there is no evidence to support the convention (generally accepted idea) that the prolonged use of laxatives causes the colonic danmage [40]. However, the long-term use of the extracts of *Ziziphus jujuba* should not be of concern in this regard. Furthermore, extracts of *Ziziphus jujuba* did not cause any change in liver or kidney function tests, nor any significant electrocardiogram (EKG) changes. Their patients reported no adverse events. However, some constipated patients need to be treated for very long periods of many years, and this study was too short to evaluate any adverse effectsa that such long-term treatment may have. From these results, the extracts of *Ziziphus jujuba* is safe and effective for chronic idiopathic constipation and can be safely recommended for at least 12 weeks of the treament of chronic idiopathic constipation. Additionally, their long-term efficacy and safety of extracts of *Ziziphus jujuba* should be tested for the confirmation of the more safety [29].

3.4. Alkaloids

3.4.1. Alkaloids from Bark of Ziziphus jujuba

In 2008, a group from the of Chembiotek Research International Pvt. Ltd, Salt Lake City, Sector V, Kolkata 700091, India found a novel sativanine-M (**10**) of 5(13)-zizyphine A type cyclic peptide alkaloids with a known nummularine-P (**11**) from *Zizyphus sativa* bark (Figure 2) [41].

In 1986, a group from theof Department of Chemistry, Gomal University, D. I. Khan, Pakistan identified sativanine-H (**12**) of 5(13)-zizyphine A type cyclic peptide alkaloids from bark of *Zizyphus sativa* (Figure 2) [42].

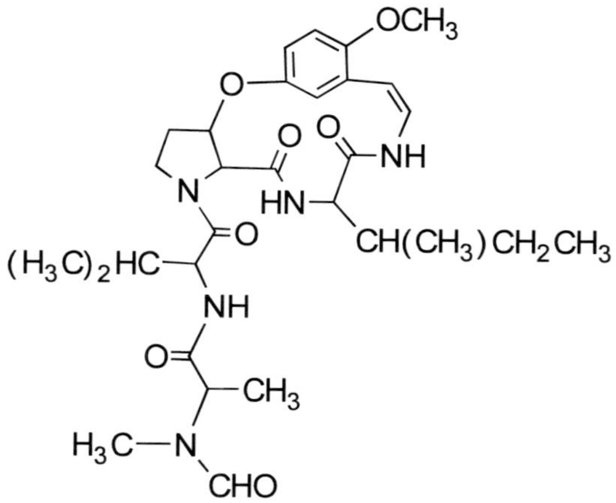

sativanine-M (**10**)

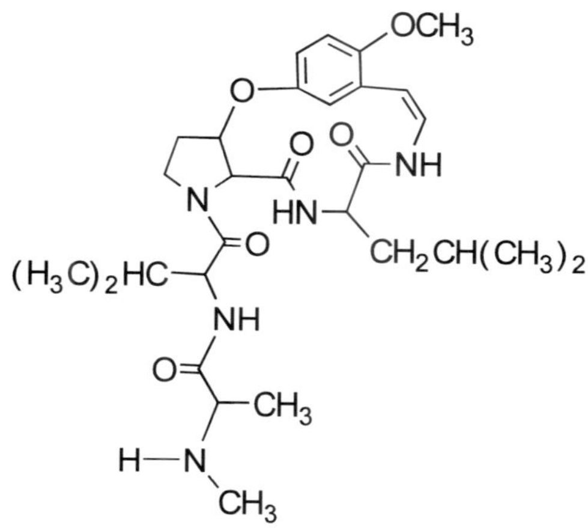

nummularine-P (**11**)

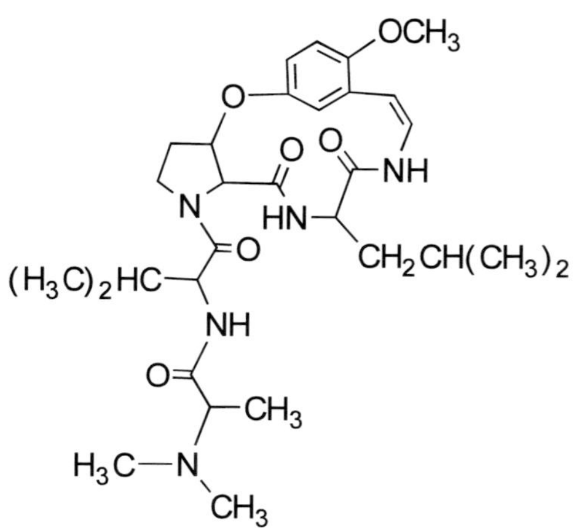

sativanine-H (**12**)

Figure 2 Sativanine-M (**10**), nummularine-P (**11**) and sativanine-H (**12**) of 5(13)-zizyphine A type cyclic peptide alkaloids from jujube bark

3.4.2. *Alkaloids from Stem Bark of Ziziphus jujuba*

In 1972, four 5(14)-amphibine B type cyclopeptide alkaloids such as amphibine-B (**13**), amphibine-C (**14**), amphibine-D (**15**) and amphibine-E (**16**) were identified from stem bark and bark of *Ziziphus amphibia* A. Cheval (Figure 3) [43].

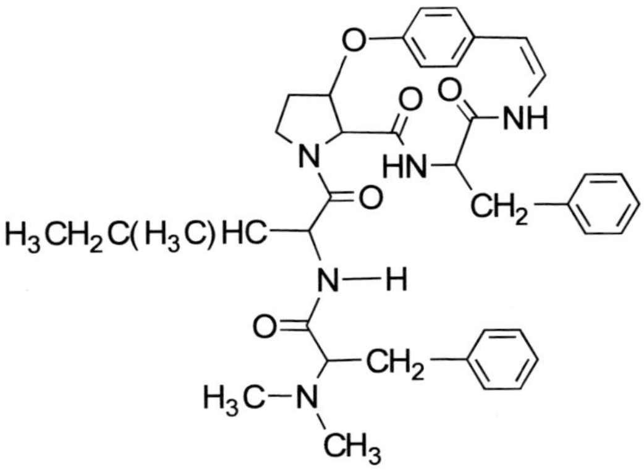

amphibine-B (**13**)

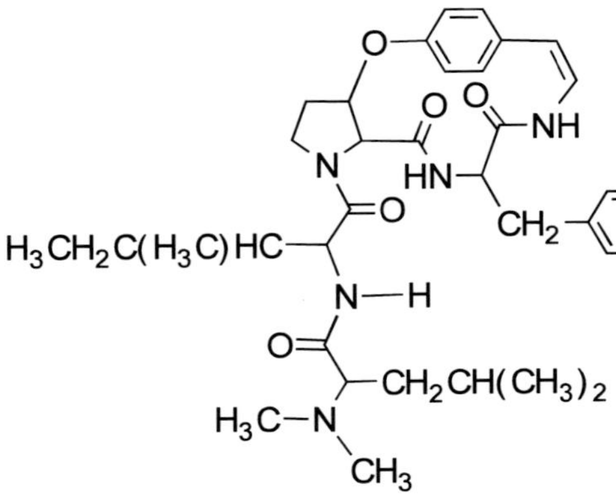

amphibine-C (**14**)

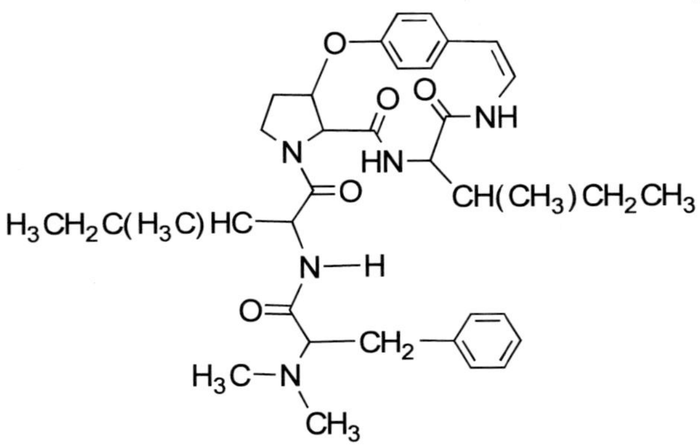

amphibine-D (**15**)

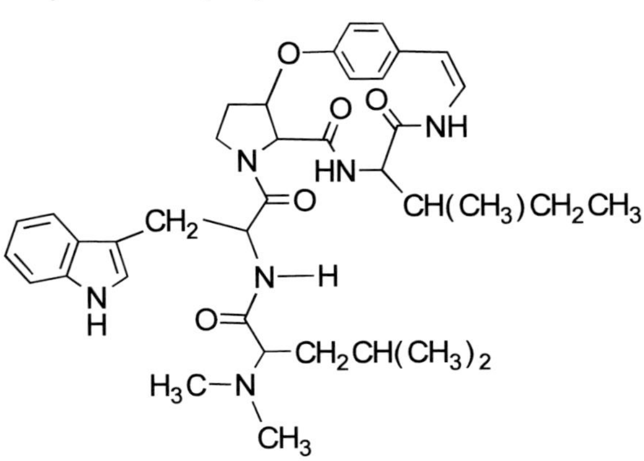

amphibine-E (**16**)

Figure 3 5(14)-Amphibine B type cyclopeptide alkaloids **such as** amphibine-B (**13**), amphibine-C (**14**), amphibine-D (**15**) and amphibine-E (**16**) from stem bark and bark of *Ziziphus amphibia* A. Cheval

In 2006, a group from theof Department of Medicinal Chemistry, Institute of Medical Sciences, Banaras Hindu University, Varanasi, 221 005, India isolated two novel 4(13)-nummularine C type cyclopeptide alkaloids such as sativanine-N (**17**) and sativanine-O (**18**) from the stem bark of *Zizyphus sativa* (Figure 4) [44].

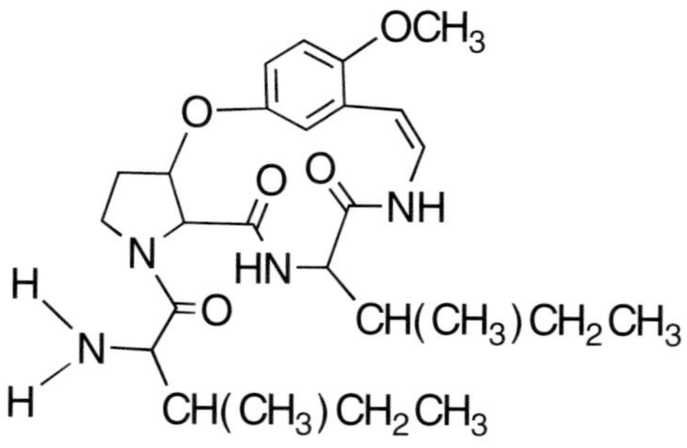

sativanine-N (**17**)

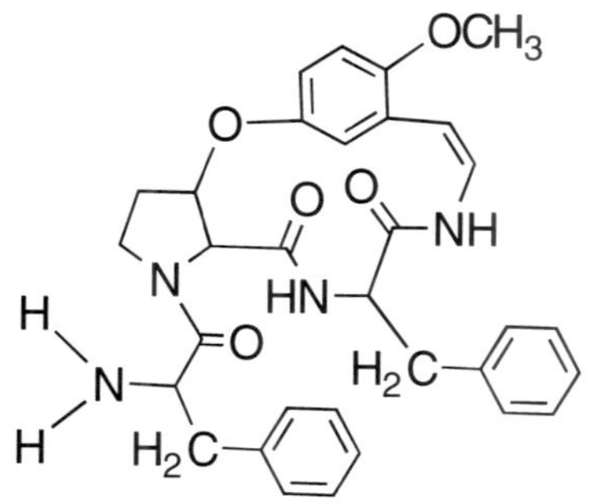

sativanine-O (**18**)

Figure 4 4(13)-Nummularine C type cyclopeptide alkaloids such as sativanine-N (**17**)) and sativanine-O (**18**) from the stem bark of *Zizyphus sativa*

3.4.3. Alkaloids from Leaves of Ziziphus jujuba

In 1977, a group from theof Institute of the Chemistry of Plant Substances, Academy of Sciences of the Uzbek SSR, Tashkent isolated a known isoquinoline alkaloid coclaurine (**19**), and quinoline alkaloids such as isoboldine (**20**), norisoboldine (**21**) and asimilobine (**22**) isolated first from the leaves of *Ziziphus jujuba*, and two novel alkaloids of a 5(13)-zizyphine A type cyclic peptide alkaloid jiziphine A (**23**) and an isoquinoline alkaloid juzirine (**24**) from jujube leaves in the period of ripening of the jujube fruits (Figure 5) [45, 46].

Here, in 2008, a group from the of College of Veterinary Medicine, Chungbuk National University, Cheongju 361-763, South Korea studied their effects of cyclopeptide alkaloid fraction of *Zizyphi Spinosi* Semen on pentobarbital-induced sleeping behaviors. Their effects of the sleeping behaviors were mediated by γ-aminobutyric acid (GABA) receptors Cl-channel activation. Additionally, cyclopeptide alkaloid fraction of *Zizyphi Spinosi* Semen in combination with a muscimol of γ-aminobutyric acid (GABA) A receptors agonist synergistically prolonged the pentobarbital-induced sleeping time. cyclopeptide alkaloid fraction of *Zizyphi Spinosi* Semen and pentobarbital increased the Cl-influx, cyclopeptide alkaloid fraction of *Zizyphi Spinosi* Semen had the similar effects with muscimol in potentiating their Cl-influx inducing effects of the low-dose pentobarbital. From these results, it is suggested that the enhancement of Cl-influx by cyclopeptide alkaloid fraction of *Zizyphi Spinosi* Semen might play an important role in the potentiation of pentobarbital-induced sleeping behaviors [47].

3.4.4. Alkaloids from Seeds of Ziziphus jujuba

In 1990, a group from the of Natural Products Research Institute, Seoul National University, Chongno-ku, Seoul, Kores isolated three 4(14)-frangulanine type cyclopeptide alkaloids such as frangufoline (sanjoinine A. **25**), sanjoinine B (**26**) and sanjoinine F (**27**), a 4(14)-pandamine type cyclopeptide alkaloids sanjoinine D (**28**), a miscellaneous type peptide alkaloid sanjoinine G2 (**29**), a 5(14)-amphibine B type cyclopeptide alkaloid amphibine-D (**15**) and a neutral cyclopeptide sanjoinenine (**30**) from *Zizyphus vulgaris* spinous seeds by column chromatography, followed by purification by preparative thin-layer chromatography (Figure 6) [48].

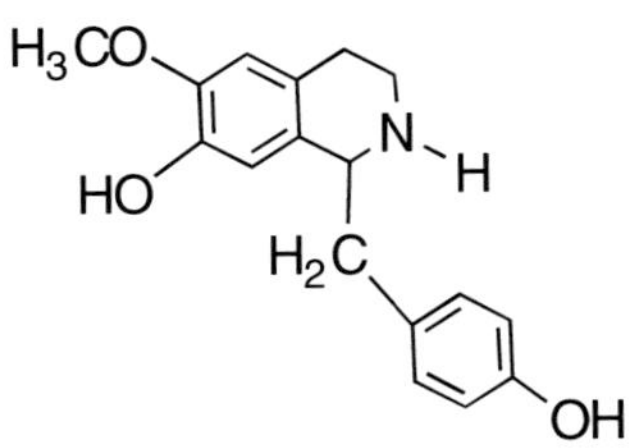

coclaurine (**19**)

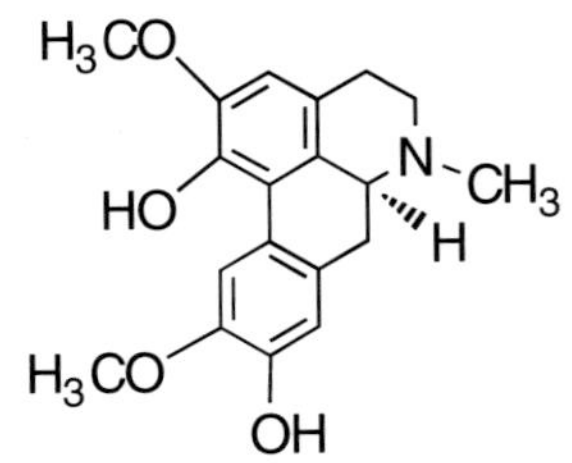

isoboldine (**20**)

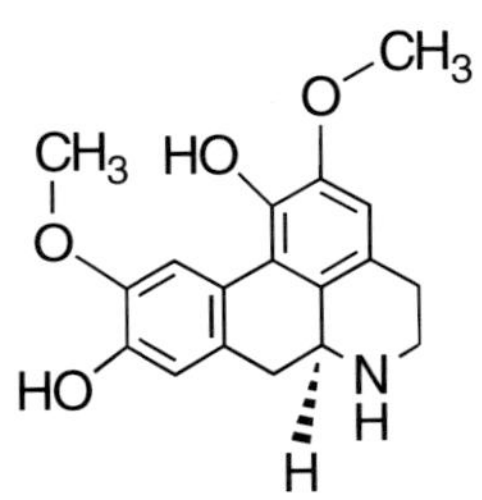

norisoboldine (**21**)

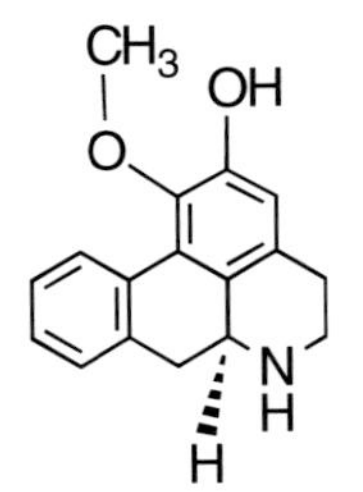

asimilobine (**22**)

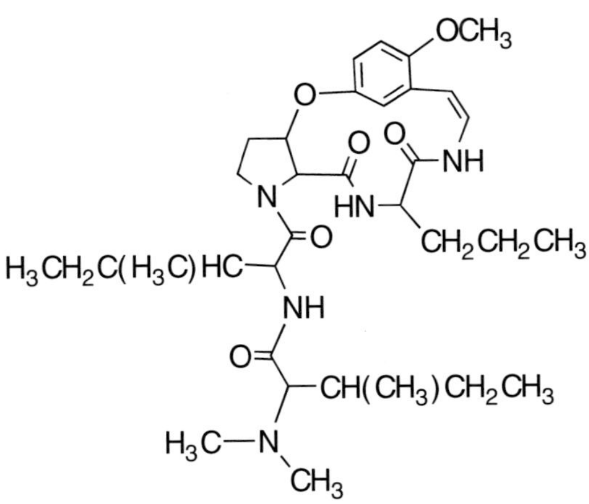

zizyphine A (**23**)

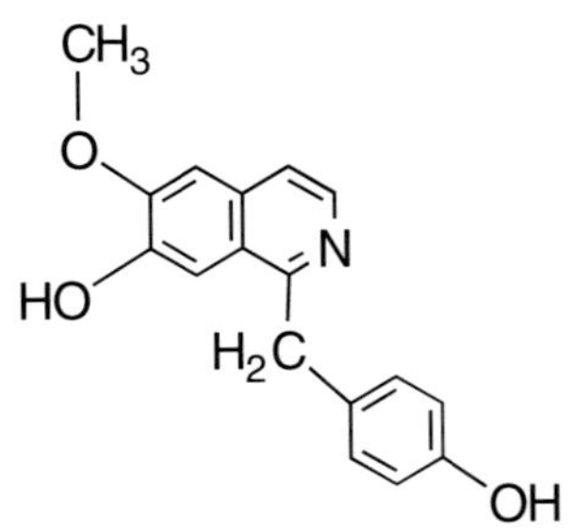

juzirine (**24**)

Figure 5 Coclaurine (**19**), isoboldine (**20**), norisoboldine (**21**), asimilobine (**22**), zizyphine A (**23**) and juzirine (**24**) from leaves of *Ziziphus jujuba*

3.4.5. Alkaloids from Root Bark of Zizyphus mauritiana

In 1996, a group from the of Laboratory of Chemistry, National Museum of Natural History, Paris, France isolated a 5(14)-amphibine B type cyclopeptide alkaloid mauritine J (**31**) from *Zizyphus mauritiana* root bark (Figure 7) [49].

3.4.6. Sedative Activity of Cyclopeptide Alkaloids

In 2001, a group from the of School of Pharmacy, College of Medicine, National Taiwan University, Taipei 100, Taiwan, Republic of China isolated three 13-membered cyclopeptide alkaloids such as paliurine G (**32**), paliurine H (**33**) and paliurine I (**34**) together with six known alkaloids such as nummularine H (**35**), daechuine-S3 (**36**), paliurine A (**37**), paliurine B (**38**), paliurine C (**39**) and paliurine F (**40**) from stems of *Zizyphus jujuba* Mill.var.ine (*Paliurus ramosissimus*) by a combination of centrifugal partition chromatography and preparative thin-layer chromatography (preparative TLC).

Among these nine cyclopeptide alkaloids, nummularine H (**35**) could shorten a barbiturate-derived drug methohexital (**41**)-induced sleeping time used as an anesthetic (Figures 8 and 9) [50].

In 2005, a group from the of Natural Products Research Institute, College of Pharmacy, Seoul National University, Seoul 110-460, Korea studied the effects of thirteeen sedative peptide alkaloids from *Zizyphus species* on the inhibitory activity against calmodulin-dependent protein kinase II.

Protein kinase II activity was assayed on the basis of its ability to activate tryptophan 5-monooxygenase (tryptophan hydroxylase 1) which an enzyme involved in the synthesis of the neurotransmitter serotonin (**42**) as its substrate in the presence of calmodulin.

Their 50% inhibitory concentration (IC_{50}) against calmodulin-dependent protein kinase II by thirteeen sedative peptide alkaloids were stronger inhibitors when compared to that (50% inhibitory concentration) of chlorpromazine (**43**) (IC_{50}, 98 μM). Among thirteeen sedative peptide alkaloids, daechuine S27 (nummularine B. **44**) (IC_{50} 2.95 μM) [51] was the potential inhibitory alkaloid which was stronger when compared to that of an anti-psychotic drug of the diphenylbutylpiperidine class pimozide (**45**) (IC_{50}, 15.0 μM), followed by sanjoinine F (**27**) (IC_{50}, 18.2 μM) and sanjoinine G2 (**29**) (IC_{50}, 19.0 μM) (Figure 10) [52].

In 2008, a group from the of College of Veterinary Medicine, Chungbuk National University, Cheongju 361-763, South Korea studied their effects on a short-acting barbiturate used as a sedative, hypnotic and anti-spasmodic pentobarbital (**46**)-induced sleeping behaviors by cyclopeptide alkaloid fraction of *Zizyphi Spinosi* Semen, and whether their effects of sleeping behaviors were mediated by γ-aminobutyric acid (GABA) receptors Cl-channel (chloride channel) activation.

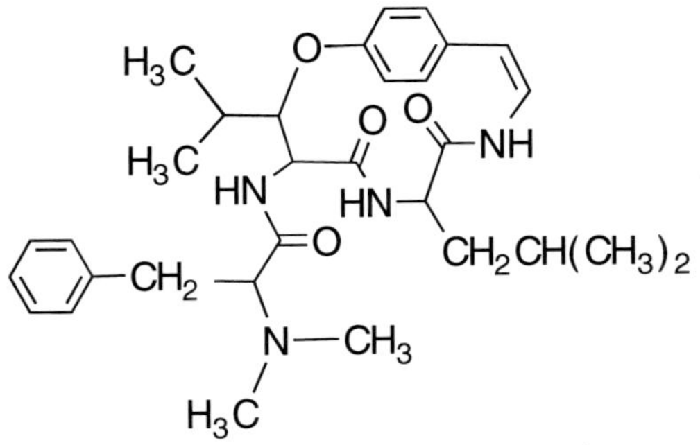

frangufoline (sanjoinine A. **25**)

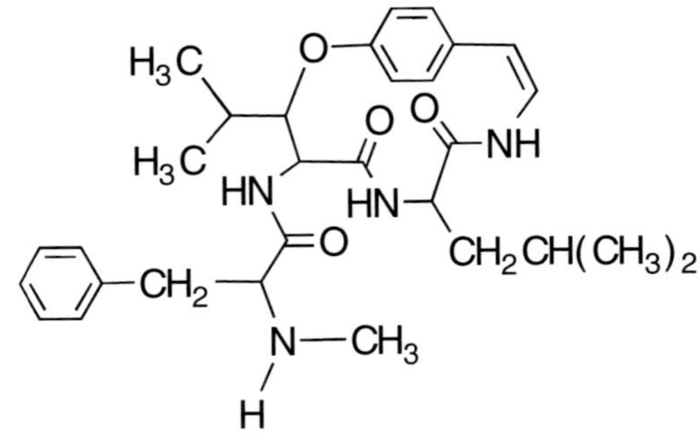

sanjoinine B (**26**)

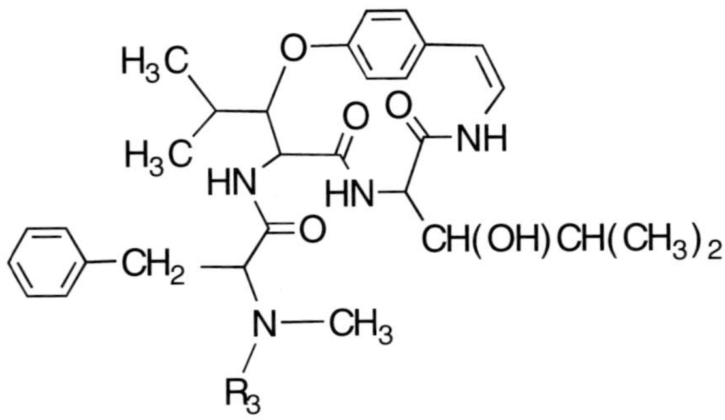

sanjoinine F (**27**)

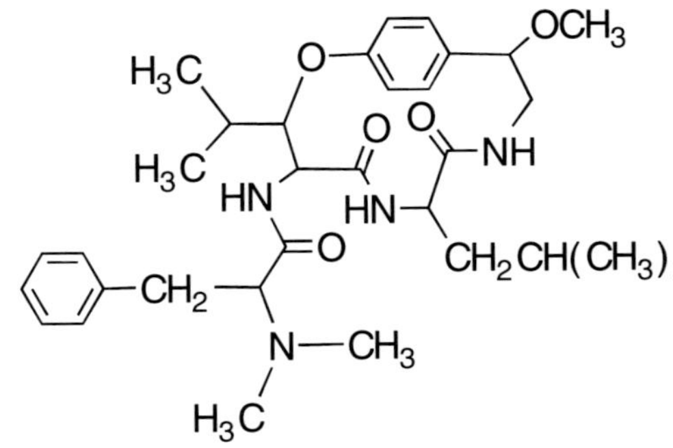

sanjoinine D (**28**)

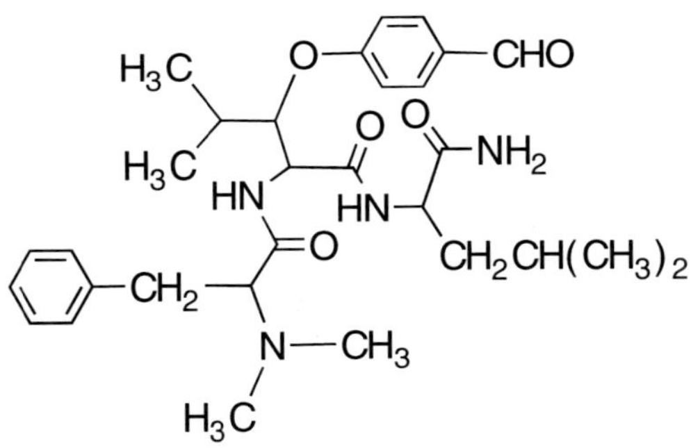

sanjoinine G2 (**29**)

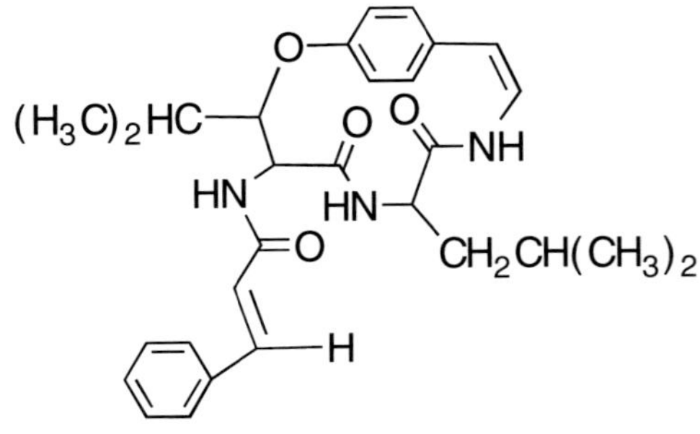

sanjoinenine (**30**)

Figure 6 Three 4(14)-frangulanine type cyclopeptide alkaloids such as frangufoline (sanjoinine A. **25**), sanjoinine B (**26**) and **sanj**oinine F (**27**), a 4(14)-pandamine type cyclopeptide alkaloid sanjoinine D (**28**), a miscellaneous type peptide alkaloid sanjoinine G2 (**29**), a 5(14)-amphibine B type cyclopeptide alkaloid amphibine-D (**15**), and a neutral cyclopeptide sanjoinenine (**30**) from *Zizyphus vulgaris* spinous seeds

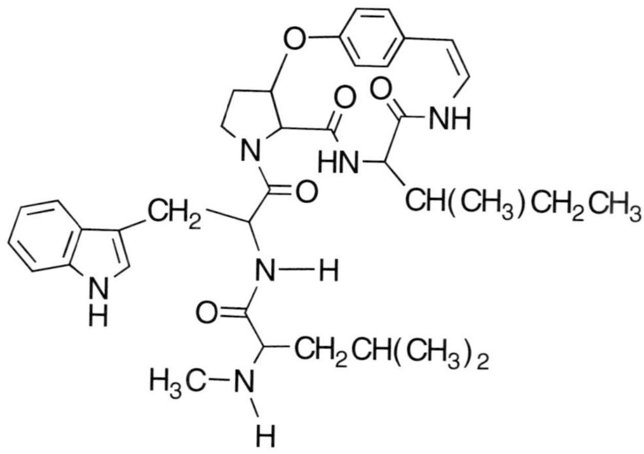

mauritine J (**31**)

Figure 7 A 5(14)-amphibine B type cyclopeptide alkaloid mauritine J (**31**) from *Zizyphus mauritiana* root bark)

As a result, cyclopeptide alkaloid fraction of *Zizyphi Spinosi* Semen shortened the sleeping onset and prolonged the sleeping time induced by pentobarbital (**46**) (42 mg/kg).

Cyclopeptide alkaloid fraction of *Zizyphi Spinosi* Semen also could significantly increase the falling asleep rate and duration of sleeping time at a sub-hypnotic dosage of pentobarbital (**46**) (28 mg/kg). Moreover, the combination of cyclopeptide alkaloid fraction of *Zizyphi Spinosi* Semen with a GABA A receptors agonist muscimol (**47**) could prolong synergistically their pentobarbital (**46**)-induced sleeping time.

Additionally, it was observed that cyclopeptide alkaloid fraction of *Zizyphi Spinosi* Semen and pentobarbital (**46**) increased the Cl^- influx through the chloride channels and cyclopeptide alkaloid fraction of *Zizyphi Spinosi* Semen showed the similar effects with muscimol (**47**) in potentiating the Cl^- influx inducing effects at the low-dose pentobarbital (**46**). From these results, it is suggested that the enhancement of the Cl^- influx by cyclopeptide alkaloid fraction of *Zizyphi Spinosi* Semen might play an important role in the potentiation of pentobarbital (**46**)-induced sleeping behaviors (Figure 10) [47].

In 1997, a group from the of Natural Products Research Institute, Seoul National University, Korea studied the enzymatic cleavage in rodents on frangufoline (sanjoinine A. **25**) of a sedative 14-membered frangulanine-type cyclopeptide alkaloid.

Frangufoline (sanjoinine A. **25**) in jujubes was converted by enzyme in rodents to a substituted linear tripeptide. Interestingly, mammalian serum has not the enzyme to catalyze the reaction [53].

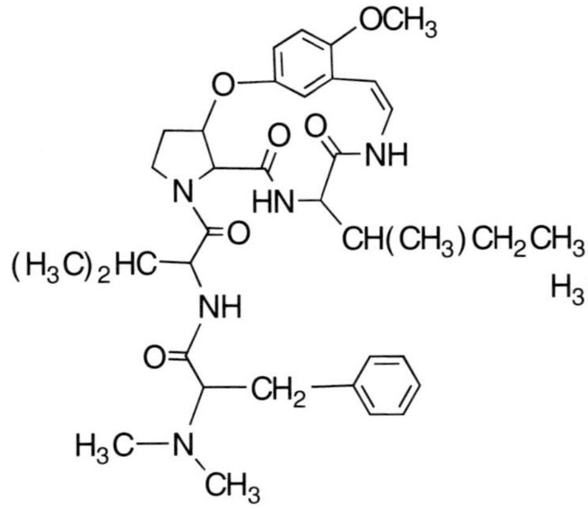

paliurine G (**32**)

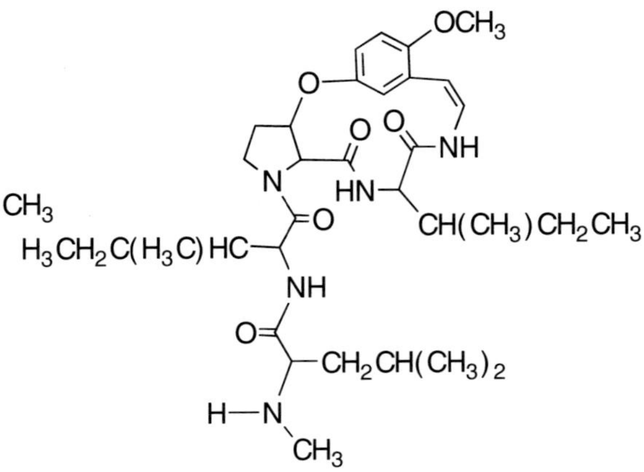

paliurine H (**33**)

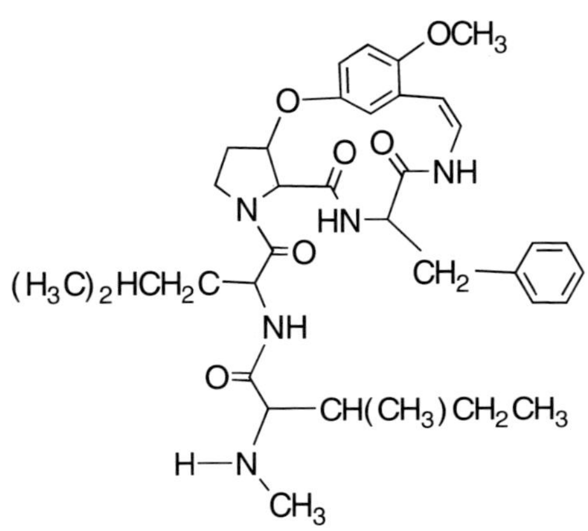

paliurine I (**34**)

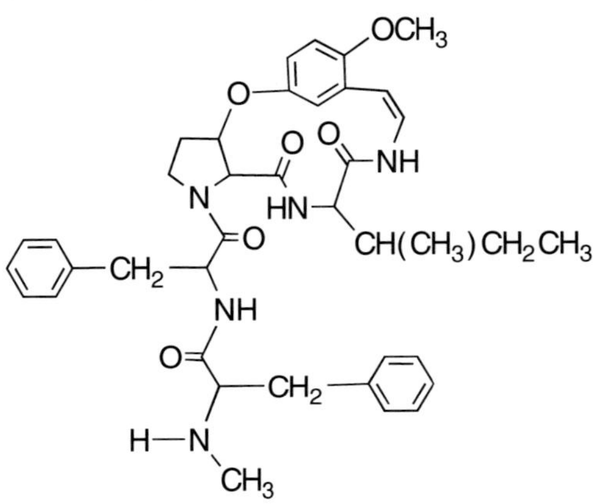

nummularine H (**35**)

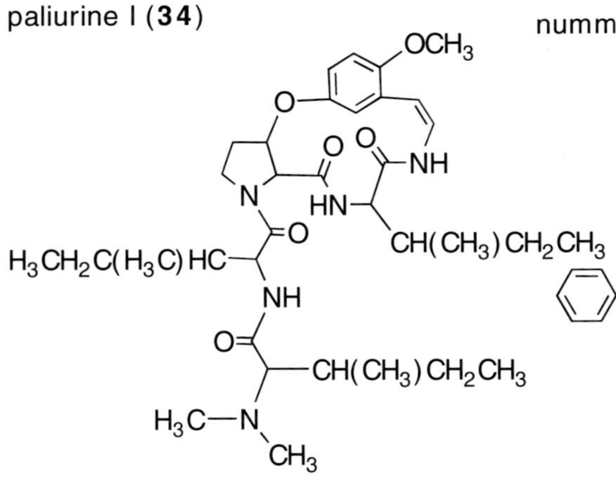

daechuine-S3 (**36**)

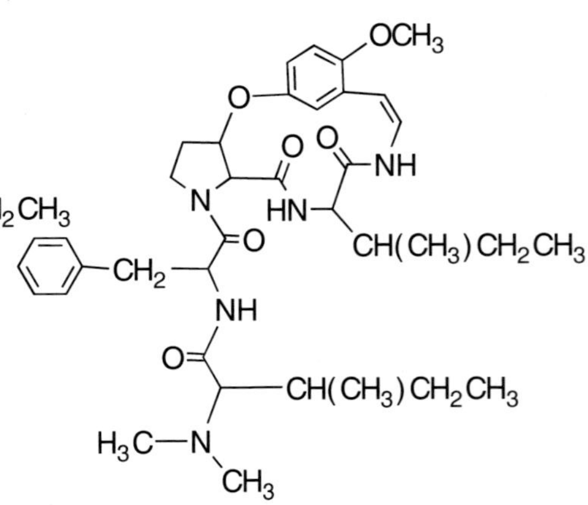

paliurine A (**37**)

Figure 8 Paliurine G (**32**), paliurine H (**33**), paliurine I (**34**), nummularine H (**35**), daechuine-S3 (**36**) and paliurine A (**37**) from stems of *Zizyphus jujuba* Mill.var.ine

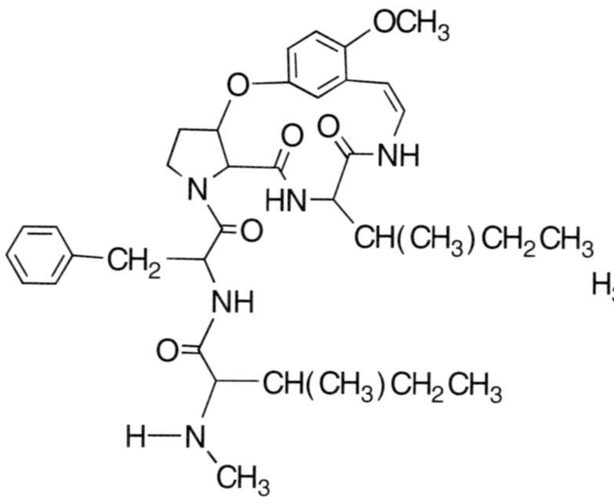

paliurine B (**38**)

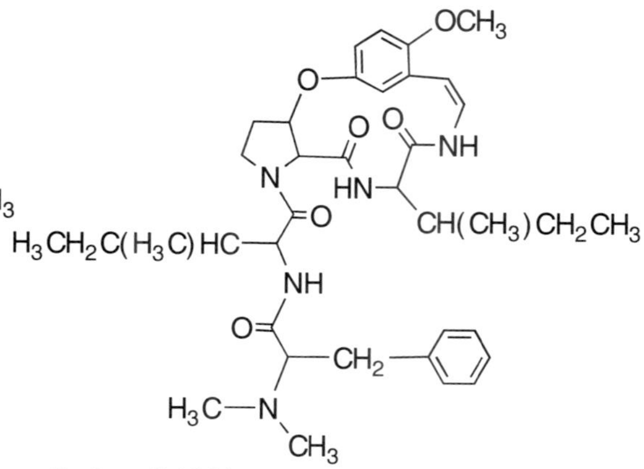

paliurine C (**39**)

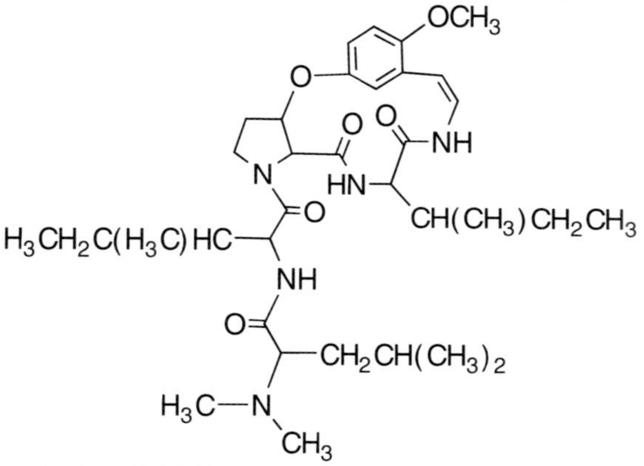

paliurine F (**40**)

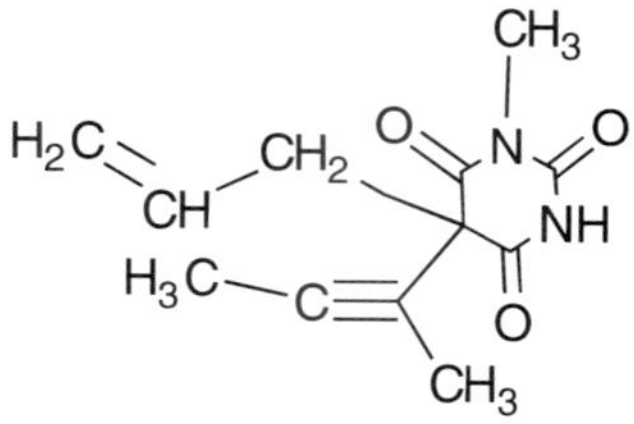

methohexital (**41**)

Figure 9 Paliurine B (**38**), paliurine C (**39**) and paliurine F (**40**) from stems of *Zizyphus jujuba* Mill.var.ine (*Paliurus ramosissimus*), and a barbiturate-derived drug methohexital (**41**)

3.5. Terpenoids (Esential Oils)

3.5.1. Terpenoids from Fruit of Ziziphus jujuba and Antutumor Activity of Jujube Triperpenoids

In 2003, a group from the of Department of Environmental Engineering, Chongiu University, Chongiu, Korea isolated eleven triperpenoic acids such as alphitolic acid (**48**), 3-*O*-(*cis*-*p*-coumaroyl)-alphitolic acid (**49**), 3-*O*-(*trans*-*p*-coumaroyl)-alphitolic acid (**50**), betulinic acid (**51**), betulonic acid (**52**) (Figure 11), 3-*O*-(*cis*-*p*-coumaroyl)-maslinic acid (**53**), 3-*O*-(*trans*-*p*-coumaroyl)-maslinic acid (**54**), oleanolic acid (**55**), oleanonic acid (**56**) (Figure 12), colubrinic acid (**57**), and zizyberenalic acid (**58**) (Figure 13) from the fruits of jujube (*Zizyphus jujuba*) and examined their cytotoxicity.

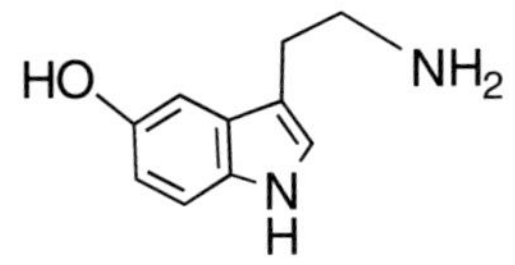

serotonin (**42**)

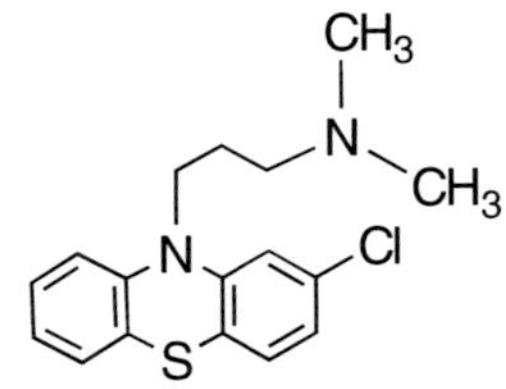

chlorpromazine (**43**)

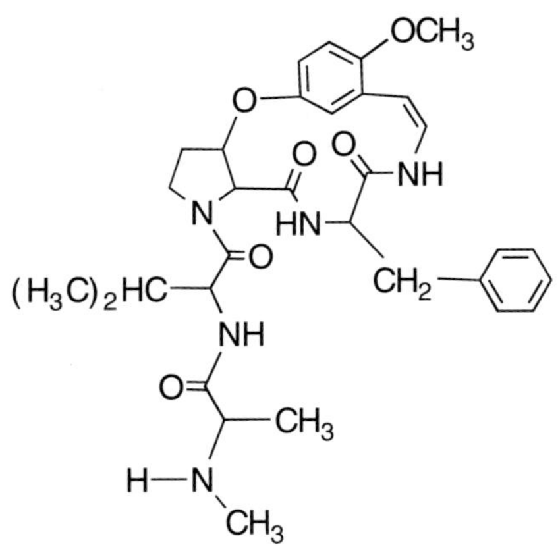

daechuine S27 (nummularine B, **44**)

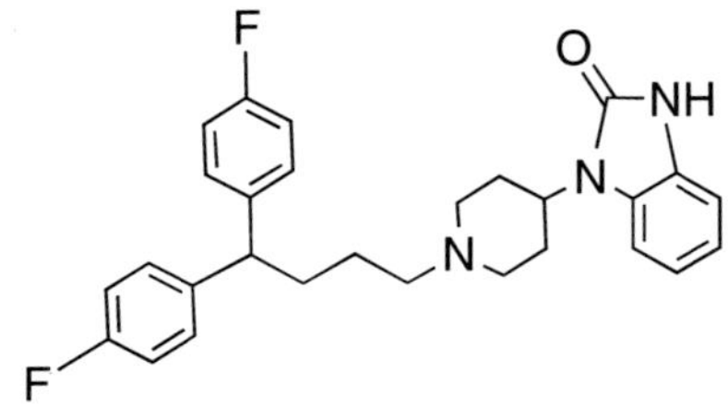

pimozide (**45**)

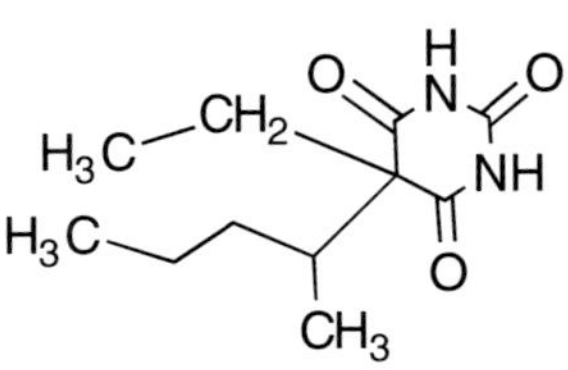

pentobarbital (**46**)

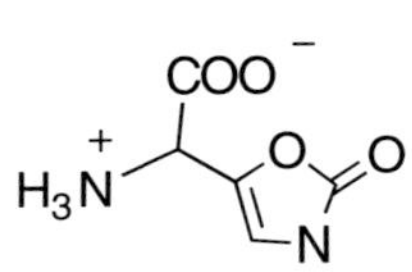

muscimol (**47**)

Figure 10 Serotonin (**42**), chlorpromazine (**43**), and daechuine S27 (nummularine B, **44**) from *Zizyphus species,* pimozide (**45**), pentobarbital (**46**), and muscimol (**47**)

These eleven triperpenoic acids were examined on their anti-tumor activity against six tumor cell strains such as K562, B16(F-10), SK-MEL-2, PC-3, LOX-IMVI and A549.

First, among eleven triperpenoic acids, oleanane type terpenoic aids such as 3-*O*-(*cis-p*-coumaroyl)-maslinic acid (**53**), 3-*O*-(*trans-p*-coumaroyl)-maslinic acid (**54**), oleanolic acid (**55**) and oleanonic acid (**56**) were no cytotoxity against these six tumor cell strains.

Second, however, lupane type triterpenes such as 3-*O*-(*cis-p*-coumaroyl)-alphitolic acid (**49**), 3-*O*-(*trans-p*-coumaroyl)-alphitolic acid (**50**), betulinic acid (**51**) and betulonic acid (**52**) had their cytotoxicity. Especially, 3-*O*-(*cis-p*-coumaroyl)-alphitolic acid (**49**) and 3-*O*-(*trans-p*-coumaroyl)-alphitolic acid (**50**) represented their significant inhibitory effects. From the result, it could suggest that the existence of coumaroyl moiety at the C-3 position of the lupane-type triterpene such as these two compounds (**49**, **50**) might play an important role in enhancing their cytotoxic activity [54].

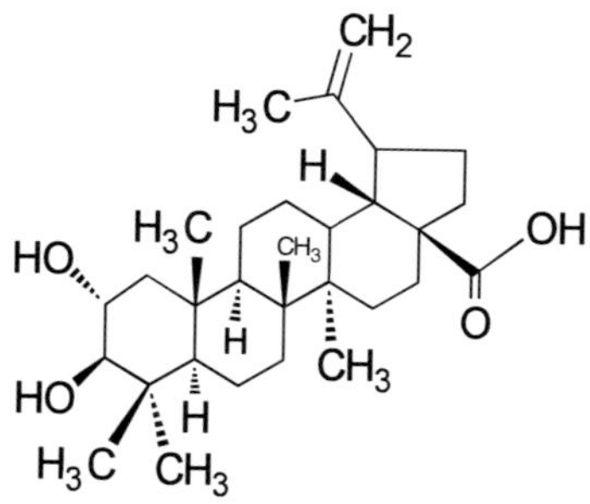

alphitolic acid (**48**)

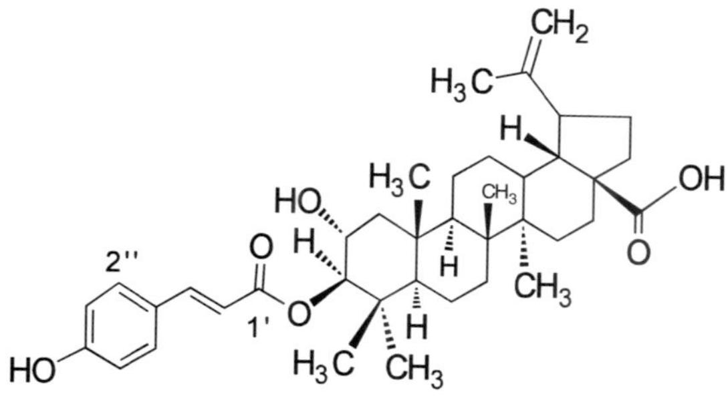

3-*O*-(*cis*-*p*-coumaroyl)-alphitolic acid (**49**)

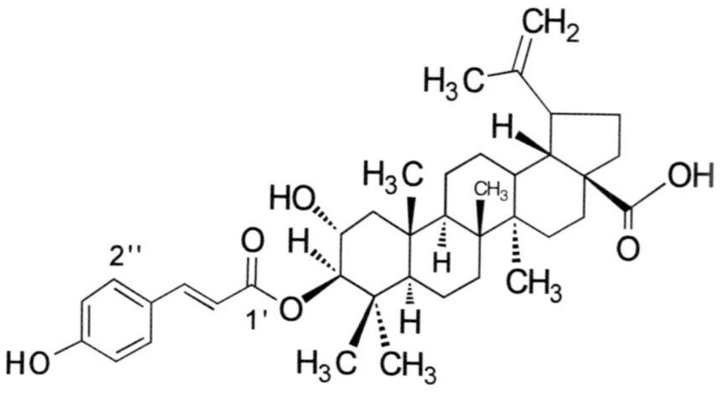

3-*O*-(*trans*-*p*-coumaroyl)-alphitolic acid (**50**)

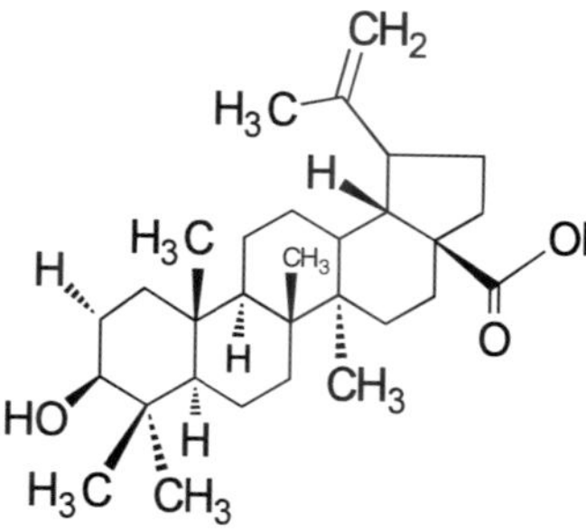

betulinic acid (**51**)

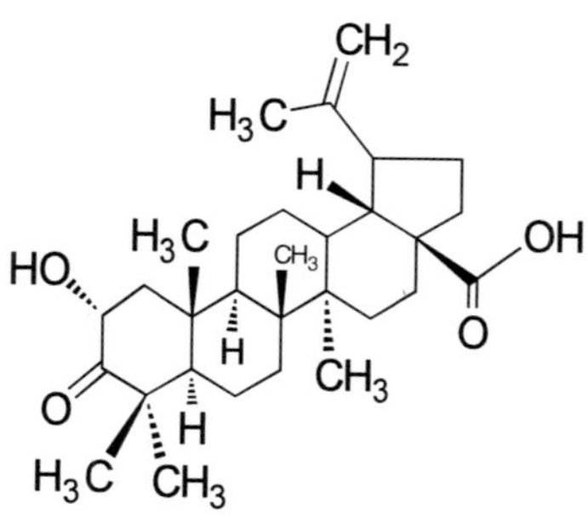

betulonic acid (**52**)

Figure 11 Five alphitolic acid (**48**), 3-*O*-(*cis*-*p*-coumaroyl)-alphitolic acid (**49**), 3-*O*-(*trans*-*p*-coumaroyl)-alphitolic acid (**50**)], betulinic acid (**51**] and betulonic acid (**52**)] of lupane type triperpenoic acids from the fruits of jujube (*Zizyphus jujuba*) etulinic acid (**51**] and betulonic acid (**52**)

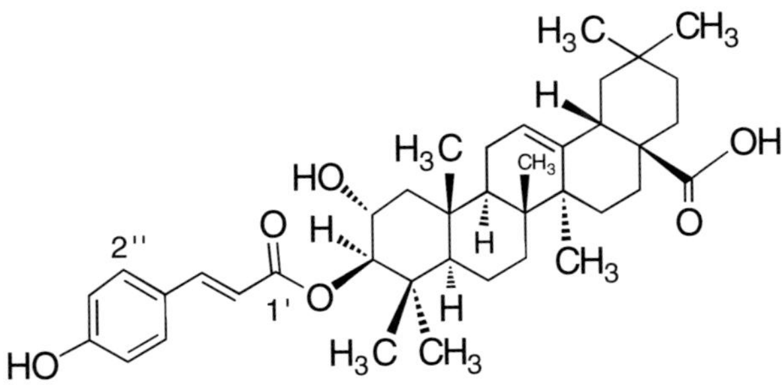

3-*O*-(*cis*-*p*-coumaroyl)-maslinic acid (**53**)]

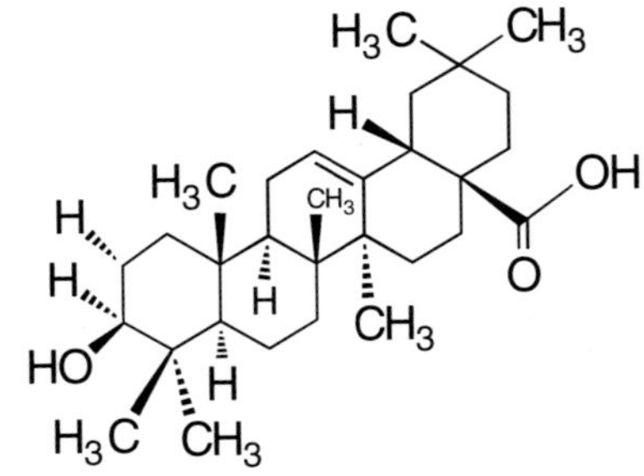

oleanolic acid (**55**)

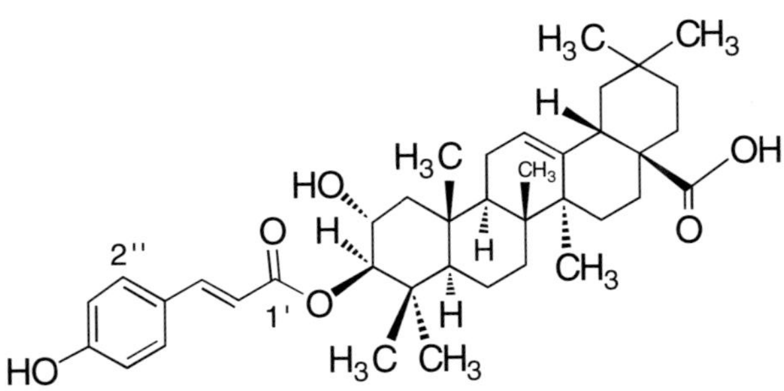

3-*O*-(*trans*-*p*-coumaroyl)-maslinic acid (**54**)]

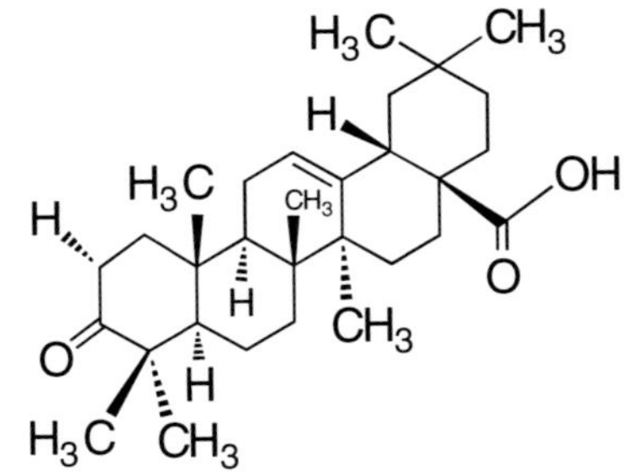

oleanonic acid (**56**)]

Figure 12 Four 3-*O*-(*cis*-*p*-coumaroyl)-maslinic acid (**53**)], 3-*O*-(*trans*-*p*-coumaroyl)-maslinic acid (**54**)], oleanolic acid (**55**)] and oleanonic acid (**56**)] of oleanane type terpenoic acids from the fruits of jujube (*Zizyphus jujuba*)

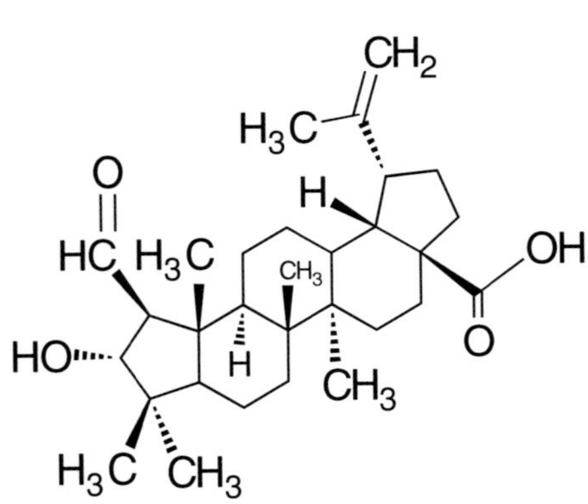

colubrinic acid (**57**)

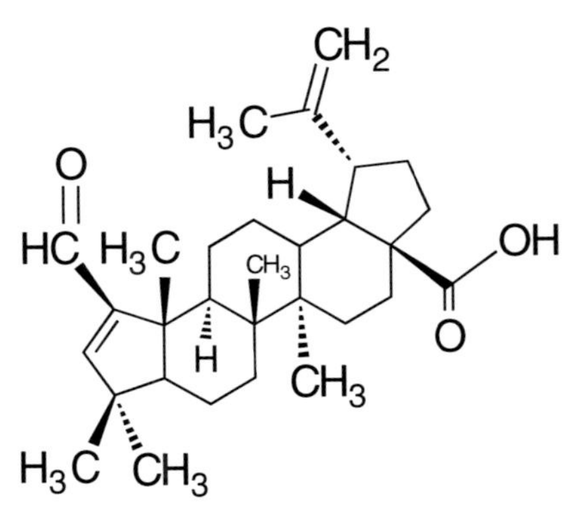

zizyberenalic acid (**58**)]

Figure 13 Colubrinic acid (**57**) and zizyberenalic acid (**58**) of two triterpenoic acids in jujube fruits

3.5.2. Triterpenes from Seeds of Zizyphus vulgaris Lamark var. spinosus Bunge

In 1940, triterpenes (lupanes) such as betulin (**59**) and betulinic acid (**60**), were identified from seeds of *Zizyphus vulgaris* Lamark var. spinosus Bunge.

Besides this, oleic acid (ω-9. **61**), linoleic acid (**62**), myristic acid (**63**), palmitic acid (**64**), stearic acid (**65**), behenic acid (**66**) and β-sitosterol (**67**) were also found from seeds of *Zizyphus vulgaris* Lamark var. spinosus Bunge (Figure 14) [55].

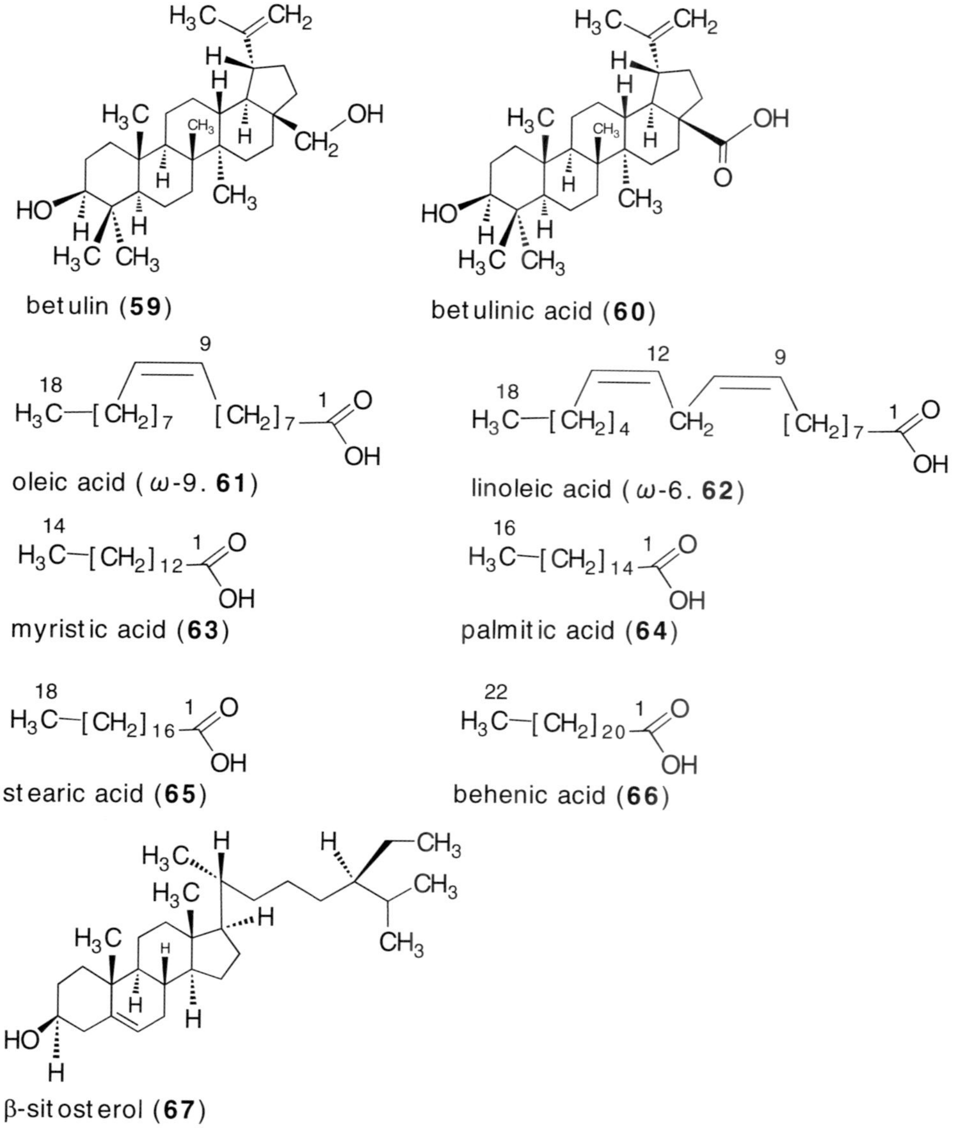

Figure 14 Betulin (**59**) and betulinic acid (**60** of two lupane types, oleic acid (ω -9. **61**), linoleic acid (ω-6. **62**), myristic acid (**63**), palmitic acid (**64**), stearic acid (**65**) and behenic acid (**66**) of six fatty acids, and β-sitosterol (**67**) of one phytosterol

3.5.3. Oleanane-type Triterpenes of Triterpenoids (3-O-cis-p-coumaroyl Maslinic Acid, 3-O-trans-p-coumaroyl Maslinic Acid, Oleanolic Acid): Erythrocyte Hemolysis Inhibition, Antitumor Effect

In 2004, a group from the of Department of Environmental Engineering, Chongiu University, Chongiu, Korea was identified anti-complements maslinic acid (**68**) derivatives of oleanane-type triterpenes such as 3-*O*-(*cis*-*p*-coumaroyl)-maslinic acid (**69**), 3-*O*-(*trans*-*p*-coumaroyl)-maslinic acid (**70**) and oleanolic acid (**71**) from the fruits of *Zizyphus jujuba* MILL. Because these oleanane-type triterpenes could inhibit the hemocytocatheresis, the components in erythrocyte could be inhibited to carry away in blood plasma. Therefore, oleanane-type triterpenes in jujube fruits might play a role to prevent the hemolytic activity of human serum against erythrocyte (Figure 15) [56].

For example, oleanolic acid (**71**) is known as one of triterpenic acids from leavefs of olive trees, and apple peel or clove buds also contains oleanolic acid (**71**).

In 2008, a group from the of Technical University of Munich, Freising of Germany separated the olive fruit extracts with maslinic acid (**68**) 73.25% and oleanolic acid (**11→71**) 25.75%. First, these two maslinic acid (**68**) and oleanolic acid (**71**) of triterpenic acids were examined on their anti-tumor effect against human HT-29 colon cancer cells and the apoptosis. As a result, oleanolic acid (**71**) could inhibit their tumor growth at the higher concentration. On the other hand, maslinic acid (**68**) could inhibit their tumor growth at the lower concentration.

Second, maslinic acid (**68**) and oleanolic acid (**71**)] were compared on their apoptosis. Oleanolic acid (**71**) without hydroxyl (OH) group at position 2 had no other activity of caspase-3 as a prime apoptosis protease. However, maslinic acid (**68**) at concentrations of 10 μmol/L, 25 μmol/L and 50 μmol/L enhanced 3, 3-5 and 5-times caspase-3 like activity, respectively when compared to the control.

From the results, it suggests that their trigger of anti-tumor activity is not oleanolic acid (**71**) in olive fruits butand is maslinic acid (**68**) (Figure 15) [57].

(memo)

Anti-complement: a substance that opposes or countracts the action of a complement.

Complement (C. alexin): Originally thea term complement wasere used to refer to the heat-labile factor in serum that causes immune cytolysis, the lysis of antibody-coated cells. However, now the term 'complement' term is used to refer to the entire functionally- related system comprising at least 20 distinct serum proteins, their cellular receptors, and related regulatory proteins that is the effector not only of immune cytolysis but also of other biologic functions including anaphylaxis, phagocytosis, opsonization, and hemolysis. Complement activation occurs by two different sequences such as the classical and alternative pathways.

3.5.4. Saponins in Seeds of Zizyphus jujuba

In 1974, jujuboside A (**72**) and jujuboside B (**73**) were identified from MeOH extract in seeds of *Zizyphus jujuba* Mill. (Figure 16) [58, 59].

3.5.5. Fatty Oil

3.5.5.1. Fatty Oils from Seeds of *Zizyphus vulgaris* Lamark var. Spinosus Bunge

In 1940, fatty acids such as oleic acid (**61**), linoleic acid (**62**), myristic acid (**63**), stearic acid (**65**), behenic acid (**66**) and β-sitosterol (**67**) were identified from seeds of *Zizyphus vulgaris* Lamark var. spinosus Bunge (Figure 14) [55].

3.5.5.2. Jujube Essential Oil: Hair Fostering (Hair Restoration) Effect

In 2010, a group from the of Department of Biotechnology, Daegu University, Kyoungsan, Kyoungbook, Republic of Korea was examined the efficacy of essential oil from seeds of *Zizyphus jujuba* for its potential role on hair growth by *in vivo* method by mice.

First, jujube essential oil was applied at three concentrations of 0.1%, 1% and 10% on the shaved skin onto the backs of BALB/c mice and monitored for 21 days. After 21 days, the length of their hairs on the administrated mice were 9.96 for 1% of jujube essential oil, and 10.02 mm for 10% of jujube essential oil, respectively, when compared to 8.94 mm of a control non-treated mice. At the same time, the weight of hair/cm^2 area of dorsal skin was measured. And their hair thickness and the hair follicles microscopically after plucking the hair immediately from the shaved area of mice were also evaluated. Among three concentrations, 1% of essential oil-treated mice was the best effective oil concentration. From these results, it could suggest that *Zizyphus* essential oil might be beneficial for the hair growth promoting activity [24].

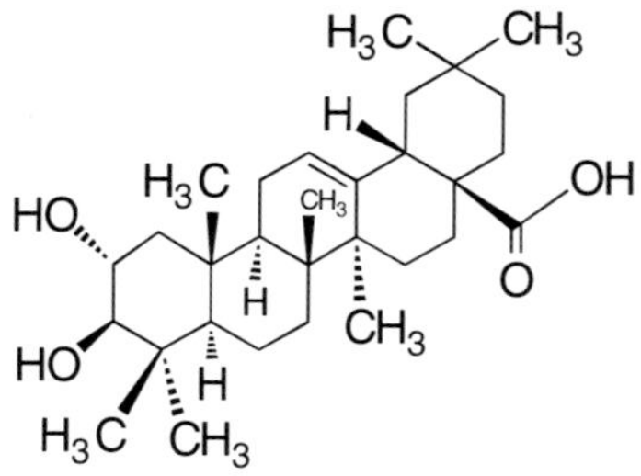

maslinic acid (**68**)

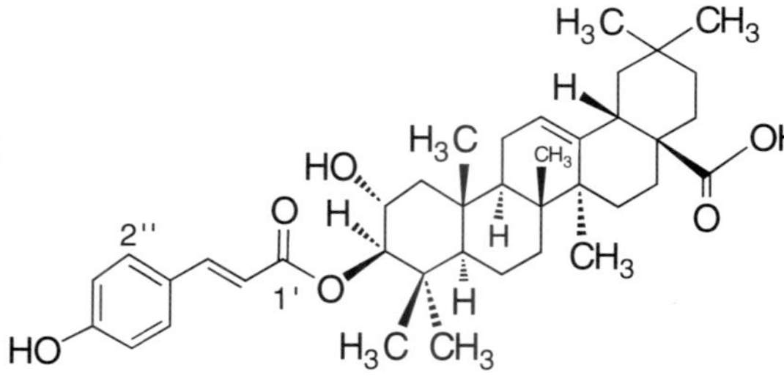

3-*O*-(*cis*-*p*-coumaroyl)-maslinic acid (**69**)]

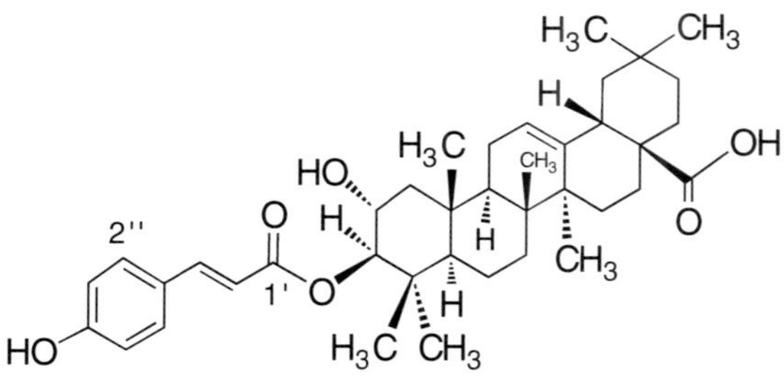

3-*O*-(*trans*-*p*-coumaroyl)-maslinic acid (**70**)

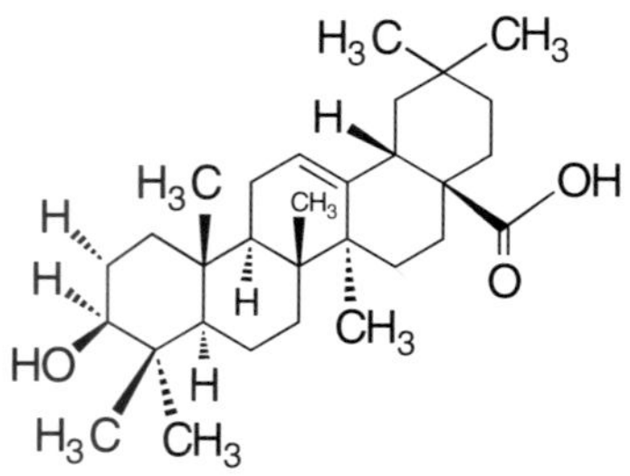

oleanolic acid (**71**)

Figure 15 Maslinic acid (**68**)], 3-*O*-(*cis*-*p*-coumaroyl)-maslinic acid (**69**), 3-*O*-(*trans*-*p*-coumaroyl)-maslinic acid (**70**)] and oleanolic acid (**71**) of oleanane-type triterpenes from jujube berry fruits

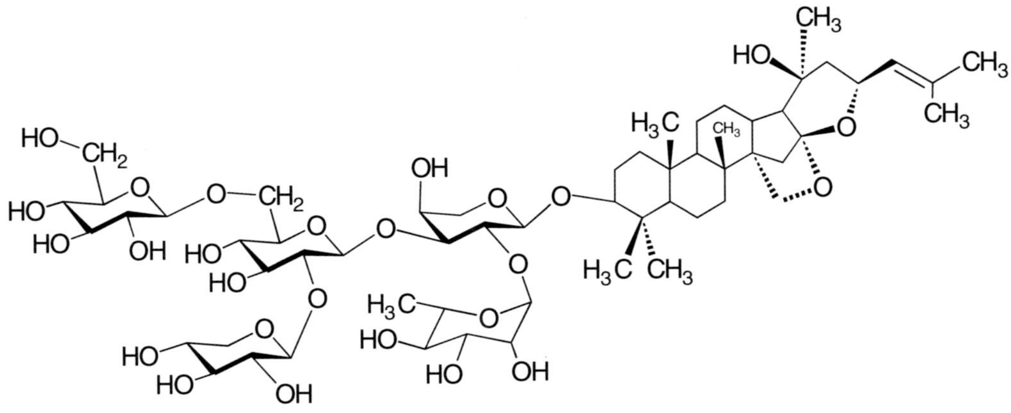

jujuboside A (**72**)

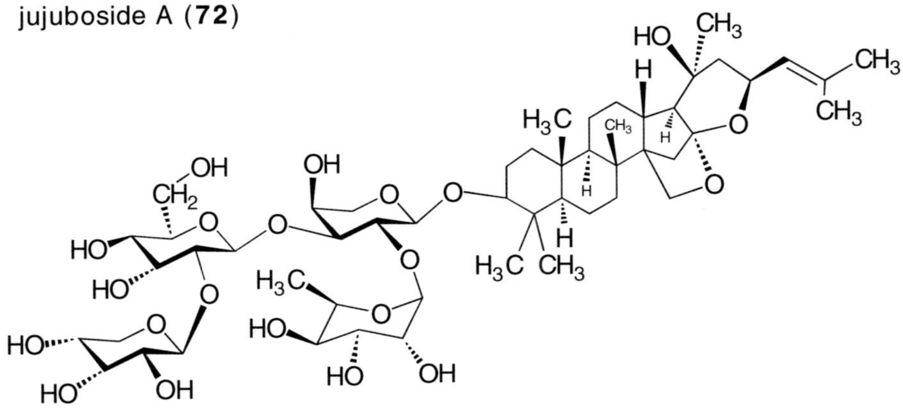

jujuboside B (**73**)

Figure 16 Jujuboside A (**72**) and jujuboside B (**73**) from seeds of Zizyphus jujuba Mill.

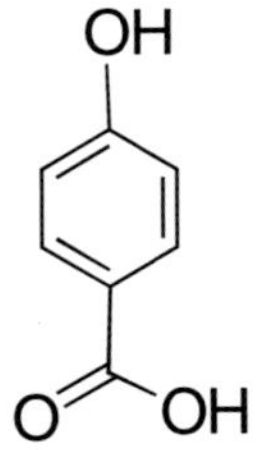

p-hydroxybenzoic acid (**74**)

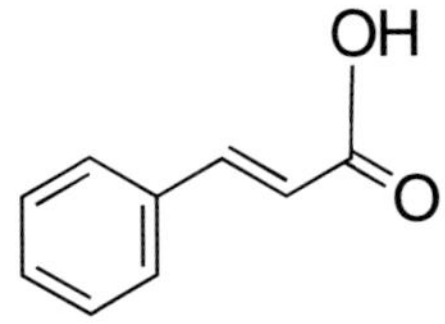

cinnamic acid (**75**)

Figure 17 *p*-Hydroxybenzoic acid (**74**) and cinnamic acid (**75**)

3.6. Flavonoids

3.6.1. Jujube Flavonoids: Antioxidative Effect

3.6.1.1. Phenolic Acids in Jujubes

In 2011, their content and antioxidant activity on phenolic acids in pulp, seed and peel of jujube were measured by high performance liquid chromatography with electrochemical detection (HPLC-ECD). Their jujubes contained the most abundant *p*-hydroxybenzoic acid (**74**) and cinnamic acid (**75**). All quantified phenolic acids were mainly present in jujube peel. Their phenolic acids in both seed and peel are present in the insoluble-bound form. On the other hand, the glycosided form and insoluble-bound phenolic acid fractions in jujube pulps had the highest total phenolic content and the most high antioxidant activity by both 1,1-diphenyl-2-picrylhydrazyl (DPPH) assay and ferric reducing antioxidant power (FRAP) assay when compared to total phenolic content and the antioxidant activity of seed and peel. By these results, it could suggest that most active phenolic acids with the antioxidant activity in different tissues of jujube might be present as the glycosided form and insoluble-bound phenol forms (Figure 17) [60].

3.6.1.2. Anthocyanins

3.6.1.2.1. Two Naringenin Type's Flavanones: Naringenin 2(S)-6,8-di-C-glucoside (6,8-di-C-glucosyl-2(S)-naringenin. 77) and Naringenin 2(R)-6,8-di-C-glucoside (6,8-di-C-glucosyl-2(R)-naringenin. 78): Antioxidative Effect

Naringenin (**76**) of flavanones has antioxidative activity such as anti-hepatotoxic and anti-peroxidative properties. In 1981, a group from the of Pharmaceutical Sciences, Kyushu University, Fukuoka, Japan isolated two novel naringenin type's glycosides of naringenin 2(*S*)-6,8-di-C-glucoside (6,8-di-C-glucosyl-2(*S*)-naringenin. **77**) and naringenin 2(*R*)-6,8-di-C-glucoside (6,8-di-C-glucosyl-2(*R*)-naringenin. **78**) from their ethanol extract of jujube fruit (*Zizyphi Fructus. Zizyphus jujuba* Mill var. inermis Bunge) (Figure 18) [61].

*3.6.1.2.2. Two Dimeric Procyanidins of Procyanidin B3 (**80**) and Procyanidin B4 (**81**) : Reversal Effects of Antibiotic Resistance*

An antibiotic drug, methicillin (**79**), belongs to the penicillin family used in the treatment of certain *Staphylococcal* infections. However, methicillin (**79**) often reveals often the resistance for *Staphylococcus aureus*. Methicillin (**79**)-ressistant *Staphylococcus aureus* (MRSA) might acquire the multidrug resistance in many cases of diseases.

In 2006, based on these problems, a group from the of Department of Pharmacognosy, Faculty of Pharmaceutical Sciences, Okayama University, Tsushima, Okayama 700-8530, Japan identified two dimeric procyanidins such as procyanidin B3 (**80**) and procyanidin B4 (**81**) from jujube fruit and studied the reversal effects of two jujube dimeric procyanidins for methicillin (**79**)-ressistant *Staphylococcus aureus* (MRSA).

First, inon an antibiotics of a form of penicillin resistant to penicillinase and effective against penicilline G (**82**)-resistant *staphylococci* oxacillin (**83**), their minimum inhibitory concentration (MIC. μg/mL) by single administration of oxacillin (**83**) as a control for four

MRSA strains was 256 for OM481, 64 for OM505, 256 for OM584 and 512 for OM623, respectively.

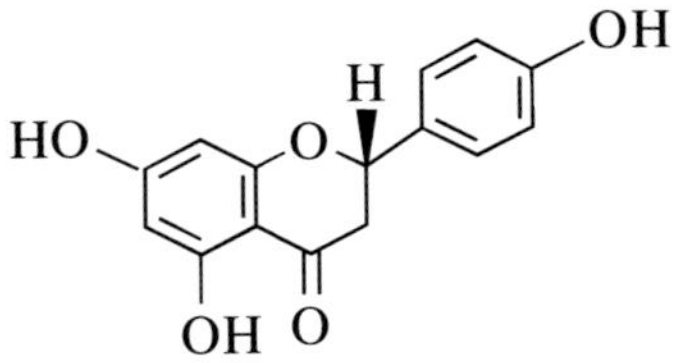

naringenin (**76**)

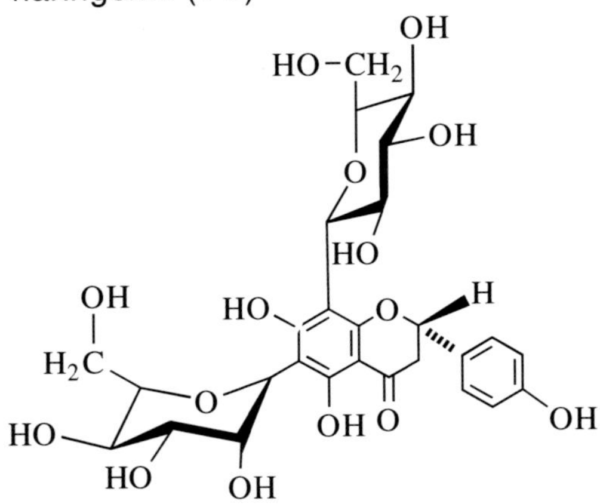

naringenin 2(*S*)-6,8-di-C-glucoside (**77**)

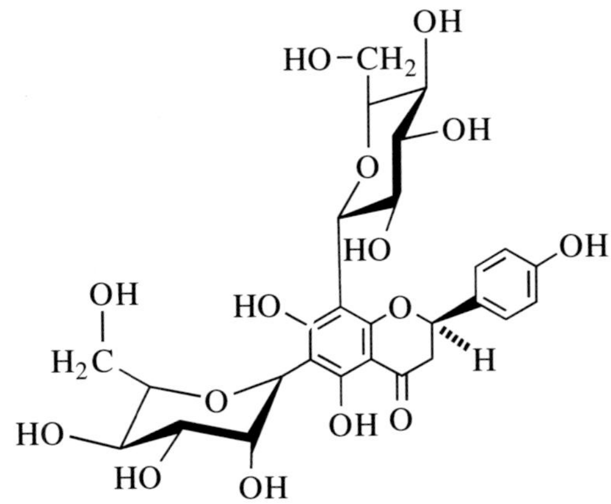

naringenin 2(*R*)-6,8-di-C-glucoside (**78**)

Figure 18 A lead compound naringenin (**76**), and two flavanones of naringenin 2(*S*)-6,8-di-C-glucoside (6,8-di-C-glucosyl-2(*S*)-naringenin. **77**) and naringenin 2(*R*)-6,8-di-C-glucoside (6,8-di-C-glucosyl-2(*R*)-naringenin. **78**) from *Zizyphi Fructus*

On procyanidin B3 (**80**), their minimum inhibitory concentration (MIC. μg/mL) by the single administration of procyanidin B3 (**80**) for four MRSA strains was 512 for OM481, 1024 for OM505, 1024 for OM584 and 1024 for OM623, respectively.

Their minimum inhibitory concentration (MIC. μg/mL) by the combined administration of oxacillin (**83**) and procyanidin B3 (**80**) for four MRSA strains was 128 for OM481, 32 for OM505, 32 for OM584 and 64 for OM623, respectively.

On procyanidin B4 (**81**), their minimum inhibitory concentration (MIC. μg/mL) by the single administration of procyanidin B4 (**81**) for four MRSA strains was 512 for OM481, 512 for OM505, 512 for OM584 and 512 for OM623, respectively.

Their minimum inhibitory concentration (MIC. μg/mL) by the combined administration of oxacillin (**83**) and procyanidin B4 (**81**) for four MRSA strains was 64 for OM481, 32 for OM505, 16 for OM584 and 32 for OM623, respectively (Figure 19) (Table 1).

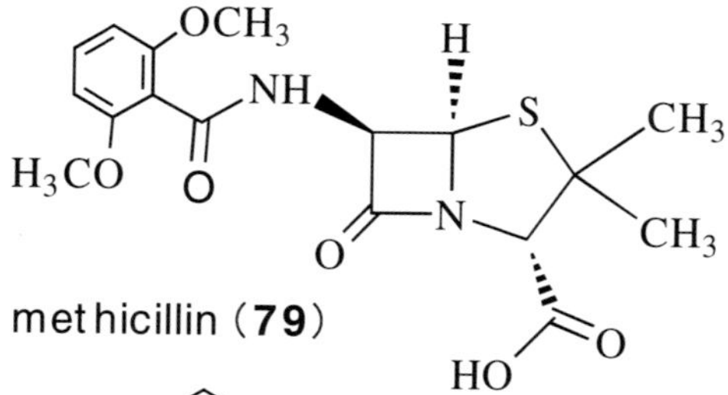

methicillin (**79**)

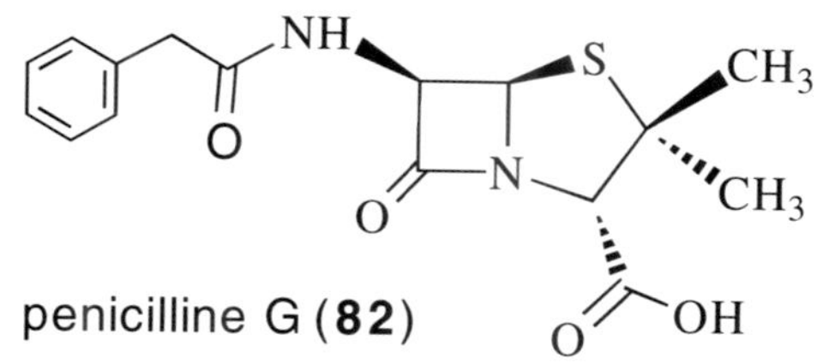

penicilline G (**82**)

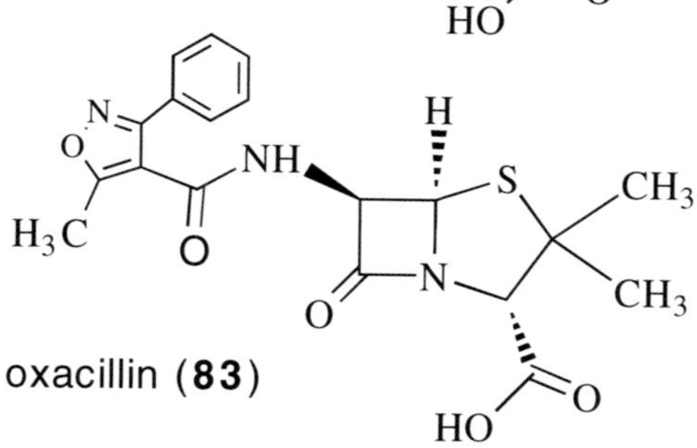

oxacillin (**83**)

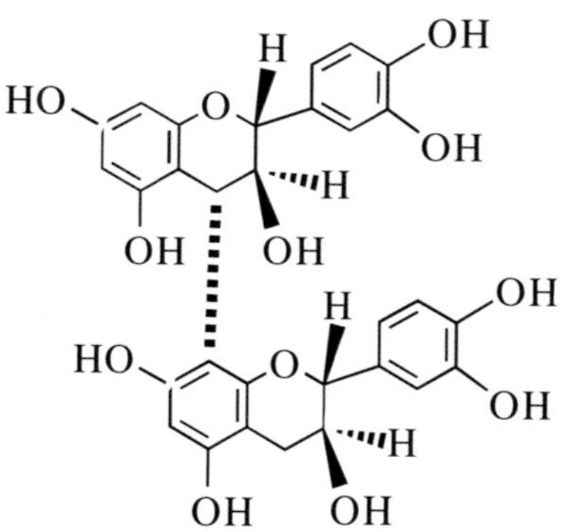

procyanidin B3 (**80**)

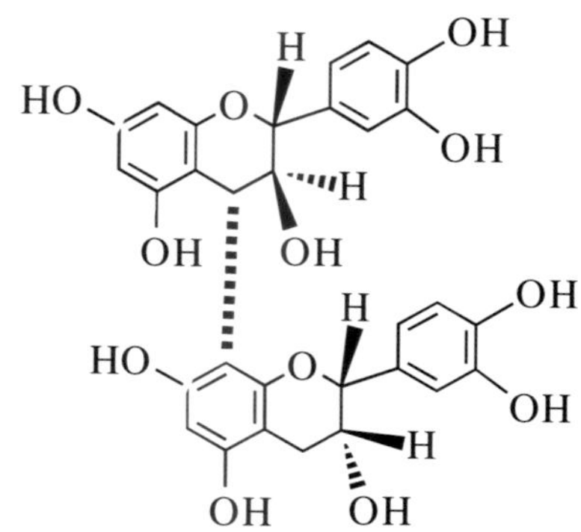

procyanidin B4 (**81**)

Figure 19 Methicillin (**79**), penicilline G (**82**) and oxacillin (**83**) of semisynthetic β-lactam antibiotic penicillins. Two dimeric procyanidins of procyanidin B3 (**80**) and procyanidin B4 (**81**) from jujube pulps

Second, on penicillin G (**82**), their minimum inhibitory concentration (MIC. μg/mL) by single administration of penicillin G (**82**) as control for four MRSA strains was 32 for OM481, 32 for OM505, 32 for OM584 and 16 for OM623, respectively.

Their minimum inhibitory concentration (MIC. μg/mL) by the combined administration of penicillin G (**82**) and procyanidin B3 (**80**) for four MRSA strains was 8 for OM481, 16 for OM505, 8 for OM584 and 8 for OM623, respectively.

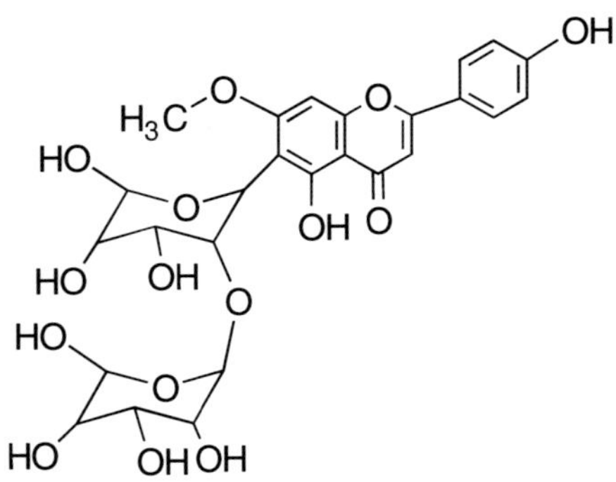

spinosin (**84**)

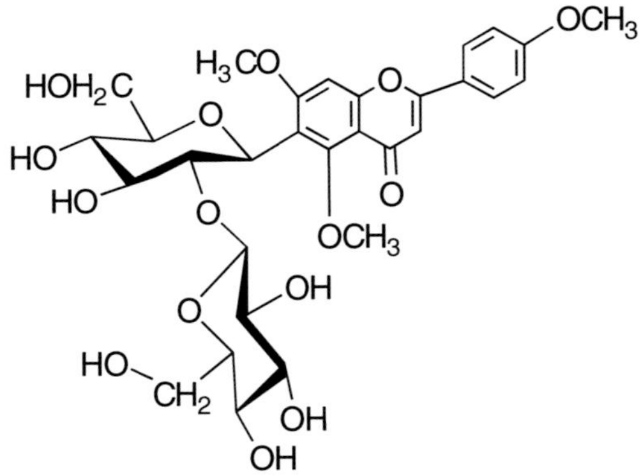

di-*O*-methylspinosin (**85**)

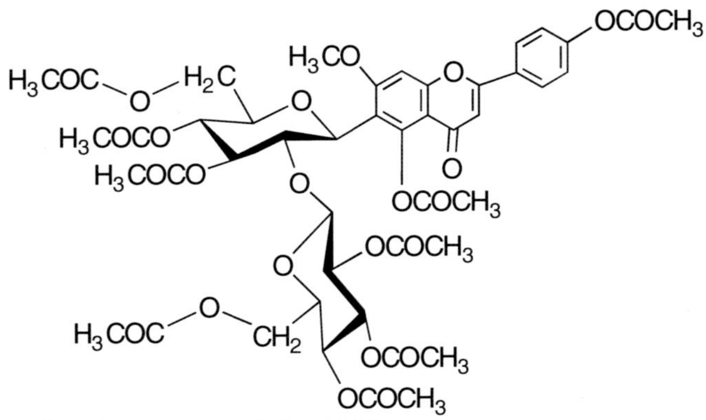

spinosin nona-acetate (**86**)

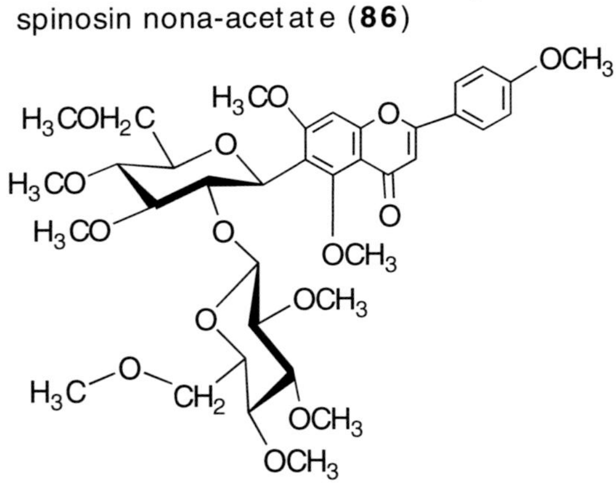

permethylspinosin (**87**)

Figure 20 Spinosin (**84**), di-*O*-methylspinosin (**85**), spinosin nona-acetate (**86**) and permethylspinosin (**87**)

Their minimum inhibitory concentration (MIC. μg/mL) by the combined administration of penicillin G (**82**) and procyanidin B4 (**81**) for four MRSA strains was 4 for OM481, 16 for OM505, 4 for OM584 and 4 for OM623, respectively (Figure 19) (Table 1) [62].

Table 1. The reversal effects of two dimeric procyanidins (80, 81) from *Zizyphus jujuba* for four methicillin (79)-ressist ant *Staphylococcus aureus* (MRSA) strains

Antibiotics, procyanidins	Minimumu inhibitory concentration (MIC) for four MRSA strains (μg/mL)			
	OM481	OM505	OM584	OM623
oxacillin (**83**) alone	256	64	256	512
oxacillin (**83**) + procyanidin B3 (**80**)(64 μg/mL)	128	32	32	64
oxacillin (**83**) + procyanidin B4 (**81**)(64 μg/mL)	64	32	16	32
penicillin G (**82**) alone	32	32	32	16
procyanidin B3 (**80**)	512	1024	1024	1024
penicillin G (**82**) + procyanidin B3 (**80**)(64 μg/mL)	8	16	8	8
procyanidin B4 (**81**)	512	512	512	512
penicillin G (**82**) + procyanidin B4 (**81**)(64 μg/mL)	4	16	4	4

3.6.1.2.3. Flavonoids in Seeds of Ziziphus Vulgaris var. spinosus

In 1979, a research group from the of Korea and Germany was identified spinosin (**84**), di-*O*-methylspinosin (**85**), spinosin nona-acetate (**86**) and permethylspinosin (**87**) from methanolic extracts of seeds of *Ziziphus vulgaris* var. spinosus (Figure 20) [63].

3.6.1.2.4. Flavonoids in Seeds of Ziziphus jujuba Miller (1)

In 1980, a research group from theof Korea and Germany was identified 6'''-sinapolyspinosin (**38**), 6'''-feruloylspinosin (**37**) and 6'''-*p*-coumaloylspinosin (**39**) from methanolic extracts of seeds of *Zizyphus jujuba* (Figure 21) [64].

3.6.1.2.5. Flavonoids in Seeds of Ziziphus jujuba Miller (2)

In 1991, a novel flavonoid 6'''-*p*-hydroxybenzoylspinosin (apigenin 6-*C*-[(6-*O*-*p*-hydroxybenzoyl) β-D-glucopyranosyl(1→2)] β-D-glucopyranoside. **91**) with spinosin (**84**), swertisin (**92**), 6'''-feruloylspinosin (**88**), 6'''-sinapolyspinosin (**89**), 6'''-*p*-coumaloylspinosin (**90**), 2''-*O*-glucosylisoswertisin (**93**) and vecenin-2 (**94**) were identified from water soluble extracts in seeds of *Ziziphus jujuba* Miller (Figure 22) [58].

3.6.1.3. Flavonoids in Korean Jujubes

In 2011, a group from theof Department of Food Service Industry and Boeun Jujube Industry Support Center, Seowon University, Cheongju-city 361-742, Republic of Korea were measured their content (g/100 g dry jujube) of flavonoids in three jujube fruits such as Boeun-deachu, Mechu and Sanzoin cultivars, and two jujube seeds such as Mechu and Sanzoin cultivars in Korea.

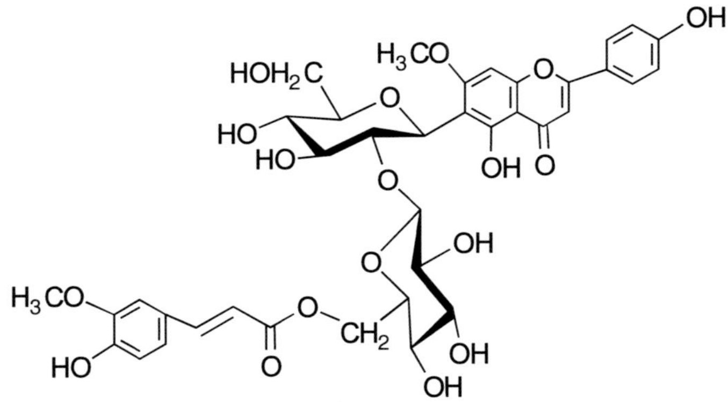

6'''-feruloyl spinosin (**88**)

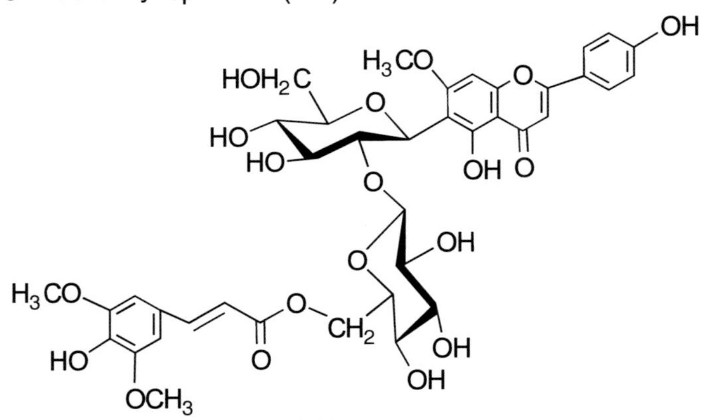

6'''-sinapolyspinosin (**89**)

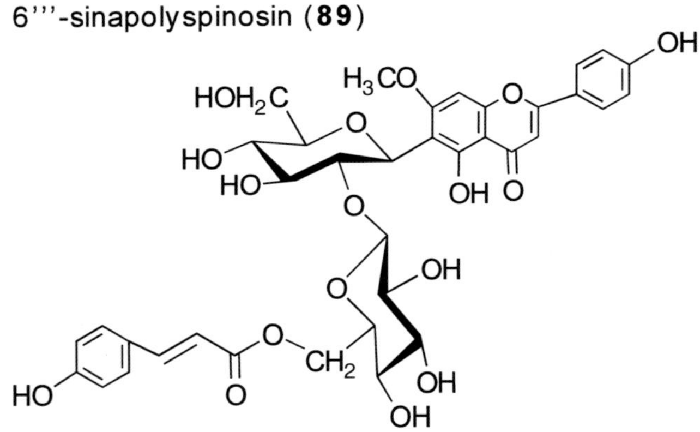

6'''-*p*-coumaloylspinosin (**90**)

Figure 21 6'''-Feruloylspinosin (**88**), 6'''-sinapolyspinosin (**89**) and 6'''-*p*-coumaloylspinosin (**90**)

1. On their Content of Flavonoids

1.1. On their Content of Flavonoids in Three Jujube Fruits such as Boeun-deachu, Mechu and Sanzoin Cultivars

First, their total flavonoid contents (g/100 g dry jujube) ranged from 0.7 to 1.8 in the pulp by high-pressure liquid chromatography.

Second, their contents (mg/100 g dry jujube) of individual flavonoids in three jujube pulps were identified by HPLC elution position, and UV/vis and mass spectra as follows: The content (mg/100 g dry jujube weight) of procyanidin B2 (**95**) was 30.7 for Boeun-deachu, 28.0 for Mechu and 14.0 for Sanzoin, respectively. The content (mg/100 g dry jujube weight) of (-)-epicatechin (**96**) was 261.0 for Boeun-deachu, 258.8 for Mechu and 352.7 for Sanzoin, respectively. The content (mg/100 g dry jujube weight) of unidentified flavonoid I (UIS-I. **97**) was 43.1 for Boeun-deachu, 46.4 for Mechu and 138.1 for Sanzoin, respectively. The content (mg/100 g dry jujube weight) of quercetin-3-*O*-rutinoside (rutin. **98**) of a glycoside of rutinose was 295.5 for Boeun-deachu, 707.5 for Mechu and 1147.0 for Sanzoin, respectively. The content (mg/100 g dry jujube weight) of quercetin-3-*O*-galactoside (hyperin. **99**) was 4.7 for Boeun-deachu, 1.4 for Mechu and 15.7 for Sanzoin, respectively. The content (mg/100 g dry jujube weight) of kaempferol-glucosyl-(1→2)rhamnoside (**100**) was 18.7 for Boeun-deachu, 32.6 for Mechu and 31.3 for Sanzoin, respectively. The content (mg/100 g dry jujube weight) of unidentified flavonoid II (UIS-II. **101**) was 36.7 for Boeun-deachu, 146.7 for Mechu and 95.1 for Sanzoin, respectively. Among the contents (mg/100 g dry jujube weight) of total 7 flavonoids on three jujube fruit extracts of Boeun-deachu, Mechu and Sanzoin cultivars, the content of Sanzoin cultivar was the highest 1744, followed by 1223 for Mechu and the lowest 690.4 of Boeun-deachu (Figure 23).

1.2. On their Content of Flavonoids in Two Jujube Seed Extracts such as Mechu and Sanzoin Cultivars

First, their total flavonoid contents (g/100 g dry jujube weight) ranged from 3.2 to 4.0 in the seeds by high-pressure liquid chromatography (HPLC).

Second, their contents (mg/100 g dry jujube weight) of individual flavonoids in two jujube seeds were identified by high-pressure liquid chromatography (HPLC) elution position, UV/vis spectra and mass spectra as follows: The content (mg/100 g dry jujube weight) of saponarin (**102**) was 170.4 for Mechu and 133.0 for Sanzoin, respectively. The content (mg/100 g dry jujube weight) of spinosin (**84**) was 1303.8 for Mechu and 2237.7 for Sanzoin, respectively. The content (mg/100 g dry jujube weight) of vitexin (**103**) was 134.0 for Mechu and 44.7 for Sanzoin, respectively. The content (mg/100 g dry jujube weight) of swertish (**104**) was 10.9 for Mechu and 14.8 for Sanzoin, respectively. The content (mg/100 g dry jujube weight) of 6'''-hydroxybenzoylspinosin (**105**) was 84.6 for Mechu and 54.0 for Sanzoin, respectively. The content (mg/100 g dry jujube weight) of 6'''-feruloylspinosin (**88**) was 1237.3 for Mechu and 1242.9 for Sanzoin, respectively. The content (mg/100 g dry jujube weight) of unidentified substance (UIS. **106**) was 256.9 for Mechu and 271.1 for Sanzoin, respectively.

Among the contents (mg/100 g dry jujube weight) of total 7 flavonoids on two jujube seed extracts of Mechu and Sanzoin cultivars, the total flavonoid contents (3998.2) of Sanzoin cultivar was higher when compared to that of 3197.9 for Mechu. Table 2 showed total phenol (TP) and two antioxidant activities (FRAP, DPPH) from pulps and seeds of *Zizyphus jujuba* when compared to butylhydroxyanisole (BHA. **107**) (Figure 24) (Table 2).

2. On Antioxidative Activity

Their antioxidative potential of both jujube fruits and jujube seeds were measured by these two assays of the ferric reducing/antioxidant power (FRAP) assay and 2,2-diphenyl-1-picrylhydrazyl (DPPH) radical scavenging assay.

First, their content values (g/100 g dry jujube weight) of total phenolic compounds of three jujube *pulps* by a modified colorimetric Folin-Ciocalteu method was 1.13 for Boeun-deachu, 1.93 for Mechu and 2.36 for Sanzoin, respectively.

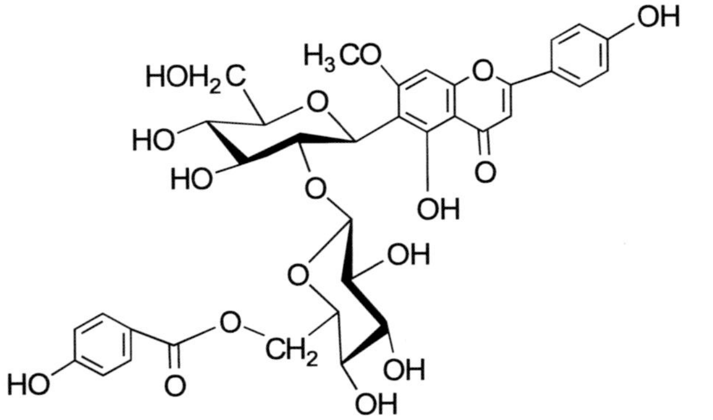

6'''-*p*-hydroxybenzoylspinosin (apigenin 6-*C*-[(6-*O*-*p*-hydroxybenzoyl) β-D-glucopyranosyl(1→2)] β-D-glucopyranoside. **91**)

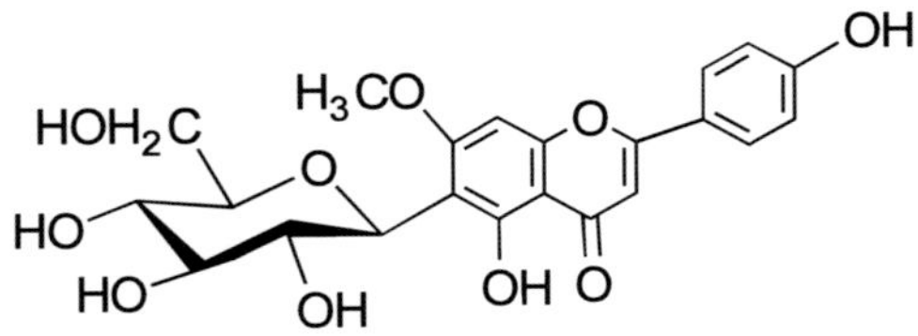

swertisin (**92**)

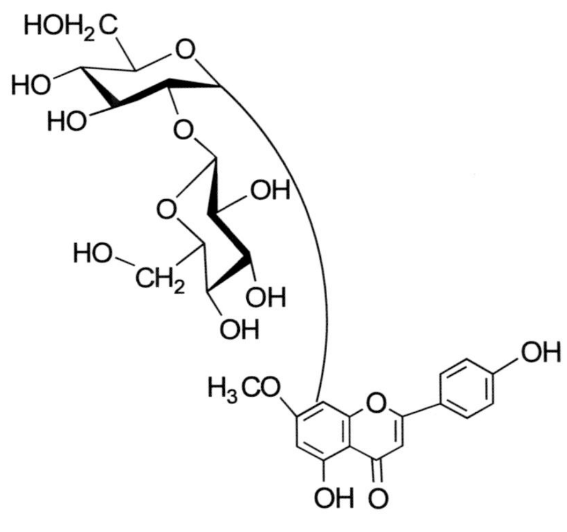

2''-*O*-glucosylisoswertisin (**93**)

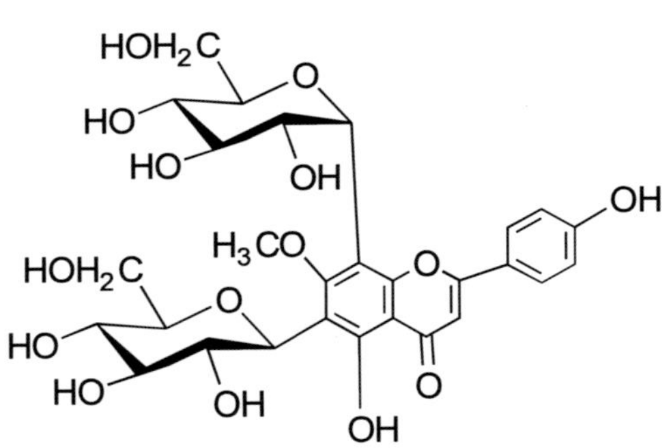

vecenin-2 (**94**)

Figure 22 6'''-*p*-Hydroxybenzoylspinosin (apigenin 6-*C*-[(6-*O*-*p*-hydroxybenzoyl) β-D-glucopyranosyl(1→2)] β-D-glucopyranoside. **91**), swertisin (**92**), 2''-*O*-glucosylisoswertisin (**93**) and vecenin-2 (di-*C*-β-D-glucopyranosyl-6,8 apigenin. **94**)

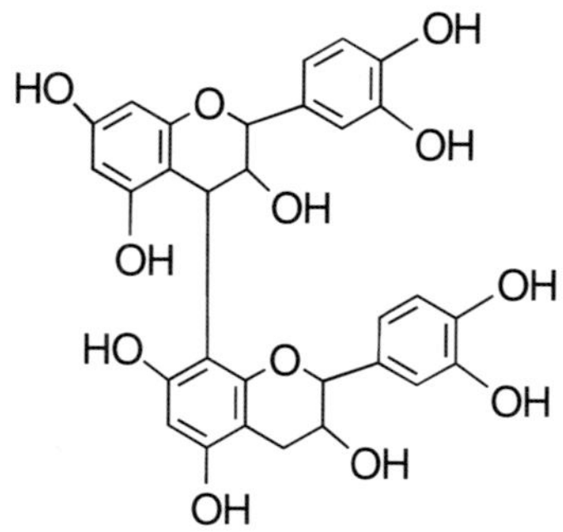

procyanidin B2 (**95**)

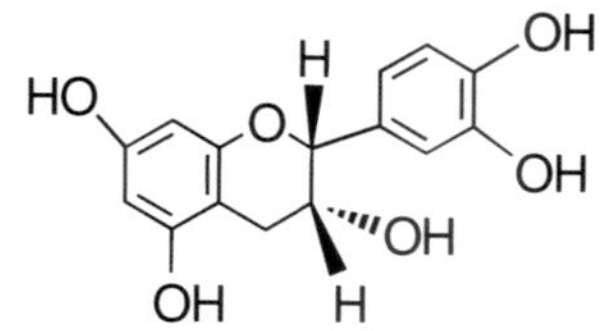

(-)-epicatechin (**96**)

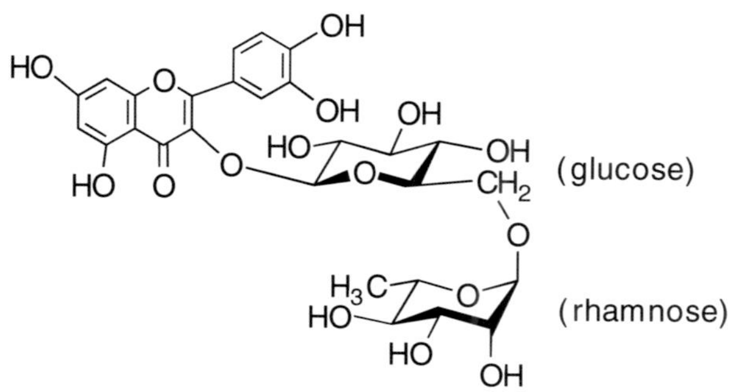

quercetin-3-*O*-rutinoside (rutin. **98**)

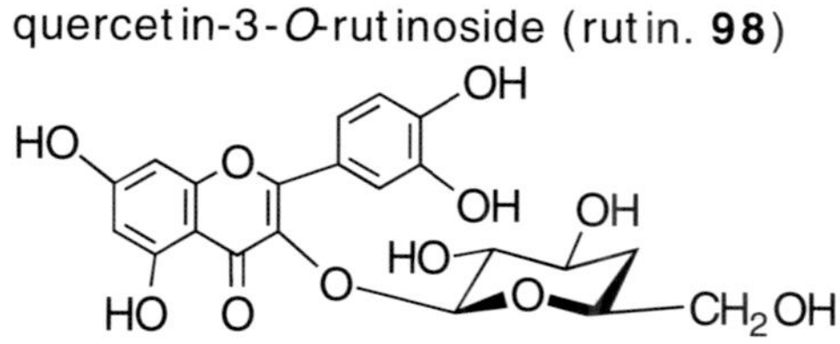

quercetin-3-*O*-galactoside (hyperin. **99**)

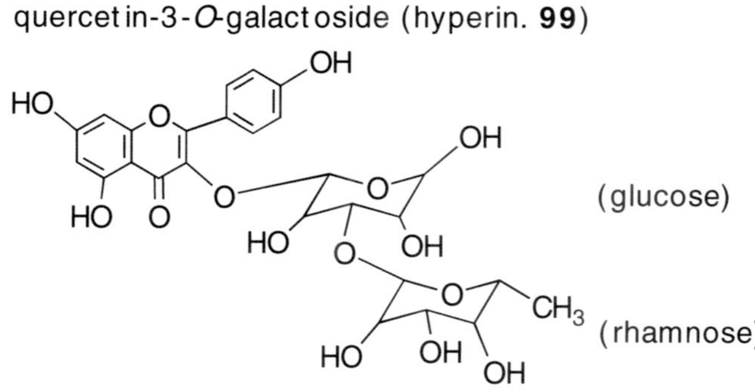

kaempferol-glucosyl-(1→2)rhamnoside (**100**)

Figure 23 Procyanidin B2 (**95**), (-)-epicatechin (**96**), quercetin-3-*O*-rutinoside (rutin. **98**), quercetin-3-*O*-galactoside (hyperin. **99**) and kaempferol-glucosyl-(1→2)rhamnoside (**100**) from jujube pulps

Their molar values (mol/100 g dry jujube weight) of total phenolic compounds of two jujube *seeds* by a modified colorimetric Folin-Ciocalteu method was 3.61 for Mechu and 4.62 for Sanzoin, respectively (Table 2).

Second, their molar value (Fe^{2+} molar value) (mol Fe^{2+}/100 g dry jujube weight) of FRAP value of three jujube *pulps* by ferric reducing/antioxidant power (FRAP) assay was 3.2 for Boeun-deachu, 9.3 for Mechu and 9.0 for Sanzoin, respectively.

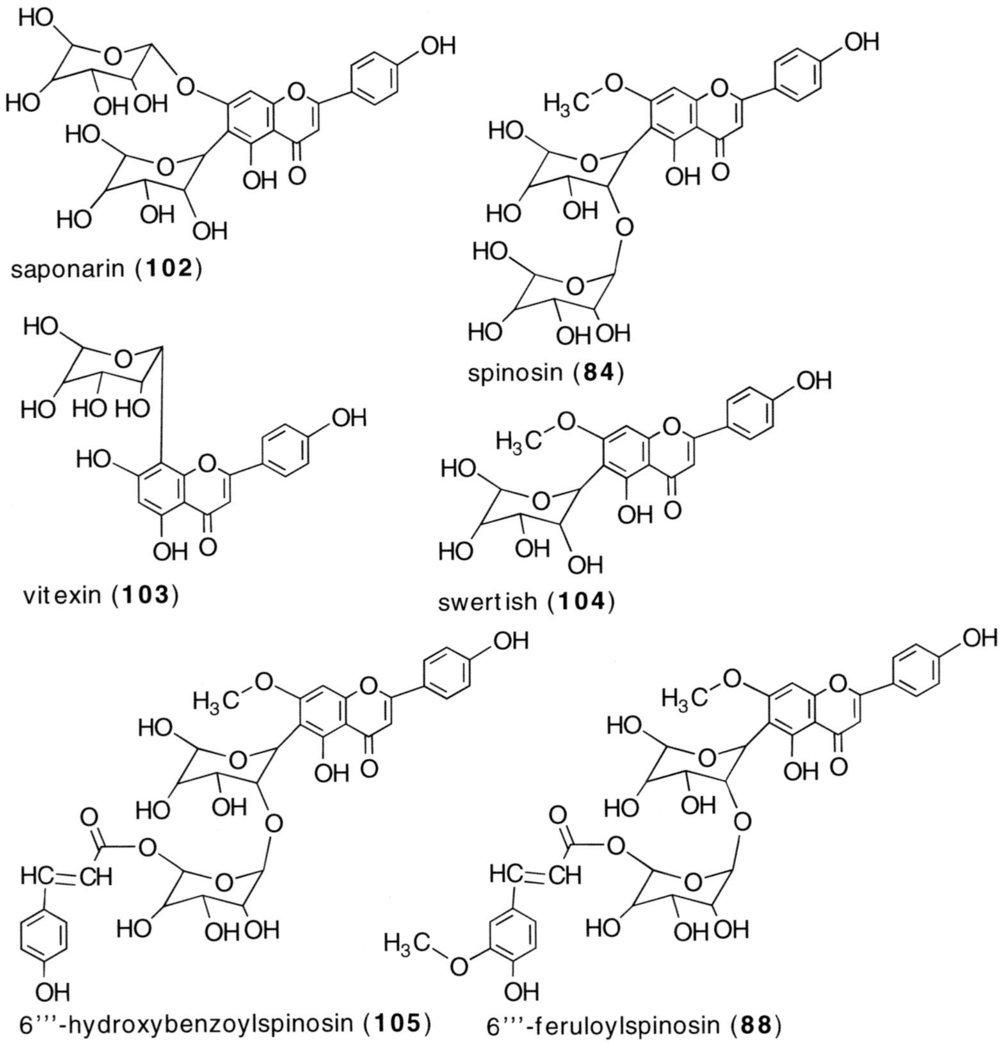

Figure 24 Saponarin (**102**), spinosin (**84**), vitexin (**103**), swertish (**104**), 6'''-hydroxybenzoylspinosin (**105**) and 6'''-feruloylspinosin (**88**) from jujube seeds

Their molar values (Fe^{2+} molar value) (mol Fe^{2+}/100 g dry jujube weight) of FRAP value of two jujube *seeds* by ferric reducing/antioxidant power (FRAP) assay was 9.9 for Mechu and 13.2 for Sanzoin, respectively.

Their values (Fe^{2+} molar value) (mol Fe^{2+}/100 g dry jujube weight) of FRAP value in three jujube fruits and two jujube seeds ranged from 3.2 with the lowest antioxidative effect for Boeun-deachu fruit to 13.2 with the highest antioxidative effect for Sanzoin seed.

Therefore, 13.2 FRAP value for Sanzoin seed was a 4.1-fold when compared to 3.2 FRAP value for Boeun-deachu fruit (Table 2).

Third, 50% inhibitory concentration (%. IC_{50}(μg/mL)) of DPPH radical scavenging effect of three jujube *pulps* by 2,2-diphenyl-1-picrylhydrazyl (DPPH) radical scavenging assay was 343.8 for Boeun-deachu, 115.6 for Mechu and 145.8 for Sanzoin, respectively, when compared to 3.95 of a standard butylhydroxyanisole (BHA. **107**).

Fifty percent50% inhibitory concentration (%. IC_{50}(μg/mL)) of DPPH radical scavenging effect of two jujube *seeds* by 2,2-diphenyl-1-picrylhydrazyl (DPPH) radical scavenging assay was 310.2 for Mechu and 195.4 for Sanzoin, respectively, when compared to 3.95 of a standard butylhydroxyanisole (BHA. **107**).

Their calculated 50% inhibitory concentration (%. IC_{50}(μg/mL)) in three jujube fruits and two jujube seeds ranged from 343.8 with the lowest antioxidative effect for Boeun-deachu pulp to 115.6 with the highest antioxidative effect for Mechu pulp. Therefore, 115.6 DPPH radical scavenging value (DPPH value) for Mechu pulp was a 2.97-fold when compared to 343.8 DPPH radical scavenging value (DPPH value) for Boeun-deachu pulp (Table 2) [65].

Table 2. Total phenol (TP) and two antioxidant activitiesw (FRAP, DPPH) from pulps and seeds of *Zizyphus jujuba*

Variety of Jujube	TP (g/100 g dry weight)	FRAP (mol Fe^{2+}/100 g dry weight)	DPPH radical scavenging effect (%. IC_{50}(μg/mL))
Jujube pups			
Boeun-deachu	1.13	3.2	343.8
Mechu	1.93	9.3	115.6
Sanzoin	2.36	9.0	145.8
Jujube seeds			
Mechu	3.61	9.9	310.2
Sanzoin	4.62	13.2	195.4
BHA (a control)			4.0

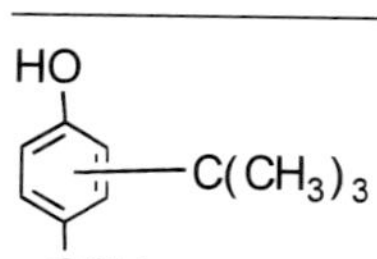

butylhydroxyanisole (BHA. **107**)

FRAP: ferric reducing/antioxidant power

DPPH: 2,2-diphenyl-1-picrylhydrazyl (DPPH) radical scavenging assay

3.6.1.4. Flavonoids in Chinese Jujubes from Slovenia

In 2008, a group from the of Department of Agronomy, Biotechnical Faculty, University of Ljubljana, Jamnikarjeva 101, SI-1000 Ljubljana, Slovenia studied their phenolics in the different varieties of Chinese jujube (*Ziziphus jujuba* Mill.) fruits.

In 2004, eight Chinese jujube (*Ziziphus jujuba* Mill.) fruits were harvested at the time of the fully mature stage in their jujube orchards of China. After that, their harvested jujubes were sundried and their phenolics in these jujubes were examined. Phenolics were extracted from the following seven varieties of *Ziziphus jujuba* Mill. such as Bianhesuanzao, Yuanlingzao, Fupingdazao, Zanhuangdazao, Zizao, Huizao and Jinsixiaozao varieties, and an acid jujube variety of *Ziziphus jujuba* Mill var. spinosa (Bunge) Hu et H. F. Chou.

Photo 6. Dried jujubes from Laoling in China. Photographed by Noboru Motohashi, 11/30/2011 Wed <492. IMGP6245(dried jujube, Laoling from China.JPG>.

Two phenolic acids such as chlorogenic acid (**7**) and caffeic acid (**8**) come from the hydroxycinnamate, and three flavonoids such as (+)-catechin (**108**), (-)-epicatechin (**96**) and quercetin-3-*O*-rutinoside (rutin. **98**) were determined from their extracts of eight dried Chinese jujube fruits by high-performance liquid chromatography (HPLC).

A difference inon their contents of phenolics was observed among the total eight jujube varieties. The fruit of acid jujube variety had the highest contents of hydroxycinnamic acids, as well as (+)-catechin (**108**), (-)-epicatechin (**96**) and quercetin-3-*O*-rutinoside (rutin. **98**) when compared to other 7 jujube varieties. On the other hand, Zizao variety had the highest (+)-catechin (**108**) content.

From these results, among eight jujube varieties, acid jujube variety might be the richest source of phenolics, while Jinsixiaozao variety was the poorest source. Generally, jujube fruits might be a more richer source of phenolics such as especially flavonoids when compared to the content of phenolics of prunes. Based on the facts, it suggests that jujube fruits as one of functional foods in the daily diets might be beneficialent for their maintenance of human health. (Photo 6) [66].

Generally, (+)-catechin (**108**) has been used as an anticoagulant, antioxidant, hepatoprotective and a substance immunostimulant which stimulates the immune response [67].

Interestingly, in 2000, (+)-catechin (**108**) was also found to be act as an infection-inhibiting factor in strawberry leaf (Figure 25) [68].

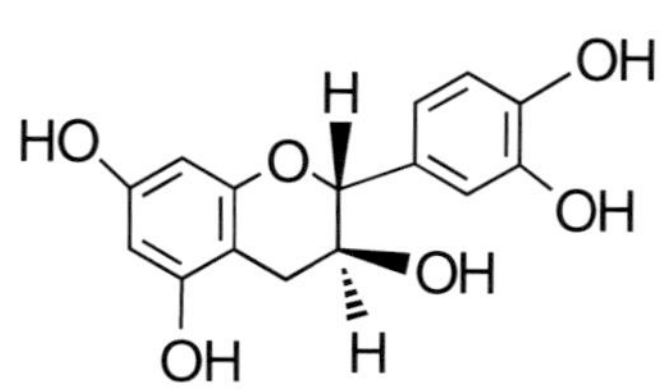

(+)-catechin (**108**)

Figure 25 (+)-Catechin (**108**) in Chinese jujube (*Ziziphus jujuba* Mill.) fruits

3.6.1.5. Flavonoids in Jujubes from Turkey

In 2010, a group from theof Suleyman Demirel University, GURBUZ Vocational School, Sutculer, Isparta, Turkey measured the content of the flavonoids on four promising jujube (*Ziziphus jujuba* Miller) selections.

Phenolic compounds, α-tocopherol (**109**), β-carotene (**110**) and fatty acid composition of four jujube selections were investigated in this study. Phenolic compounds, α-tocopherol (**109**) and β-carotene (**110**) were analyzed with an HPLC device equipped with a diode array detector, and the fatty acid methyl esters (**111**) were analyzed with QP 5050 GC/MS equipped with a CP-Wax 52 CB column. Among the jujube selections, considerable differences in phenolic and fatty acid contents were found.

First, (+)-catechin (**108**), caffeic acid (**8**) of phenolic acids, *p*-coumaric acid (**112**), ferulic acid (**113**) of plant polyphenols, quercetin-3-*O*-rutinoside (rutin. **98**), apigenin-7-*O*-glucoside (**114**), eriodictyol (**115**) of a bitter-masking flavanone, quercetin (**116**), *p*-hydroxybenzoic acid (**74**), chlorogenic acid (**7**) and syringic acid (**117**) of a naturally occurring *O*-methylated trihydroxybenzoic acid were isolated from *leaves.* Rutin content was very high in the leaves of all the jujube selections, ranging from 269.0 to 367.90 mg/100 g, followed by apigenin-7-*O*-glucoside (**114**) (22.90-49.38 mg/100 g) and eriodictyol (**115**) (5.06-6.27 mg/100 g).

Second, seven phenolic compounds such as (+)-catechin (**108**), caffeic acid (**8**), (-)-epicatechin (**96**), ferulic acid (**113**), quercetin-3-*O*-rutinoside (rutin. **98**), *p*-hydroxybenzoic acid (**74**) and chlorogenic acid (**7**) were isolated from *fruits* of jujube selections. (+)-Catechin (**108**) level ranged from 2.46 to 3.74 mg/100 g, and quercetin-3-*O*-rutinoside (rutin. **98**) level ranged from 0.88 to 3.60 mg/100 g for fruits. Predominant phenolics were quercetin-3-*O*-

rutinoside (rutin. **98**) and apigenin-7-*O*-glucoside (**114**) for *leaves*. Predominant phenolics were (+)-catechin (**108**) and quercetin-3-*O*-rutinoside (rutin. **98**) for *fruits*. Jujube leaves contained higher amounts of phenolic compounds when compared to amounts of phenolic compounds of jujube fruits.

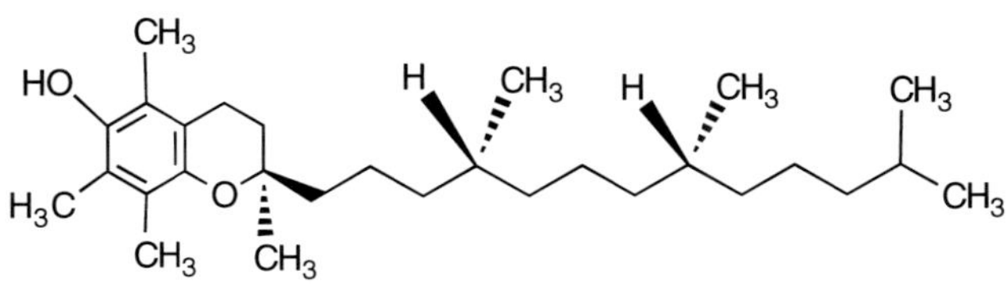

α-tocopherol (**109**)

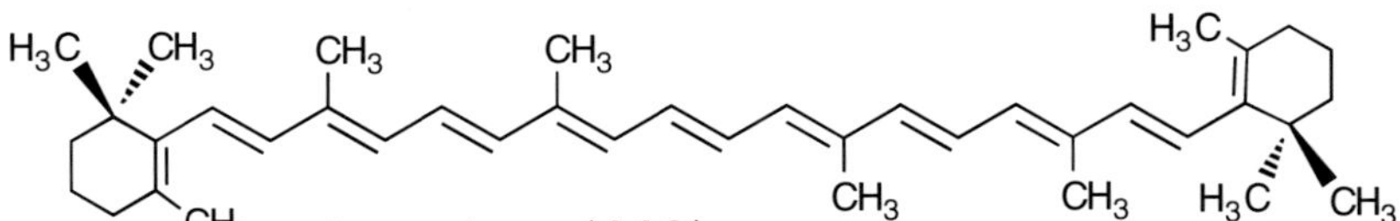

β-carotene (**110**)

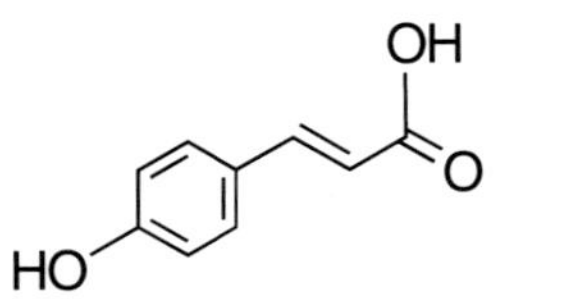

p-coumaric acid (**112**)

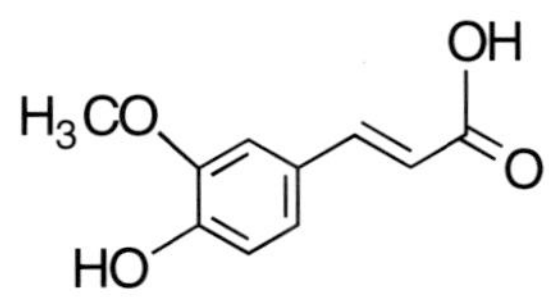

ferulic acid (**113**)

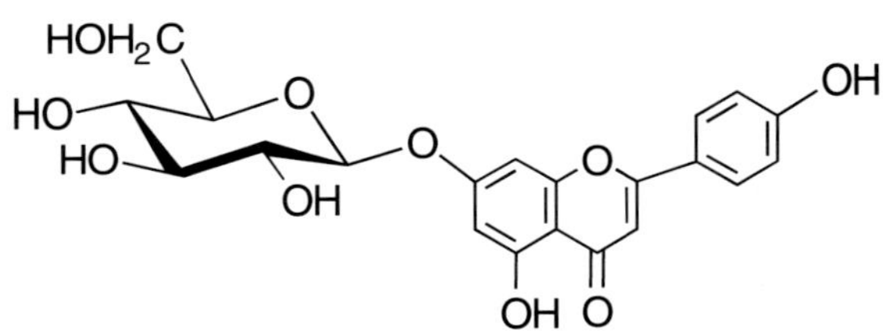

apigenin 7-*O*-glucoside (**114**)

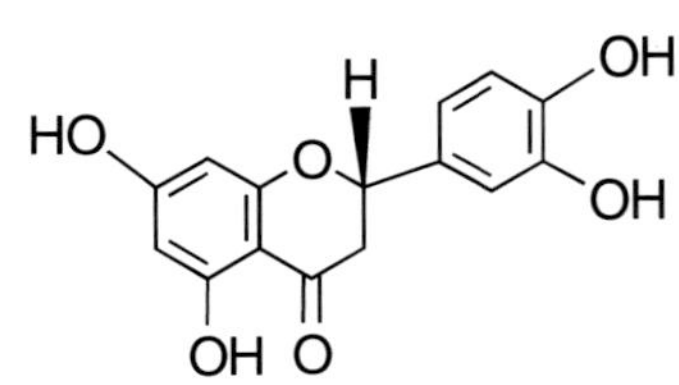

eriodictyol (**115**)

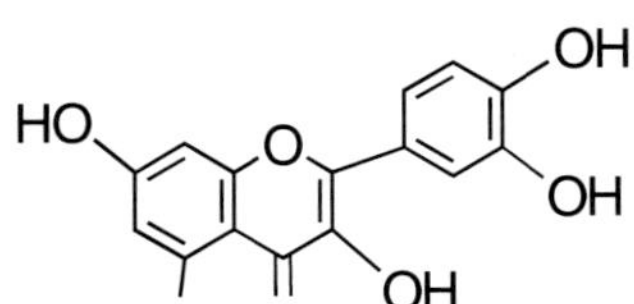

quercetin (**116**)

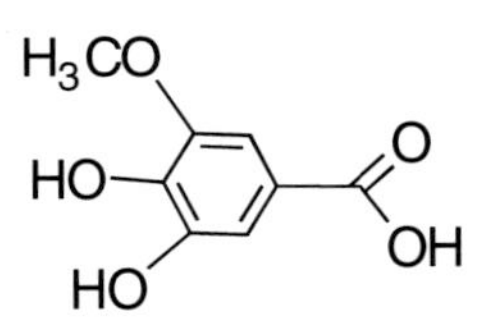

syringic acid (**117**)

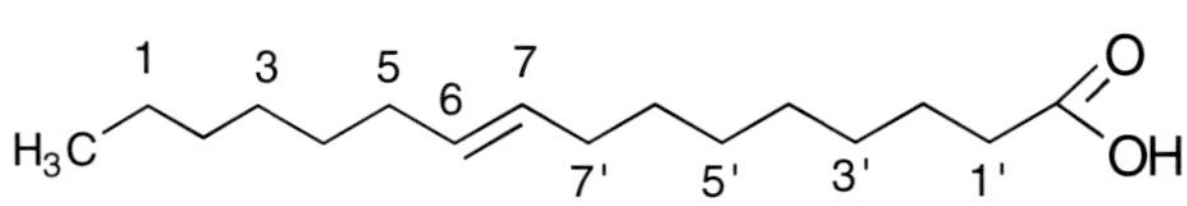

palmitoleic acid (**118**)

Figure 26 α-Tocopherol (**109**), *β*-carotene (**110**), *p*-coumaric acid (**112**), ferulic acid (**113**), apigenin 7-*O*-glucoside (**114**), eriodictyol (**115**), quercetin (**116**), syringic acid (**117**) and palmitoleic acid (**118**).

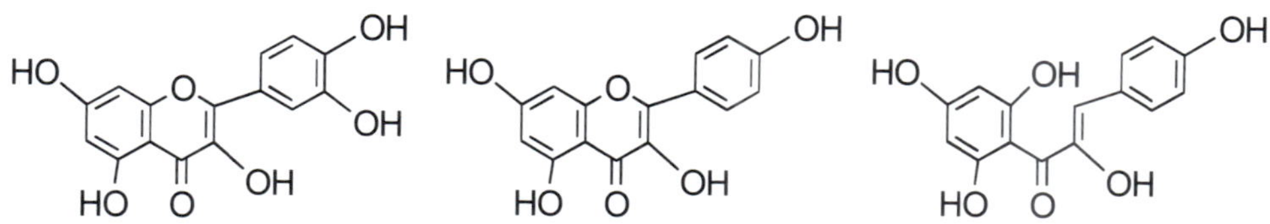

quercetin (**116**) kaempferol (**119**) phloretin (**130**)

Figure 27 Two quercetin (**116**) and kaempferol (**119**) of flavonoids, and phloretin (**120**) of chalcones from jujube pulps

Third, α-tocopherol (**109**) was only detected in fruits of selections 20-Ç-22 and 20-Ç-52 (0.04 and 0.07 mg/100 g, respectively).

Fourth, *β-carotene (110) in jujube fruits* was significantly higher in 20-Ç-22 (35.0 μg/100 g) when compared to β-carotene (**110**) in three other selections.

Fifth, lipid content of the fresh fruits ranged from 0.06% to 0.10% among the four jujube selections. The predominant fatty acids in all four jujube selections were oleic acid (ω-9. **61**), linoleic acid (ω-6. **62**), palmitic acid (**64**) and palmitoleic acid (**118**) of a monounsaturated fatty acid, having 16 carbon atoms and one double bond, found in the adipose tissue of all animals. Unsaturated fatty acids comprised 68.54–72.44% of the total fat in jujube fruit (Figure 26) [69].

Photo 7. Jujube tea from Laoling in China. Photographed by Noboru Motohashi, 11/30/2011 Wed <492. IMGP6247(jujube, tea, Laoling in China>.

3.6.1.6. Flavonoids in Jujubes from Italy

In 2009, a group from the of Dipartimento di Chimica Bioorganica e Biofarmacia, Università di Pisa, via Bonanno 33, 56126 Pisa, Italy examined their flavonoids in two ripe edible jujube fruits such as *Zizyphus jujuba* Miller (syn. *Z. sativa* Gaertner, *Z. vulgaris* Lam.) and Christ's thorn jujube *Zizyphus spina-christi* (L.) Willd by a high performance liquid chromatography (HPLC)/electrospray ionization mass spectrometry (ESI-MS) analyses. Two quercetin (**116**), kaempferol (**119**) derivatives of flavonoids, and one phloretin (**120**) derivative of chalcones were identified from both methanol extracts of jujubes (Figure 27) [70].

Generally, kaempferol (ケンペロール. **29→119**) has been used as 5-lipoxygenase-inhibitor, antioxidant, antihistaminic and hypotensive [67].

3.7. Precautions and Adverse Reactions

There are no knownir dose, health hazard and side effects ofby jujube- prescribed treatment have been no known by now [32].

3.8. Jujube Fruit Tea

In China and the USA and other countries of Asia, jujube fruit tea (Photo 7) has been effectively used for their improvementst of the diverse disease symptoms such as anxiety, insomnia, ulcers, loss of appetite, inflammation, diarrhea, dry skin and much more [71].

REFERENCES

[1] Sturrock, D. *Sturrock's Fruits for Southern Florida.* 146, 1980, Horticultural Books, Inc., Stuart, Florida, USA.

[2] Mehra, K.L., History of the use of jujube (*Zizyphus* Species) in ancient India. *Indian J Hort* 24, 33-37, 1967.

[3] Tomoda, M., Asakura, H., Iida, A. Water-soluble carbohydrates of *Zizyphi frucyus*. I. A comparative study on Japanese and Chinese *Zizyphi fructu. Syoyakugaku Zasshi* 23(2), 45-48, 1969.

[4] Catapano, A.L., Maggi, F.M., Tragni, E. Low density lipoprotein oxidation, antioxidants, and atherosclerosis. *Curr Opin Cardiol* 15(5), 355-363, 2000.

[5] Eder, K., Flader, D., Hirche, F., Brandsch, C. Excess dietary vitamin E lowers the activities of antioxidative enzymes in erythrocytes of rats fed salmon oil. *J Nutr* 132(11), 3400-3404, 2002.

[6] Libby, P. Inflammation in atherosclerosis. *Nature* 420(6917), 868-874, 2002.

[7] Ko, S.H., Choi, S.W., Ye, S.K., Yoo, A.S., Kim, H.S., Chung, M.H. Comparison of anti-oxidant activities of seventy herbs that have been used in Korean traditional medicine. *Nutrition Research and Practice* 2(3), 143-151, 2008.

[8] Kuliev, A.A., Guseinova, N.K. The content of vitamin C, B1, B2 and E in some fruits. *Referativnyi Zhurnal* 2, 69-73, 1974.

[9] Tomoda, M., Shimuju, N., Gonda, R. Pectic substances II. The location of O- acetyl groups and the Smith degradation of *Ziziphus* Pectin A. *Chem Pharm Bull* 33(9), 4017-4020, 1985.

[10] Peng, W.H., Hsieh, M.T., Lee, Y.S., Lin, Y.C., Liao, J. Anxiolytic effect of seed of *Ziziphus jujuba* in mouse models of anxiety. *J Ethnopharmacol* 72(3), 435-441, 2000.

[11] Na, M, An, R, Lee, S, Hong, N, Yoo, J, Lee, C, Kim, J, Bae, K. Screening of crude drugs for antioxidative activity. *Korean J Pharmacognosy* 32, 108-115, 2001.

[12] Otsuka, H, Ogihara, Y, Shibata, S. Isolation of coclaurine from *Zizyphus jujuba* by droplet counter-current chromatography. *Phytochemistry* 13, 2016, 1974.

[13] Gong Cheng, Yanjing Bai, Yuying Zhao, Jing Tao, Yi Liu, Guangzhong Tu, Libin Ma, Ning Liao and Xiaojie Xu. Flavonoids from *Ziziphus jujuba* Mill var. spinasa. *Tetrahedron* 56, 8915-8920, 2000.

[14] Su, B.N., Cuendet, M., Farnsworth, N.R., Fong, H.H., Pezzuto, J.M., Kinghorn, A.D. Activity-guided fractionation of the seeds of *Ziziphus jujuba* using a cyclooxygenase-2 inhibitory assay. *Planta Med* 68(12), 1125-1128, 2002.

[15] Suksamrarn, S, Panseeta, P, Kunchanawatta, S, Distaporn, T, Ruktasing, S, Suksamrarn, A. Ceanothane- and lupane-type triterpenes with antiplasmodial and antimycobacterial activities from *Ziziphus cambodiana*. *Chem Pharm Bull (Tokyo)* 54(4), 535-537, 2006.

[16] Lee, S.M., Park, J.G., Lee, Y.H., Lee, C.G., Min, B.S., Kim, J.H., Lee, H.K. Anti-complementary activity of triterpenoides from fruits of *Zizyphus jujuba*. *Biol Pharm Bull* 27(11), 1883-1886, 2004.

[17] Al-Reza, S.M., Bajpai, V.K., Kang, S.C. Antioxidant and anti-listerial effects of seed essential oil and organic extracts from *Zizyphus jujuba*. *Food Chem Toxicol* 47(9), 2374-2380, 2009.

[18] Huang, X.,, Kojima-Yuasa, A., Norikura, T., Kennedy, D.O., Hasuma, T., Matsui-Yuasa, I. Mechanism of the anti-cancer activity of *Zizyphus jujuba* in HepG2 cells. *Am J Chin Med* 35(3), 517-532, 2007.

[19] Lee, S.M., Min, B.S., Lee, C.G., Kim, K.S., Kho, Y.H. Cytotoxic triterpenoids from the fruits of *Zizyphus jujuba*. *Planta Med* 69(11), 1051-1054, 2003.

[20] Jiang, J.G., Huang, X.J., Chen, J. Separation and purification of saponins from Semen *Ziziphus jujuba* and their sedative and hypnotic effects. *J Pharm Pharmacol* 59(8), 1175-1180, 2007.

[21] Jiang, J.G., Huang, X.J., Chen, J., Lin, Q.S. Comparison of the sedative and hypnotic effects of flavonoids, saponins, and polysaccharides extracted from Semen *Ziziphus jujube*. *Nat Prod Res* 21(4), 310-320, 2007.

[22] Kubota, H, Morii, R, Kojima-Yuasa, A, Huang, X, Yano, Y, Matsui-Yuasa, I. Effect of *Zizyphus jujuba* extract on the inhibition of adipogenesis in 3T3-L1 preadipocytes. *Am J Chin Med* 37(3), 597-608, 2009.

[23] Al-Reza, S.M., Yoon, J.I., Kim, H.J., Kim, J.S., Kang, S.C. Anti-inflammatory activity of seed essential oil from *Zizyphus jujuba*. *Food Chem Toxicol* 48(2), 639-643, 2010.

[24] Yoon, J.I., Al-Reza, S.M., Kang, S.C. Hair growth promoting effect of *Zizyphus jujuba* essential oil. *Food Chem Toxicol* 48(5), 1350-1354, 2010.

[25] Li, J., Shan, L., Liu, Y., Fan, L., Ai, L. Screening of a functional polysaccharide from *Zizyphus Jujuba* cv. Jinsixiaozao and its property. *Int J Biol Macromol* 49(3), 255-259, 2011.

[26] Chang, S.C., Hsu, B.Y., Chen, B.H. Structural characterization of polysaccharides from *Zizyphus jujuba* and evaluation of antioxidant activity. *Int J Biol Macromol* 47(4), 445-453, 2010.

[27] Pahuja, M., Mehla, J., Reeta, K.H., Joshi, S., Gupta, Y.K. Hydroalcoholic extract of *Zizyphus jujuba* ameliorates seizures, oxidative stress, and cognitive impairment in experimental models of epilepsy in rats. *Epilepsy Behav* 21(4), 356-363, 2011.

[28] Zhang, H., Jiang, L., Ye, S., Ye, Y., Ren, F. Systematic evaluation of antioxidant capacities of the ethanolic extract of different tissues of jujube (*Ziziphus jujuba* Mill.) from China. *Food Chem Toxicol* 48(6), 1461-1465, 2010.

[29] Naftali, T., Feingelernt, H., Lesin, Y., Rauchwarger, A., Konikoff, F.M. *Ziziphus jujuba* extract for the treatment of chronic idiopathic constipation: a controlled clinical trial. *Digestion* 78(4), 224-228, 2008.

[30] Rahmatullah, M., Jahan, R., Khatun, M.A., Jahan, F.I., Azad, A.K., Bashar, A.B.M.A., Miajee, Z.U.M.E.U., Ahsan, S., Nahar, N., Ahmad, I., Chowdhury, M.H. A Pharmacological Evaluation of Medicinal Plants used by Folk Medicinal Practitioners of Station Purbo Para Village of Jamalpur Sadar Upazila in Jamalpur district, Bangladesh. *Am Eurasian J Sustain Agric* 4(2),170-195, 2010.

[31] Dai Yin-fang, Liu Cheng-jim. *Fruits as Medicine.* 49a-50a, 1999, Pelanduk Publications (M) Sdn. Bhd., Malaysia.

[32] PDR for Herbal Medicines. 2nd ed., 439b, 2000.

[33] Yan, J., Li, J., Zhao, H., Chen, N., Cao, J., Jiang, W. Effects of oligochitosan on post-harvest *Alternaria* rot, storage quality, and defense responses in Chinese jujube (*Zizyphus jujuba* Mill. cv. Dongzao) fruit. *J Food Prot* 74(5), 783-788, 2011.

[34] Reuter, G. Present and future of probiotics in Germany and in central Europe. *Biosci Microflora* 16, 43-51, 1997.

[35] Rui-Lin Huang, Zhe-Yuan Deng, Cheng-bo Yang, Yu-Long Yin, Ming Yong Xie, Guo-Yao Wu, Tie-Jun Li, Li-Li Li, Zhi-Ru Tang, Ping Kang, Zheng-Ping Hou, Dun Deng, Hua Xiang, Xiang Feng Kong, Yu-Ming Guo. Dietary oligochitosan supplementation enhances immune status of broilers. *J Sci Food Agric* 67(1), 153-159, 2007.

[36] Albersheim, P. *Plant Carbohydrate Biochemistry,* In Pridham JB (ed). Academic Press, London and New York, 1974, pp.150-156.

[37] Shimizu, N., Tomoda, M. Pectic substances. I. The major pectin from the fruits of *Zizyphus jujuba* MILLER var. inermis REHD. *Chem Pharm Bul*l 31(2), 499-506, 1983.

[38] Gunness, P., Gidley, M.J. Mechanisms underlying the cholesterol-lowering properties of soluble dietary fibre polysaccharides. *Food Funct* 1(2), 149-155, 2010.

[39] Hosseini, E., Grootaert, C., Verstraete, W., Van de Wiele, T. Propionate as a health-promoting microbial metabolite in the human gut. *Nutr Rev* 69(5), 245-258, 2011.

[40] Müller-Lissner, S.A., Kamm, M.A., Scarpignato, C., Wald, A. Myths and misconceptions about chronic constipation. *Am J Gastroenterol* 100(1), 232-242, 2005.

[41] Pandey, M.B., Singh, A.K., Singh, V.P., Pandey, V.B. Cyclopeptide alkaloids from Zizyphus sativa bark. *Nat Prod Res* 22(3), 219-221, 2008.

[42] Shah, A.H., Miana, G.A., Devi, S., Pandey, V.B. Sativanine-H: A new alkaloid from the bark of *Zizyphus sativa*. *Planta Med* (6), 500-501, 1986.

[43] Tschesche, R., Kaussmann, E.U., Fehlhaber, H.W. Alkaloids from Rhamnaceae. 13. Amphibine-B, -C, -D, and -E, four peptide alkaloids from *Ziziphus amphibia* A. Cheval. *Chem Ber* 105(9), 3094-3105, 1972.

[44] Singh, S., Pandey, M.B., Singh, J.P., Pandey, V.B. Peptide alkaloids from *Zizyphus sativa* bark. *J Asian Nat Prod Res* 8(8), 733-737, 2006.

[45] Ziyaev, R., Irgashev, T., Israilov, I.A., Abdullaev, N.D., Yunusov, M.S., Yunusov, S.Y. Alkaloids of *Ziziphus jujuba*. Structure of iusiphine and iusirine. *Khimiya Prirodnykh Soedinenii.* 2, 239-243, 1977.

[46] Ziyaev, R., Irgashev, T., Israilov, I.A., Abdullaev, N.D., Yunusov, M.S., Yunusov, S.Y. Alkaloids of *Ziziphus jujuba*. Alkaloids of *Ziziphus jujuba* the structure of juziphine and juzirine. *Chem Nat Comp* 13(2), 204-207, 1977.

[47] Ma, Y., Han, H., Nam, S.Y., Kim, Y.B., Hong, J.T., Yun, Y.P., Oh, K.W. Cyclopeptide alkaloid fraction from *Zizyphi Spinosi* Semen enhances pentobarbital-induced sleeping behaviors. *J Ethnopharmacol* 117(2), 318-324, 2008.

[48] Han, B.H., Park, M.H., Han, Y.N. Cyclic peptide and peptide alkaloids from seeds of *Ziziphus vulgaris*. *Phytochemistry* 29(10), 3315-3319, 1990.

[49] Jossang, A., Zahir, A., Diakite, D. Mauritine, J., aA cyclopeptide alkaloid from *Zizyphus mauritiana*. *Phytochemistry* 42(2), 565-567, 1996.

[50] Lee, S.S., Su, W.C., Liu, K.C. Cyclopeptide alkaloids from stems of *Paliurus ramossisimus*. *Phytochemistry* 58(8), 1271-1276, 2001.

[51] Han, B.H., Park, M.H., Park, J.H. Chemical and pharmacological studies on sedative cyclopeptide alkaloids in some Rhamnaceae plants. *Pure Appl Chem* 61(3), 443-448, 1989.

[52] Han, Y.N., Hwang, K.H., Han, B.H. Inhibition of calmodulin-dependent protein kinase II by cyclic and linear peptide alkaloids from *Zizyphus* species. *Arch Pharm Res* 28(2), 159-163, 2005.

[53] Suh, D.Y., Kim, Y.C., Kang, Y.H., Han, Y.N., Han, B.H. Metabolic cleavage of frangufoline in rodents: *in vitro* and *in vivo* study. *J Nat Prod* 60(3), 265-269, 1997.

[54] Lee, S.M., Min, B.S., Lee, C.G., Kim, K.S., Kho, Y.H. Cytotoxic triterpenoids from the fruits of *Zizyphus jujuba*. *Planta Med* 69(11), 1051-1054, 2003.

[55] Kawaguchi, T., Kin, T. The components in jujube seeds. *Yakugaku Zasshi* 60, 343-352, 1940.

[56] Lee, S.M., Park, J.G., Lee, Y.H., Lee, C.G., Min, B.S., Kim, J.H., Lee, H.K. Anti-complementary activity of triterpenoides from fruits of *Zizyphus jujuba*. *Biol Pharm Bull* 27(11), 1883-1886, 2004.

[57] Juan, M.E., Planas, J.M., Ruiz-Gutierrez, V., Daniel, H., Wenzel, U. Antiproliferative and apoptosis-inducing effects of maslinic and oleanolic acids, two pentacyclic triterpenes from olives, on HT-29 colon cancer cells. *Br J Nutr* 100(1), 36-43, 2008.

[58] Tanaka, Y., Sanada, S. Studies on the constituents of *Ziziphus jujuba* Miller. *Syoyakugaku Zasshi* 45(2), 148-152, 1991.

[59] Kawai, K., Akiyama, T., Ogihara, Y., Shibata, S. A new sapogenin in the saponins of *Zizyphus jujuba*, *Hovenia dulcis* and *Bacopa monniera*. *Phytochemistry* 13, 2829-2832, 1974.

[60] Wang, B.N., Liu, H.F., Zheng, J.B., Fan, M.T., Cao, W. Distribution of phenolic acids in different tissues of jujube and their antioxidant activity. *J Agric Food Chem* 59(4), 1288-1292, 2011.

[61] Okamura, N., Yagi, A., Nishioka, I. Studies on the constituents of *Zizyphi Fructus*. V. Structures of glycosides of benzyl alcohol, vomifoliol and naringenin. *Chem Pharm Bull* 29, 3507-3514, 1981.

[62] Hatano, T., Kusuda, M., Inada, K., Ogawa, T.O., Shiota, S., Tsuchiya, T., Yoshida, T. Effects of tannins and related polyphenols on methicillin-resistant *Staphylococcus aureus*. *Phytochemistry* 66(17), 2047-2055, 2005.

[63] Woo, W.S., Kang, S.S., Shim, S.H., Wagner, H., Chari, V.M., Seligmann, O., Obermeier, G. The structure of spinosin (2''-*O*-β-glucosylswertisin) from *Zizyphus vulgaris* var. spinosus. *Phytochemistry* 18(2), 353-355, 1979.

[64] Woo, W.S., Kang, S.S., Wagner, H., Seligmann, O., Chari, V.M. Acylated flavone-*C*-glycosides from the seeds of *Zizyphus jujuba*. *Phytochemistry* 19, 2791-2793, 1980.

[65] Choi, S.H., Ahn, J.B., Kozukue, N., Levin, C.E., Friedman, M. Distribution of free amino acids, flavonoids, total phenolics, and antioxidative activities of Jujube (*Ziziphus jujuba*) fruits and seeds harvested from plants grown in Korea. *J Agric Food Chem* 59(12), 6594-6604, 2011.

[66] Hudina, M., Liu, M., Veberic, R., Stampar, F., Colaric, M. Phenolic compounds in the fruit of different varieties of Chinese jujube (*Ziziphus jujuba* Mill.). *J Hortic Sci Biotechnol* 83(3), 305-308, 2008.

[67] Gary Null. The Clinician's Handbook of Natural Healing. 676, 1997.

[68] Yamamoto, M., Nakatsuka, S., Otani, H., Kohmoto, K., Nishimura, S. (+)-catechin acts as an infection-inhibiting factor in strawberry leaf. *Phytopathology* 90(6), 595-600, 2000.

[69] San, B., Yildirim, A.N. Phenolic, alpha-tocopherol, beta-carotene and fatty acid composition of four promising jujube (*Ziziphus jujuba* Miller) selections. *J Food Compos Anal* 23, 706-710, 2010.

[70] Pawlowska, A.M., Camangi, F., Bader, A., Braca, A. Flavonoids of *Zizyphus jujuba* L. and *Zizyphus spina-christi* (L.) Willd (Rhamnaceae) fruits. *Food Chem* 112, 858-862, 2009.

[71] http://www.zooscape.com/cgi-bin/maitred/GreenCanyon/questc10 1240/r01.

In: Anthocyanins: Structure, Biosynthesis and Health Benefits ISBN: 978-1-62257-329-5
Editor: Noboru Motohashi

Chapter 10

STRUCTURAL DIVERSITY OF ANTHOCYANINS IN FRUITS

***Vlasios Goulas*[1], *Ariel R. Vicente*[2,3] *and George A. Manganaris*[1*]**

[1]Cyprus University of Technology, Department of Agricultural Sciences, Biotechnology and Food Science, Lemesos, Cyprus

[2]Centro de Investigación y Desarrollo en Criotecnología de Alimentos, Facultad de Ciencias Exactas, CONICET-UNLP, La Plata, Argentina

[3]Cátedra de Agroindustrias, Facultad de Ciencias Agrarias y Forestales Universidad Nacional de La Plata, Argentina

ABSTRACT

Anthocyanins are water-soluble pigments responsible for blue, purple and red color of many fruits such as berries, grapes, cherries, pomegranates, plums, apples, and some citrus and tropical fruits. Anthocyanins have ecological functions as attractants to pollinators and/or predators,predators; provide protection against excessive radiation, act as antimicrobials and most promptly they have high antioxidant potencycapacity. Chemically, anthocyanins belong to a parent class of molecules, the flavonoids, and are glycosiydes containing a sugar moiety and an aglycone unit (the anthocyanidin) which is derived from the flavylium ion. More than 635 different anthocyanins and 23 anthocyanidins by now have been already described. The structural differences among anthocyanins are related to the number of hydroxyl or methoxyl groups in the anthocyanidin skeleton, the position and the number of bonded sugar residues as well as by the aliphatic or aromatic carboxylates bonded to them. The distribution of the six most common anthocyanidins in fruits is: cyanidin 30%, delphinidin 22%, pelargonidin 18%, peonidin 7.5%, malvidin 7.5% and petunidin 5%. The conformation of anthocyanins has great impact in their color, stability, copigmentat co-pigmentation capacity and antioxidant properties. The most common glycoside derivatives are 3-monosides, 3-biosides, 3,5 and 3,7-diglucosides. The 3-glucosyl derivatives of the three non-methylated anthocyanidins (cyanidin, delphinidin, pelargononidin) are the most common

[*] Corresponding author: george.manganaris@cut.ac.cy.

in nature. The content and diversity of anthocyanins in fruits are affected by genetic factors, environmental conditions and agricultural practices. Harvest maturity, manipulation, storage conditions (temperature, humidity), postharvest treatments as well as processing can deeply affect the steady state content of anthocyanins in fruits. In the current chapter, the distribution of anthocyanins in fruits is also described. In addition, the effect of pre- and postharvest conditions and/or treatments on their accumulation and fate is analyzed.

1. Functions of Anthocyanins

The term anthocyanin is derived from the Greek words *anthos* and *cyan* that mean flower and blue, respectively [1]. Anthocyanins are good absorbers of visible light, thus appearing as colored substances. Their peaks of radiation absorption fall between 270-280 nm and 460-550 nm in the ultraviolet (UV) and visible regions [2]. Anthocyanins are among the most important groups of visible plant pigments, together with carotenoids and chlorophylls [3]. Anthocyanins are predominant abundant in many reproductive organs, but may also accumulate in leaves, stems, and roots [4]. By giving pigmentation to flowers and to fruits, anthocyanins play a definite role in the pollination process and seed dispersal. Consequently they and hence they have considerable relevance in the interaction of plants with other organisms [5]. Given their strong absorption of light it has been shown that they anthocyanins can exert an important role in the protection against excessive UV radiation [6].

Anthocyanins have been suggested to control pathogens and act also as antioxidants, protecting cells against the deleterious oxidative reactions [7,8]. The *in vitro* antioxidant capacity of anthocyanins is in great part associated with their ability to donate an H atom from an aromatic hydroxyl group to the free radicals, and to delocalize an unpaired electrons. Anthocyanins have been also suggested to have anti-carcinogenic and anti-inflammatory activities, thus participating in the prevention of cardiovascular diseases, obesity, and diabetes. Most recognized putative health-promoting effects are related to their antioxidant properties [4]. The increased awareness and interest about the potential impact of foods on health has generated renewed the interest in the study of anthocyanins [4].

Given their widespread distribution, anthocyanins are common antioxidants in the human diet. Their relative abundance and widespread distribution in diversefferent plant materials results in leads to a high dietary consumption of anthocyanins when compared to other flavonoids [4]. The daily intake of anthocyanins ranges from 3 to 215 mg (3 to 47 mg of the corresponding anthocyanidins) [9]. In the US, a daily consumption of 12.5 mg has been estimated [10]. After ingestion anthocyanins are incorporated to some extent [11]. Some works suggest that absorption of anthocyanins takes place in the stomach and small intestine. Compared to isoflavones, which represents the most bioavailable group of flavonoids, anthocyanins are poorly absorbed. While their uptake into the epithelial tissues seems to be quite efficient, transportation into circulation and tissue distribution is limited. Significant proportion of anthocyanins could be degraded *in vivo* into phenolic acids and an aldehydes [12].

2. Structure of Anthocyanins

Anthocyanins are glycosides containing a sugar moiety and an aglycone flavonoid unit, known as anthocyanidin [13]. Like all flavonoids, the anthocyanidin has two aromatic rings (A and B-rings) associated *via* C-C bonds through a third containing oxygen heterocyclic ring (C-ring) (Figure 1). The main structural difference of anthocyanidins with other flavonoids such as flavones, flavonols, flavanones and flavan-3-ols relates to the degree of oxidation of the C ring. In anthocyanidins, the aglycon part (anthocyanidin) is a positively charged oxonium compound [14], structurally based on the flavylium ion (2-phenylbenzopyrilium) [15], (Figure 2). Owing to their polarity, anthocyanins are readily soluble in water. Glycosylation increases water solubility, whereas acylation reduces it. For analytical purposes, anthocyanins are normally extracted from plant materials by using methanol that contains small amounts of hydrochloric or formic acids.

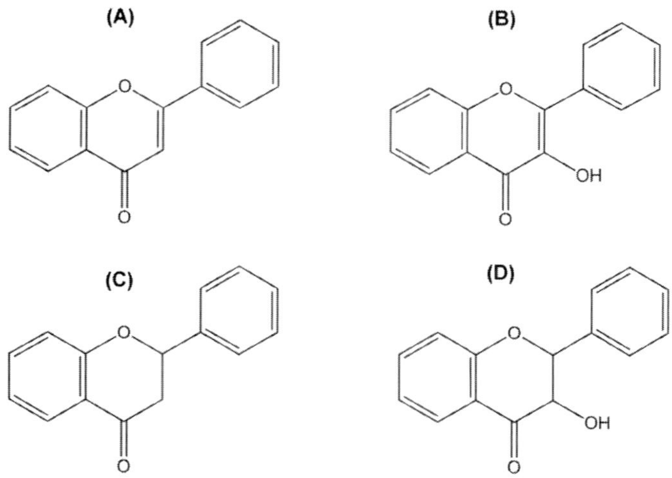

Figure 1. General structure of flavones (A), flavonols (B), flavanones (C) and flavanols (D).

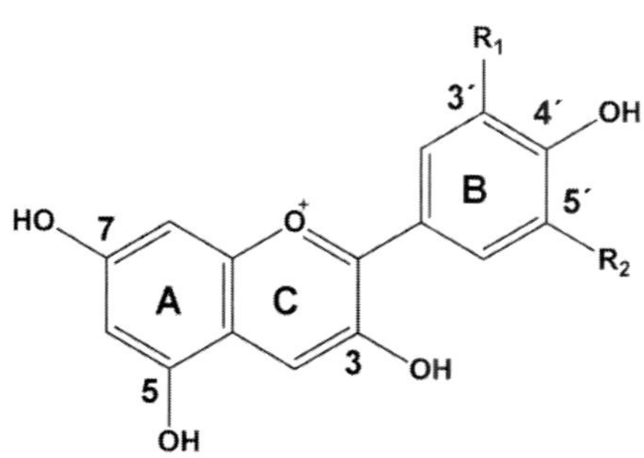

Anthocyanidin	R_1	R_2
Pelargonidin	H	H
Cyanidin	OH	H
Delphinidin	OH	OH
Peonidin	OCH_3	H
Petunidin	OH	OCH_3
Malvidin	OCH_3	OCH_3

Figure 2. Structures of most common anthocyanidins in fruits.

The great diversity of anthocyanins found in nature makes them a very complex and interesting group. More than 635 different anthocyanins, with a variety of building blocks, have been reported [4]. The distinction between individual anthocyanins is related to the anthocyanidin, which differs mainly on the number of hydroxyl groups and the methoxylation degree of the B ring [16]. Six anthocyanidins are most commonly found in higher plants: pelargonidin (Pg), peonidin (Pn), cyanidin (Cy), malvidin (Mv), petunidin (Pt) and delphinidin (Dp) (Figure 2) [10, 17]. Their distribution in fruits is: Cy 30%, Dp 22%, Pg 18%, Pn 7.5%, Mv 7.5% and Pt 5% [18]. Anthocyanin diversity could increase based on the nature and the number of sugars bonded to the specific anthocyanidin. Sugars may be present as mono-, di- or trisaccharides. Glucose, galactose, rhamnose and arabinose are the sugars most commonly present in anthocyanins. The sugar residues are commonly linked in position 3 of the aglycone through α– or β–linkages. Additional sugar moieties may be linked at positions 5 and 7, and more rarely at 3′ and 5′. The most widespread glycoside derivatives in nature are 3-monosides, 3-biosides, 3,5 and 3,7-diglycosides [3,19]. Ninety percent of anthocyanins contain glucose as the only accompanying sugar [18].

Acylating agents, including cinnamic acids such as caffeic, *p*-coumaric, ferulic and sinapic acid may be further associated with the sugars [19]. The presence of aliphatic acids (*e.g.* acetic, malic, malonic, oxalic and succinic acids) has been also mentioned [15, 20, 21].

3. Factors Affecting the Color of Anthocyanins

3.1. Anthocyanin Structure

The structure of anthocyanidins determines fruit color, producing shades from orange-red (pelargonidin) to blue-violet (delphinidin) at non-physiological pH 1. As a general rule, hydroxylation induces a shift of the visible max toward greater wavelengths (bathochromic effect, also known as blueing effect) [22]. On the other hand, mMethylation of hydroxyl groups causes a reverse trend. Consequently, the anthocyanidins with more hydroxyls are more bluish, while those with more methoxyl show intense red hues [23]. Glycosylation at C-3 comes along with a reduction of maximum absorption wavelength. The type and the number of glycosidic units at C-3 do not change the color significantly, but additional glycosylation at C-5 induces a slight shift to red- purple.

3.32. Copigmentat Co-pigmentation

The interaction with other colorless organic compounds or metallic ions to form molecular or complex associations (known as copigmentat co-pigmentation), can affect color and increase its intensity [24]. Copigmentat co-pigmentation may occur through intra- or intermolecular interactions. In the first case, acyl residues are covalently linked to the glycosyl moieties. The acylation with hydroxycinnamic acids causes the bathochromic effect and also increases the absorptivity (hyperchromic effect) [15, 19, 25]. Copigmentat Co-pigmentation increases with the degree of methoxylation and glycosylation of the anthocyanin chromophore.

Intermolecular copigmentat co-pigmentation occurs through interactions of the anthocyanins with colorless compounds through weak hydrophobic forces. This type of co-pigmentation also displays bathochromic and hyperchromic effects. Copigmentat co-pigmentation is thought to play a key role in the changes in wine color during maturation and aging [24].

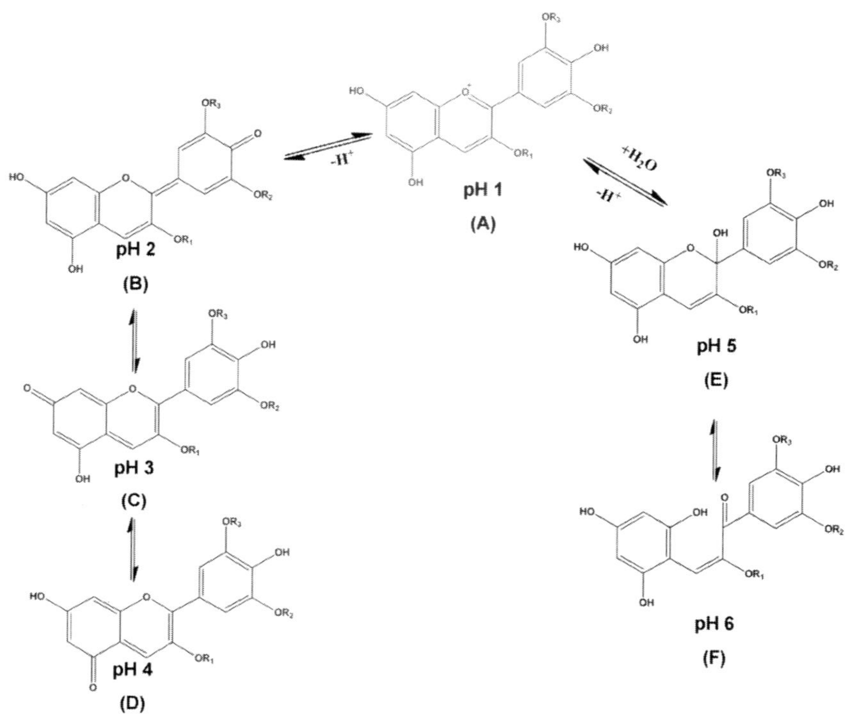

Figure 3. Chemical transformations of anthocyanins at different pH values.

3.43. pH

Color expression, is highly affected by the pH. At pH 1, the flavylium cation (red color) is the major species and contributes to purple and red colors (Figure 3A). At pH values between 2 and 4, the quinoidal blue species are the predominant (Figure 3B-D), while colourless carbinol pseudobase (Figure 3E) or chalcone (Figure 3F) can be detected at pH values between 5 and 6. At higher pH values, the degradation of anthocyanins may occur depending on their substitution pattern [17, 19, 21, 26, 27].

4. BIOSYNTHESIS OF ANTHOCYANINS

The biosynthetic pathway of anthocyanins is one of the most extensively studied routes of plant metabolism. It takes place in the cytosol but the pigments finally accumulate in the vacuole. Anthocyanins could be either found dissolved or in discrete regions of the vacuoles as inclusions or globules called anthocyanoplasts [28].

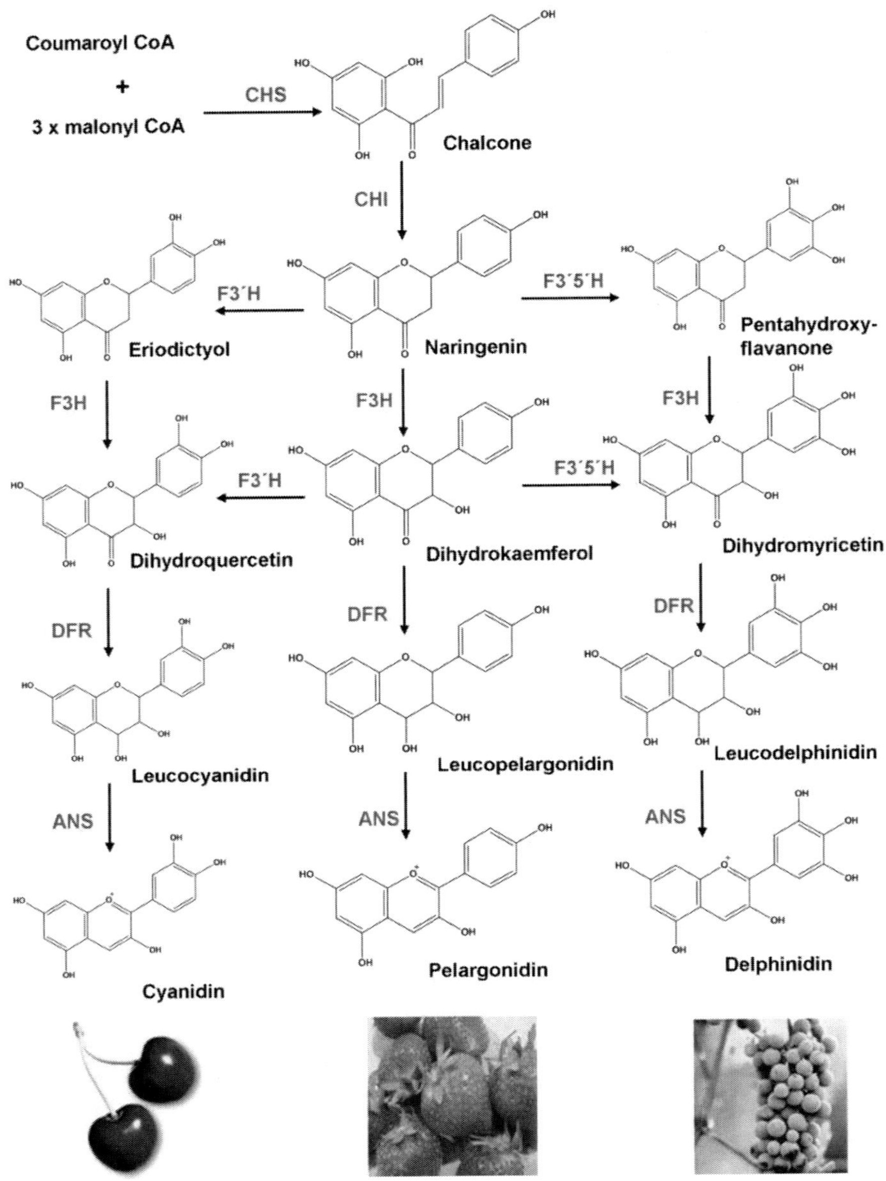

Figure 4. The basic upstream flavonoid pathway for the biosynthesis of anthocyanidins in fruits.

Most enzymes and genes involved in anthocyanin biosynthetic pathway have been characterized over the last three decades [29, 30]. Anthocyanins may be synthesized through the shikimic acid route to produce the amino acid phenylalanine or *via* the acetate-malonate pathway. Anthocyanin synthase (ANS) has been considered as the first key enzyme that could lead the flavonoid flux into anthocyanin branch, though it also plays significant role in the biosynthesis of proanthocyanidins [31, 32, 33]. Overall, the whole pathway can be subdivided into the basic flavonoid upstream pathway (Figure 3) and the specific anthocyanin downstream branch (Figure 4).

4.1. The Basic Flavonoid Upstream Pathway of Anthocyanin Biosynthesis

Chalcone synthase (CHS) catalyzes the first step in the biosynthesis of anthocyanins. It performs sequential condensations with *p*-coumaroyl-CoA and three acetate units from malonyl-CoA to produce a naringin chalcone. However, this intermediate product is used for the formation of the other phenolic compounds [34, 35]. Then, naringenin chalcone is stereospecificially converted to the flavone (2*S*)-naringenin by chalcone isomerase (CHI). The CHI action induces the formation almost absolutely of the biologically active (2*S*)-flavone [36].

The B ring of the naringenin may be further hydroxylated by flavonoid 3´- hydroxylase (F3´H) or flavonoid 3´, 5´- hydroxylase (F3´5´H) to produce eriodictyol or pentahydroxyflavanone, respectively. All of these three (2*S*)-flavanones can be converted to the corresponding dihydroflavonols by the catalysis of flavanone-3-hydroxylase (F3H). In addition, the dihydrokaemferol, oxidized product of naringenin, is used as substrate for F3´H and F3´5´H in order to produce the corresponding dihydroflavonols, dihydroquercetin and dihydromyricetin, respectively. These dihydroflavonols can be reduced to leucoanthocyanidins by the action of dihydroflavanol 4- reductase (DRF). Finally, the colorless leucoanthocyanidins can be converted to the corresponding colored anthocyanidins by ANS [5, 30, 36].

4.2. The Specific Pathway for Anthocyanin Modification

Free anthocyanidins are rapidly glycosylated, methylated and acylated after biosynthesis (Figure 5). Glycosylation increases the hydrophilicity and the stability of anthocyanins and is carried out by the 3-*O*-glycosyltranferases (3-GT). The 3-*O*-glycosylation is almost always a perquisite for further modification, such as additional glycosylation, methylation and acylation. Glycosylation at position 5 results from the action of specigfic 5-*O*-glycosyltranferases (5-GT) [29, 32].

Methylation of hydroxyl groups of anthocyanidins is a common modification that affects color and stability of anthocyanins. Methoxyl groups are found in three main anthocyanidins (peonidin, petunidin and malvidin) and lead to more stable compounds. The methylation at position C-3´ or both at the positions C-3´ and C-5´ on the B-ring of anthocyanins requires *S*-adenosyl-L-methionine (SAM) as methyl group donnor and *O*-methyltranferases (OMT). The hydroxyl group at the position C-4 is rarely methylated [29, 35].

Acylation is one of the most interesting modifications of anthocyanins. The addition of aromatic or aliphatic compounds at the C6´´ position of sugar residues is further increases anthocyanin structural diversity. Acylation protects anthocyanins from quick discoloration in neutral or weakly acidic solution. The reaction is is catalyzed by the anthocyanin acyltranferases (ACT) which show high substrate specificity for both anthocyanin acyl group acceptors and donors [32].

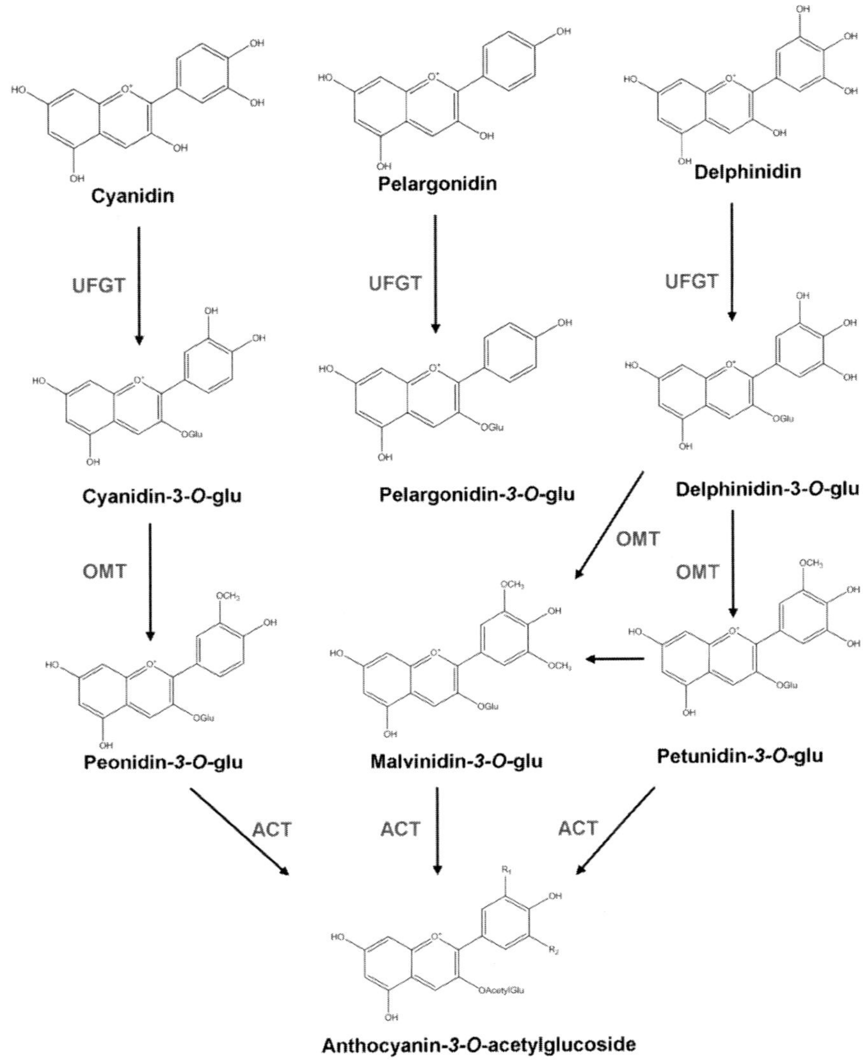

Figure 5. The specific pathway for the anthocyanin modification.

5. Fruit Anthocyanins

The systematic study of anthocyanins has highlighted both their quantitative and structural differences in fruits (Tables 1 and 2). Anthocyanin content range from 0 undetectable levels in various fruit species (Table 1) to 611 mg/100 g edible fresh weight in bilberries (Table 1) [37]. Although anthocyanins have been used in various products, the major demand comes from the beverage industry [4]. For commercial purposes, anthocyanins have been extracted from grapes, berries, and apples, among other fruits [38]. Dark-colored vegetables such as eggplant, purple corn, red sweet potato, red cabbage, red onion and radishes are also rich in anthocyanins.

Table 1. Anthocyanidin content in common fruits (Adapted from reference [47])

Fruit	Content (mg/100g FW)
Acai	53.6
Acerola	22.6
Apple Fuji	0.7
Apple Gala	1.1
Apple Golden	0.0
Apple Granny Smith	0.0
Apple Red delicious	3.8
Avocado	0.3
Blackberry	90.6
Blueberry	141.0
Cherry	27.7
Cranberry	85.5
Currant black	154.8
Currant red	75.0
Currant white	0.0
Eldberry	485.3
Grape Concord	65.6
Grape red	44.0
Grape green	0.0
Kiwifruit	0.0
Melon	0.0
Nectarine and peach	1.8
Pear	12.2
Pineapple	0.0
Plum red	6.98
Plum black	39.7
Plum yellow	0.3
Raspberry	40.9
Strawberry	23.8
Watermelon	0.0

The anthocyanin profile is also quite different among fruit species (Table 2). This difference has been very useful for chemotaxonomic studies and to detect adulterations [39]. The number of anthocyanins varies from few in peaches and nectarines, plums, cherries, strawberries, apples, pears and blackberries to thirty one in 'Concord' grapes. As previously mentioned, the most commonly found anthocyanidins in fruits are cyanidin (Cy), delphinidin (Dp), petunidin (Pt), pelargonidin (Pg), peonidin (Pn) and malvinidin (Mv) (Table 2). The glycoside derivatives of the three non-methylated anthocyanidins (Cy, Dp and Pg) are the most common in nature, being found in 69% of the fruits [38]. Cyanidin is the most widely distributed as aglycone, present in almost each fruit, but its concentrations may vary significantly (1.5 and 485 mg/100 g fresh weight in strawberry and eldberry, respectively). Wild and cultivated blueberries are a good source of malvinidin. High content of delphinidin is found in black currant, while peonidin is present in various grapes and berries with the highest amount in cranberry [10].

Table 2. Anthocyanidin distribution in common fruits

Fruits	Cy	Dp	Pt	Pg	Pn	Mv	References
Acai	+	-	-	+	+	-	[130, 131]
Apples	+	-	-	-	+	-	[10, 40, 41]
Blackberry	+	+	+	+	+	+	[10, 132]
Blueberry	+	+	+	-	+	+	[10, 37, 48]
Cherry, sweet	+	-	-	+	+	-	[10, 37]
Cranberry	+	+	+	+	+	+	[10, 45].
Currant, black	+	+	+	+	+	-	[10, 45, 46]
Fig	+	-	-	+	-	-	[133, 134]
Gooseberry	+	-	-	-	+	-	[10, 44]
Grape, red	+	+	+	-	+	+	[68, 135]
Nectarine	+	-	-	-	-	-	[10, 49]
Peach	+	-	-	-	-	-	[10, 49]
Plum	+	-	-	-	+	-	[10, 49, 51]
Pomegranate	+	+	-	+	-	-	[55, 136]
Raspberry	+	-	-	+	+	-	[10, 37, 45]
Strawberry	+	-	+	+	-	-	[10, 137, 138]

Cy: cyanidin, Dp: delphinidin, Pt: petulidin, Pg: pelargonidin, Pn: peonidin, and Mv: malvinidin.

The major form of anthocyanins in most fruits is the monoglycoside accounting for 70-100% of the total anthocyanin content in common fruits. The corresponding contribution of diglycoside anthocyanins is significantly lower and highly variable for the species considered. Sweet cherry and black currant are exceptions, characterized by a higher proportion of diglycoside anthocyanins (94% and 76%, respectively). The content of triglycoside anthocyanins is significantly limited in fruits and their presence has been reported only in red currant, raspberry and elderberry (6%) [10]. The acylated anthocyanins are generally minor in most fruits, with percentages ranging from 0 to 6%, except for blackberries and blueberries, where the total acylated content could be 15% of the total content [10]. Here, the present review briefly we describes some features of the major anthocyanins found in common fruits:

5.1. Apple

Apple anthocyanins are usually a mix of two different cyanidin glycosides, of which cyanidin-3-*O*-galactoside is more common than cyanidin-3-*O*-glycoside [40, 41]. The amount of cyanidin-3-*O*-galactoside and cyanidin-3-*O*-glycoside ranged from 10 to 550 mg/kg on a dry weight basis for red apple cultivars, while the corresponding content in green-colored cultivars was between 10 and 30 mg/g dry weight [40]. Vrhovsek and co-workers reported the presence of four different cyanidin glycosides, additionally cyanidin-3-*O*-arabinoside and cyanidin-3-*O*-xyloside [42]; however, these compounds were found at low concentration is some red peel-colored cultivars.

5.2. Black Currant

Anthocyanins are the major group of phenolic compounds in black currants acounting for ~80% of the total phenolic content [43]. The chromatographic separation of anthocyanin from six black currant cultivars showed that these currants share a similar anthocyanin profile. A total of fourteen delphinidin-, cyanidin-, malvinidin-, petunidin-, and peonidin-based anthocyanins have been identified across all black currant cultivars [44, 45]. Diglycoside anthocyanins such as delphinidin-3-*O*-rutinoside and cyanidin-3-*O*-rutinose represent 76% of total in currants [10]. Other common anthocyanins found in black currants include delphinidin-3-*O*-glucoside and *p*-coumaroyl derivatives of the glycosides of petunidin and peonidin [45, 46].

5.3. Blueberry

The most abundant anthocyanins in blueberries are the derivatives of both malvinidin and delphinidin [47]. The study of anthocyanin profiles of sixteen blueberry genotypes showed a great diversity of anthocyanins among *Vaccinium* species since at least 25 different anthocyanins have been identified. Sixteen anthocyanins are monoglycosides of five different anthocyanidins, while the other nine anthocyanins are derivatives resulting from acylation with acetic acid [48].

5.4. Peach and Nectarine

Peach and nectarine usually share identical anthocyanin profiles. The anthocyanin composition of these stone fruits is quite simple as only two cyanidin monoglycosides have been detected. The main pigment is cyanidin-3-*O*-glycoside, while cyanidin-3-*O*-rutinoside is found at lower concentration [49, 50].

5.5. Plum

Cyanidin and peonidin derivatives are mainly found in plum. The major anthocyanins in plums are cyanidin-3-*O*-rutinoside and cyanidin-3-*O*-glucoside, while the corresponding peonidin derivatives are detected at lower concentrations [51]. The presence of cyanidin galactoside, xyloside, malonyl-glycoside and (6´´-acetoyl)-glycoside has been also reported [49, 52]. The same anthocyanins were present in most cultivars but with wide variation in terms of quantityquantitative changes has been monitored [52]. Intriguingly,R red and black plums demonstrated a similar anthocyanin profile [50].

5.6. Pomegranate

Pomegranate usually contains 3-glycosides and 3,5-diglycosides of delphinidin, cyanidin and pelargonidin [53]. A study of the anthocyanin profile of pomegranate cultivars showed in all cases the same six pigments. These were identified as delphinidin-3-*O*-glycoside, delphinidin-3,5-*O*-diglycoside, cyanidin-3-*O*-glycoside, cyanidin-3,5-*O*-diglycoside, pelargonidin-3-*O*-glucoside and pelargonidin-3,5-*O*-diglucoside [54]. Similar anthocyanin profiles have been also reported in Spanish and Italian cultivars with differences in the relative amounts of the individual anthocyanins [54, 55].

5.7. Red Grapes

Red grape anthocyannins have been characterized in detail [56]. The anthocyanin profile of grapes is complex and varies according to the cultivar considered. The chromatographic study of red grapes revealed at least 31 different anthocyanins [50, 57]. The most common anthocyanins include cyanidin, delphinidin, peonidin and malvinidin 3-*O*-monoglycosides along with the corresponding acetyl, *p*-coumaroyl and caffeoyl derivatives [57, 58]. Malvidin derivatives (acylated and non-acylated) are the main anthocyanins present in the skin of red grapes [56]. Delphinidin is the most abundant anthocyanidin in 'Concord' grapes [47]. Wu and co-workers reported that anthocyanin composition of red grapes includes only monoglycoside anthocyanins and that the acylated anthocyannins account for 22% of the total content [11].

5.8. Strawberry

The major anthocyanins of strawberries are 3-glycoside, 3-arabinoside, 3-(6´´-malonylglycoside) and 3-(6´´rhamnosylglycoside) of pelargonidin and cyanidin [50, 59]. Pelargonidin derivatives are the most represented anthocyanins in all strawberry cultivars [60]. Qualitative and quantitative differences have been observed among strawberry genotypes. The acylated anthocyanins range between 1-9%, while the diglycoside anthocyanins do not exceed the 20% of the total content [10].

5.9. Sweet Cherry

Five anthocyanins have been identified in different sweet cherry cultivars. The most abundant anthocyanin is almost always cyanidin-3-*O*-rutinoside that contributes to the characteristic dark red-purple color of sweet cherries, followed by cyanidin-3-*O*-glycoside in all studied cultivars [61, 62]. Peonidin pelargonidin and malvinidin have been detected at lower amounts [63, 64].

6. Factors Affecting the Anthocyanin Profile of Fruits

The content and structural diversity of anthocyanins in fruits depends on the cultivar, cultural practices, environmental factors, developmental stage, phytohormones, postharvest treatments and storage conditions.

6.1. Cultivar

Anthocyanin content and composition show significant variations among cultivars of the same species. Since each cultivar displays different anthocyanin content looking for improved quality, including desirable anthocyanin profile has been an important goal of many fruit breeding programs. The anthocyanin composition of selected blackberry cultivars showed three different anthocyanin patterns [65, 66]. The first pattern was found in eight of the eleven studied cultivars and it contained cyanidin-3-*O*-glucoside, as the major constituent, cyanidin-3-*O*-rutinoside as a second major constituent, traces of cyanidin-xyloside, cyanidin-3-*O*-glucoside acylated with malonic acid and/or cyanidin-3-*O*-dioxalylglucoside. The second pattern displayed similarities with the first pattern but it contained lower amounts of cyanidin-3-*O*-rutinoside higher amounts of cyanidin-3-*O*-xyloside. The third pattern was similar with the first one but presented a higher proportion of cyanidin-3-*O*-dioxalylglucoside and the absence of cyanidin-3-*O*-rutinoside. The anthocyanin composition of red grapes has been also thoroughly studied [67, 68]. The anthocyanin composition of fifteen Spanish red grapes cultivars showed distinctive anthocyanin profiles [39].

6.2. Cultural Practices

Various studies have analyzed the effect of cultural practices on fruit pigmentation [69]. Regulated deficit irrigation was found to increase the anthocyanin content in the peel of peaches. Additionally, crop load affected the anthocyanin profile [70]. Some authors have found evidence suggesting that organic products might accumulate higher levels of antioxidants than the conventionally produced commodities [71, 72, 73]. Organically grown plants are usually exposed to different stress condition, which could in turn induce the biosynthesis of anthocyanins. However, there are some reports showing no differences or even lower levels of anthocyanins in organically grown fruit. Grapes from conventional agriculture presented higher contents of delphinidin, petunidin, malvinidin and acylated malvinidin glycosides than organically-grown fruit [74]. Therefore, it is not possible to conclude that organically-grown fruits have higher antioxidant potency capacity when compared to the conventionally grown fruits [75].

Application of exogenous phytohormones can greatly affect the expression of genes involved in the biosynthesis of anthocyanin, and consequently modulate their accumulation. Abscisic acid spray could significantly increase anthocyanin content of harvested grapes without negatively affecting other quality attributes [76]. The use of salicylic acid stimulates anthocyanin biosynthesis in grape berries by inducing the expression of genes encoding phenylalanine-amonia lyase, a key enzyme for the phenylpropanoid metabolism [32].

6.3. Environmental Factors

Anthocyanin biosynthesis is usually governed by environmental factors such as sunlight exposure, UV radiation and temperature. These variables may determine the content and the composition of fruit anthocyanins by affecting both the expression of structural and regulatory genes [32]. It has been demonstrated that preharvest exposure to UV irradiation can activate anthocyanin biosynthetic pathway. Sun -exposed fruit has been associated with higher anthocyanin content [77]. In grape, improvement of light penetration into the canopy enhanced anthocyanin and soluble phenol levels [78]. Low light intensity reduced color development in red apple cultivars, a common situation when fruit are located well within the tree canopy [41]. In pear, fruit bagging promoted degradation of chlorophyll and prevented anthocyanin synthesis [79]. The effect was associated with the interception of UV rather than with differences in fruit temperature [79]. The anthocyanin content of grapes increased with UV-B treatment [80]. UV radiation has been shown to induce both phenylalanine ammonia-lyase (PAL) and chalcone synthase (CHS), catalyzing the first committed step in flavonoid biosynthesis [81]. The MYB-mediated transcription factors are regulatory factors involved in sunlight induction of anthocyanin synthesis and seem to be conserved in various species [82, 83]. An MYB transcription factor from snapdragon was expressed using the fruit-specific E8 promoter specifically in tomato fruit, causing elevated levels of anthocyanin compounds [84].

Temperature may also affect anthocyanin biosynthesis. Moderate temperatures favor anthocyanin biosynthesis, whereas high temperatures over 35 °C are associated with decreased biosynthesis and even with their degradation. In grapes, high temperatures during night at early stages of ripening results in poor production of anthocyanins [32].

6.4. Developmental Stage

Ripening of fruits results from a coordinated set of biochemical and physicochemical changes such as the modification of the cell wall structure, the accumulation of carbohydrates and pigments leading to a soft sweet and appealing fruit to predators. Anthocyanin accumulation during fruit ripening follows three stages: (i) a phase of rapid increase in the concentration, (ii) a transitional phase characterized by an accumulation at a lower rate, leading to a maximum concentration when the fruit reaches organoleptic maturity and (iii) a phase of decreasing concentration at overripe stages [85].

6.5. Fruit Manipulation and Postharvest Environment

Wounding on fruit peel may favor the loss of anthocyanins. Through the disruption of cellular compartimentalization, the physical barriers preventing oxidation of phenolics by polyphenol oxidases (PPOs) and peroxidases (PODs) are eliminated. In addition physical damage increases leakage, raises oxygen partial pressure, exhacerbates the production of free radicals. Careful handling and minimization of bruising is recommended to prevent losses of anthocyanins in fresh fruits. Postharvest treatments and storage conditions can have affect fruit anthocyanin content. The pigments may be subject to degradation, but *de novo* biosynthesis may also occur.

a. *Storage temperature:* Under At moderate temperatures anthocyanin biosynthesis continues during fruit storage [86]. Anthocyanins may be also synthesized in response to stressful events, such as low temperatures. Chilling increases PAL activity and triggers antocyananin biosynthesis, causing pulp bleeding in plum [87, 88]. Anthocyanins turnover could result from the hydrolysis by anthocyanases or by coupled oxidation in the presence of polyphenols and PPO or POD [89, 90]. Direct degradation of anthocyanins by reaction with hydrogen peroxide (H_2O_2) or hydroxyl radicals (OH•) has been also reported [91].
b. *Control and modified atmospheres:* The effect of control and modified atmospheres depend on the species considered, the gas concentration and especially on the time frame evaluated. By delaying ripening, atmospheres rich in CO_2 and reduced in O_2 could decrease the accumulation of anthocyanins [92]. Elevated CO_2 atmospheres can minimize decay and maintain fruit firmness, but increase the pH and adversely affect the internal flesh color of strawberries [93]. Under long term storage, control and modified atmospheres have been shown to result in higher contentsresultes in fruit with higher of athocyanin contents probably due to a reduction of oxidative reactions, involving the oxidation of phenolic compound [94].
c. *Other physical treatments:* Prestorage hot water treatments reduced anthocyanin synthesis and retain overall quality of white asparagus spears [95]. In strawberry, exposure to hot air (45 °C, 3h) also delayed the accumulation of pelargonidin 3-*O*-glucoside. Such treatments inhibited protein synthesis and decreased the rate of most ripening-associated processes [96].

   UV-C illumination as a postharvest treatment has proven to be beneficial in delaying postharvest fruit senescence and reducing decay in various fruits and vegetables [97]. It has been reported that anthocyanin content might be reduced [98, 99], remain unaffected [100, 101], or increase [102] in UV-C treated produce. This might depend on the developmental stage at which the tissue was treated and on the UV-C dose received [97]. Besides the dose applied, radiation intensity may affect the final outcome of UV treatments [103, 104, 105].
d. *Chemical treatments:* Methyl jasmonate (MeJA) was effective in inhibiting pericarp browning and in maintaining the contents of anthocyanins in lychee [106]. and strawberry. However, no effects changes were observed on the anthocyanin profile [107]. A combination of MeJA and ethanol had an additive effect in reducing postharvest decay and improving antioxidant capacity in Chinese bayberries [108].

Oxalic acid, salicylic and methyl-salycilic acid (MeSa) were effective to delayed ripening and maintained higher levels of anthocyanins after long-term low-temperature storage in cherry fruit [109]. MeSa also increased antioxidants in pomegranates [110]. Benzo-thiadiazole-7-carbothioic acid *S*-methyl ester (BTH) at 0.2 g/L induced anthocyanin biosynthesis in strawberry [111]. Pretreatment with ethylene of 'Aleatico' grape increased weight loss during subsequent dehydration and increased the accumulation of anthocyanins and polyphenols [112]. Wine from grapes treated with ethylene revealed higher concentrations of phenols and anthocyanins in wine, although aromatic compounds were lost, decreasing their aroma profiles [113].

7. Stability of Anthocyanins during Processing-Storage

The aglycone form of anthocyanins is rarely found in nature because of its poor stability. Isolated anthocyanins are highly instable and very susceptible to degradation [114].The stability of anthocyanins could be affected by various factors [115, 116]. Glycosylation at position 3 results in higher stability and increased solubility. Interestingly, 5-glycosylation of cyaniding 3-glucoside increased instability at most pH values. In contrast, aromatic acylation resulted in a dramatic increase in color stability, at different pHs from 3 to 7.3 [117]. Acylated anthocyanins are usually used as food colorants because of their higher stability over non-acylated compounds [114].

Self -association of anthocyanins or with other compounds usually results in increased stability [19]. Interactions between anthocyanin molecules are thought to occur through hydrophobic forces that allow their vertical stacking. Several studies have determined that color and stability of anthocyanins could be increased by interaction with other molecules [118]. This is particularly important in the wine and juice industry. Various phenolic compounds have been shown to exerted positive effects in anthocyanin stability [115]. Flavone-*C*-glycosides and flavonols have been shown to be stabilizing agents [116]. Copigmentat Co-pigmentation with caffeic acid significantly increased the stability of anthocyanins. Metals can interact with some anthocyanins and can be stabilizing agents. One of the pre-requisite anthocyanindin-metal interactioninteractions is to have more than one free hydroxyl group in the chromane ring capable of metal chelation [119]. Complexes of anthocyanidins with aluminum tin, copper and iron have been identified [120].

Anthocyanin extracts could be preserved better at higher concentrations [114]. At low pH, anthocyanins are more stable than in alkaline solutions. The acid lowers the solution's pH value and prevents the degradation of the non-acylated anthocyanin pigments. Depending on the pH values, anthocyanins could exist as flavylium anions, hemiketals, quinoid bases, and chalcones. At pH values over 8, anthocyanins degrade to gallic acid, protocatechuic acid, and 4-hydroxybenzoic acid for Dp, Pg, and Cn, respectively [121].

Light and temperature are other factors that can impact anthocyanin stability. Increasing High proccesing temperatures results in higher large losses of anthocyanin [122]. Pigments are store better in dark than under light conditions [123]. Loss of cyanidin 3-*O*-glucoside after 3 months of storage in the dark was about 54% lower than after 3 months of storage in the sun [124].

Sulfites can cause discoloration of anthocyanin by formation of sulfur derivatives [19]. Hydrogen sulfite (H_2S) reacts with anthocyanins causing a nucleophilic attack in the molecule and forming colorless adducts [125]. At low levels of sulfite, the reaction could be reversed by reducing the pH or through moderate heating. At low concentration, sugars can favor degradation [126]. However, 20% sugar had a protective effect on the color of berry extracts. Anthocyanins could also be degraded in the presence of ascorbic acid (AsA) [127]. This can has been shown to occur by through production of free radicals or by direct reaction of AsA with the anthocyanin [128, 129]. Most processed fruit-derived products show significantly lower content of anthocyanins than the corresponding nonun-processed counterparts (Table 3). Freezing or freeze drying appears to be among the best processing methods to prevent anthocyanin degradation.

Table 3. Anthocyanidin content in some fruit-derived processed foods (Adapted from reference [47])

Food	Content (mg/100g)
Jam strawberry	0.31
Jelly grape	0.24
Juice grape juice	53.6
Juice strawberry	22.6
Juice orange blood	0.7
Prunes	0.8
Raisins	0.1
Strawberry frozen	12.3
Wine	0-116.4

REFERENCES

[1] Valls J, Millan S, Marti MP, Borras E, Arola L. Advanced separation methods of food anthocyanins, isoflavones and flavonols. *Journal of Chromatography A* 1216, 7143-7172, 2009.

[2] Eder R. Pigments. Handbook of food analysis. NY, USA: Marcel Dekker Ed.; 2000.

[3] Kong JM, Chia LS, Goh NK, Chia TF, Brouillard R. Analysis and biological activities of anthocyanins. *Phytochemistry* 64, 923-933, 2003.

[4] He JA, Giusti MM. Anthocyanins: Natural colorants with health-promoting properties. *Annual Reviews of Food Science and Technology* 1, 163-186, 2010.

[5] Jaakola L, Maata K, Pirttila AM, Torronen R, Karenlampi S, Hohtola A. Expression of genes involved in anthocyanin biosynthesis in relation to anthocyanin, proanthocyanidin, and flavonol levels during bilberry fruit development. *Plant Physiology* 130, 729-739, 2002.

[6] Mazza, G, Miniati, E. Anthocyanins in fruits, vegetables, and grains. Boca Raton, FL USA: CRC Press; 1993.

[7] Larson RA. The antioxidants of higher plants. *Phytochemistry* 27, 969-978, 1988.

[8] Lewis NG. *Plant phenolics. In Antioxidants in higher plants*. Boca Raton, FL, USA: CRC Press, 1993.

[9] Drossard, C, Alexy U, Bolzenius K, Kunz C, Kersting M. Anthocyanins in the diet of infants and toddlers: intake, sources and trends. *European Journal of Nutrition,* 50, 705-711,, 2011, DOI 10.1007/s00394-011-0184-5.

[10] Wu X, Cao G, Prior RL. Absorption and metabolism of anthocyanins in elderly women after consumption of elderberry or blueberry. *Journal of Nutrition* 132, 1865-1871, 2002.

[11] Wu X, Beecher GR, Holden JM, Haytowitz DB, Gebhardt SE, Prior RL. Concentrations of anthocyanins in common foods in the United States and estimation of normal consumption. *Journal of Agricultural and Food Chemistry,* 54, 4069-4075, 2006.

[12] Woodward G, Kroon P, Cassidy A, Kay C. Anthocyanin stability and recovery: Implications for the analysis of clinical and experimental samples. *Journal of Agricultural and Food Chemistry* 57, 5271-5278, 2009.

[13] de Rijke E, Out P, Niessen WMA, Ariese F, Gooijer C, Brinkman UAT. Analytical separation and detection methods for flavonoids. *Journal of Chromatography,* 1112, 31-63, 2006.

[14] Walle T. Absorption metabolism of flavonoids. *Free Radical Biology and Medicine* 36, 829-837, 2004.

[15] Pascual-Teresa SD, Sanchez-Ballesta MT. Anthocyanins: from plants to health. *Phytochemistry Reviews* 7, 281-299, 2008.

[16] Andersen ØM. *Encyclopedia of life sciences.* Nature Publishing Group, MacMillan Reference Ltd, London 2001. http://www.els.net.

[17] Estevez L, Mosquera RA. Conformational and substitution effects on the electron distribution in a series of anthocyanidins. *Journal of Physical Chemistry A* 113, 9908-9919, 2009.

[18] Andersen ØM, Jordheim M. *The anthocyanins. In Flavonoids: Chemistry, Biochemistry and Applications.* ed. ØM Andersen, KR Markham, 2006. pp. 471–552. Boca Raton, FL, CRC Press, 2006.

[19] Castañeda-Ovando A, Pacheco-Hernandez MDL, Paez-Hernandez ME, Rodriguez JA, Galan-Vidal CA. Chemical studies of anthocyanins: a review. *Food Chemistry* 113, 859-871, 2009.

[20] Clifford, MN. Anthocyanins- nature, occurrence and dietary burden. *Journal of the Science of Food and Agriculture* 80, 1063-1072, 2000.

[21] Krenn L, Steitz M, Schlicht C, Kurth H, Gaedcke F. Anthocyanin and proanthocyanidin- rich extracts of berries in food supplements – analysis with problems. *Pharmazie* 62, 803-812, 2007.

[22] Gómez-Míguez M, González-Manzano S, Escribano-Bailón MT, Heredia FJ, Santos-Buelga, C. Influence of different phenolic copigments on the color of malvidin 3-glucoside. *Journal of Agricultural and Food Chemistry* 54, 5422-5429, 2006.

[23] Delgado-Vargas F, Paredes-López O. Natural colorants for food and nutraceutical uses. Boca Ratón, Florida, USA: CRC Press; 2003.

[24] Boulton R. The copigmentat co-pigmentation of anthocyanins and its role in the color of red wine: A critical review. *American Journal of Enology and Viticulture*, 52, 67-87, 2001.

[25] Stintzing FC, Carle R. Functional properties of anthocyanins and betalains in plants, foods, and in human nutrition. *Trends in Food Science and Technology* 15, 19-38, 2004.

[26] Pina F, Melo, MJ, Maestri M, Passaniti P, Camaioni N, Balzani V. Photo- and pH-induced transformations of flavylium cation: "write- lock- read- unlock- erase" cycles. *European Journal of Organic Chemistry* 11, 3199-3207, 1999.

[27] Ito F, Tanaka N, Katsuki A, Fujii T. Why do flavylium salts show so various colors in solution? Effect of concentration and water on the flavylium's color changes. *Journal of Photochemistry and Photobiology A: Chemistry* 150, 153-157, 2002.

[28] Nozue M, Yasuda H. Occurrence of anthocyanoplasts in cell suspension cultures of sweet potato. *Plant Cell Reports* 4, 252-255, 1985.

[29] Springob K, Nakajima JI., Yamazaki M, Saito K. Recent advances in the biosynthesis and accumulation of anthocyanins. *Natural Product Reports,* 2003, 20, 288-303, 2003.

[30] Petroni K, Tonelli C. Recent advances on the regulation of anthocyanin synthesis in reproductive organs. *Plant Science* 181, 219-229, 2011.

[31] Cheng L, Xu Y, Grotewold E, Jin Z, Wu F, Zhao D. Characterization of anthocyanidin synthase (ANS) gene and anthocyanidin in rare medicinal plant *Saussurea medusa. Plant Cell, Tissue and Organ Culture* 89, 63-73, 2007

[32] He F, Mu L, Yan GL, Liang NN, Pan QH, Wang J, Reeves MJ, Duan, CQ. Biosynthesis of anthocyanins and their regulation in colored grapes. *Molecules* 15, 9057-9091, 2010.

[33] Wang H, Wang W, Li H, Zhang P, Zhan J, Hyang W. Expression and tissue and subcellular localization of anthocyanidin synthase (ANS) in grapevine. *Protoplasma* 248, 267-279, 2011.

[34] Tanaka Y, Ohmiya A. Seeing is believing: engineering anthocyanin and carotenoids biosynthetic pathways. *Current Opinion in Biotechnology* 19, 190-197, 2008.

[35] McCallum S, Woodhead M, Hacket CA, Kassim A, Paterson A, Graham J. Genetic and environmental effects influencing fruit colour and QTL analysis in raspberry. *Theoretical and Applied Genetics* 121, 611-627, 2010.

[36] Tanaka Y, Sasaki N, Ohmiya A. Biosynthesis of plant pigments: anthocyanins, betalains and carotenoids. *The Plant Journal* 54, 733-749, 2008.

[37] Koronen JM, Happonen AM, Mattila PH, Torronen AR. Contents of anthocyanins and ellagitannins in selected foods consumed in Finland. *Journal of Agricultural and Food Chemistry* 55, 1612-1619, 2007.

[38] Dey PM, Harborne JB. *1. Plant phenolics methods in plant biochemistry.* 2nd printing. London: Academic Press Limited; 1993.

[39] Garcia-Benitez E, Revilla E, Cabello F. Anthocyanin pattern of several red grapes cultivars and wines made from them. *European Food Research and Technology* 215, 32-37, 2002.

[40] Wojdylo A, Oszmianski J, Laskowski P. Polyphenolic compounds and antioxidant activity of new and old apple varieties. *Journal of Agricultural and Food Chemistry* 56, 6520-6530, 2008.

[41] Jakopic J, Stampar F, Veberic R. The influence of exposure to light on the phenolic content of 'Fuji' apple. *Scientia Horticulture* 123, 234-239, 2009.

[42] Vrhovsek, U, Rigo, A, Tonon, D, Mattivi, F. Quantitation of polyphenols in different apple varieties. *Journal of Agricultural and Food Chemistry* 52, 6532-6538, 2004.

[43] Anttonen MJ, Karjalainen RO. High-Perfomance Liquid Chromatography analysis of black currant (*Ribes nigrum* L.) fruit phenolics grown either conventionally or organically. *Journal of Agricultural and Food Chemistry* 54, 7530-7538, 2006.

[44] Wu X, Gu L, Prior RL, McKay S. Characterization of anthocyanins and proanthocyanidins in some cultivars of *Ribes, Aronia,* and *Sambucus* and their antioxidant capacity. *Journal of Agricultural and Food Chemistry* 52, 7846-4856, 2004.

[45] Borges G, Degevene A, Mullen W, Crozier A. Identification of flavonoid and phenolic antioxidants in black currants, blueberries, raspberries, red currants, and cranberries. *Journal of Agricultural and Food Chemistry* 58, 3901-3909, 2010.

[46] Gavrilova V, Kajdzanoska M, Gjamovski V, Stefova M. Separation, characterization and quantification of phenolic compounds in blueberries and red and black currants by HPLC-DAD-ESI-MS^n. *Journal of Agricultural and Food Chemistry* 59, 4009-4018, 2011.

[47] Bhagwat S, Haytowitz, DB, Holden, JM. USDA database for the flavonoid content of selected foods. Release 3. Nutrient Data Laboratory Beltsville Human Nutrition Research Center Agricultural Research Service U.S. Department of Agriculture, 2011.

[48] Kalt W, McDonald JE, Ricker RD, Lu X. Anthocyanin content and profile within and among blueberry species. *Canadian Journal of Plant Science* 79, 617-623, 1999.

[49] Tomás-Barberán FA, Gil MI, Cremin P, Waterhouse AL, Hess-Pierce B, Kader AA. HPLC-DAD-ESIMS analysis of phenolic compounds in nectarines, peaches and plums. *Journal of Agricultural and Food Chemistry* 49, 4748-4760, 2001.

[50] Wu X, Prior RL. Systematic identification and characterization of anthocyanins by HPLC-ESI-MS/MS in common foods in the United States: fruits and berries. *Journal of Agricultural and Food Chemistry* 53, 2589-2599, 2005.

[51] Slimestad R, Vangdal E, Brede C. Analysis of phenolic compounds in six Noewegian plum cultivars (*Prunus domestica* L.). *Journal of Agriculture and Food Chemistry* 57, 11370-11375, 2009.

[52] Usenik V, Stampar F, Veberic R. Anthocyanins and fruit colour in plums (*Prunus domestica* L.) during ripening. *Food Chemistry* 114, 529-534, 2009.

[53] Gil MI, Tomás-Barberán FA, Hess-Pierce B, Holcroft DM, Kader AA. Antioxidant activity of pomegranate juice and its relationship with phenolic composition and processing. *Journal of Agricultural and Food Chemistry* 48, 4581-4589, 2000.

[54] Gil MI, Cherif J, Ayed N, Artés F, Tomás-Barberán FA. Influence of cultivar, maturity stage and geographical location on the juice pigmentation of Tunisian pomegranates. *Zeitschrift für Lebensmittel-Untersuchung und -Forschung* 201, 361-364, 1995.

[55] Mousavinejad G, Emam-Djomeh Z, Rezaei K, Khodapaeast MHH. Identification and quantification of phenolic compounds and their effects on antioxidant activity in pomegranate juices of eight Iranian cultivars. *Food Chemistry,* 115, 1274-1278, 2009.

[56] Guerrero RF, Loazid A, Palma M, Puertas B, Gonzalez-Barrio R, Gil-Izquierdo A, Garcia-Barroso C, Cantos-Villar E. Phenolic characterization of red grapes autochthonous to Andalucia. *Food Chemistry* 112, 949-955, 2009.

[57] Pomar F, Novo M, Masa A. Varietal differences among the anthocyanin profiles of 50 red table grape cultivars studied by high performance liquid chromatography. *Journal of Chromatography A* 1094, 34-41, 2005.

[58] Jin ZM, He JJ, Bi HQ, Cui XY, Duan CQ. Phenolic compound profiles in berry skins from nine red wine grape cultivars in Northwest China. *Molecules* 14, 4922-4935, 2009.

[59] Andersen OM, Fossen T. Torskangerpoll K, Fossen A. Hauge U. Anthocyanin from strawberry (*Fragaria ananassa*) with the novel aglycone, 5-carboxylpyranopelargonidin. *Phytochemistry* 65, 405-410, 2004.

[60] Tulipani S, Mezzetti B, Capocasa F, Bompadre S, Beekwilder J, Ric De Vos CH, Capanoglu E, Bovy A, Battino M. Antioxidants, phenolic compounds, and nutritional

quality of different strawberry genotypes. *Journal of Agricultural and Food Chemistry* 56, 696-704, 2008.

[61] Usenik V, Fabcic J, Stampar F. Sugars, organic acids, phenolic composition and antioxidant activity of sweet cherry (*Prunus avium* L.). *Food Chemistry,* 107, 185-192, 2008.

[62] Serradilla MJ, Lozano M, Bernalte MJ, Ayuso MC, Lopez-Corrales M, González-Gómez D. Physicochemical and bioactive properties evolution during ripening of 'Ambrunes' sweet cherry cultivar. *LWT-Food Science and Technology* 44, 199-205, 2011.

[63] González-Gómez D, Lozano M, Fernandez-Leon MF, Bernalte MJ, Ayuso MC, Rodriguez AB. Sweet cherry phytochemicals: identification and characterization by HPLC-DAD/ESI-MS in six sweet cherry cultivars grown in Valle del Jerte (Spain). *Journal of Food Composition and Analysis* 23, 533-539, 2010.

[64] Usenik V, Fajt N, Mikulic-Petkovsek M, Slatanar A, Stampar F, Veberic R. Sweet cherry pomological and biochemical characteristics influenced by rootstock. *Journal of Agricultural and Food Chemistry* 58, 4928-4933, 2010.

[65] Siriwoharn T, Wrolstad RE, Finn CE, Pereira CB. Influence of cultivar, maturity, and sampling on blackberry (*Rubus* hybrid) anthocyanins, polyphenolics, and antioxidant properties. *Journal of Agricultural and Food Chemistry* 52, 8021-8030, 2004.

[66] Sapers GM, Hicks KB, Burgher AM, Hargrave SL, Sondey SM, Bilyk A. Anthocyanin pattern in ripening thornless blackberries. *Journal of the American Society of Horticulture Science* 111, 945-950,1986.

[67] Ortega-Regules A, Romero-Cascales I, López-Roca JM, Ros-García JM, Gómez-Plaza E. Anthocyanin fingerprint of grapes: environmental and genetic variations. *Journal of the Science of Food and Agriculture* 86, 1460-1467, 2006.

[68] Figueiredo-Gonzalez M, Martinez-Carballo E, Canchi-Grande B, Santiago GL, Martinez MC, Simal-Gandara J. Pattern recognition of three *Vitis vinifera* L. red grapes varieties based on anthocyanin and flavonols profiles, with correlations between their biosynthesis pathways. *Food Chemistry* 130, 9-19, 2012.

[69] Kays SJ. Preharvest factors affecting appearance. *Postharvest Biology and Technology* 15, 233-247, 1999.

[70] Buendia B, Allende A, Nicolas E, Alarcon JJ, Gil MI. Effect of regulated deficit irrigation and crop load on the antioxidant compounds of peaches. *Journal of Agricultural and Food Chemistry* 56, 3601-3608, 2008.

[71] Woese K, Lange C, Boess D, Bogel KW A comparison of organically and conventionally grown foods-results of a review of the relevant literature. *Journal of the Science of Food and Agriculture* 74, 281-293, 1997.

[72] Weibel F, Bickel R, Leuthold S, Alfoldi T. Are organically grown apples tastier and healthier? A comparative field study using conventional and alternative methods to measure fruit quality. *Acta Horticulturae* 57, 417-426, 2000.

[73] Asami DK Hong YH, Barrett DM, Mitchell AE. A comparison of the total phenolic and ascorbic acid contents of freeze-dried and air-dried marionberry, strawberry and corn grown using conventional, organic and sustainable agricultural practices. *Journal of Agricultural and Food Chemistry* 51, 1237–1241, 2003.

[74] Vian MA, Tomao V, Coulomb PO, Lacombe JM, Dangles O. Comparison of the anthocyanin composition during ripening of Syrah grapes grown using organic or

conventional agriculture practices. *Journal of Agricultural and Food Chemistry* 54, 5230-5235, 2006.

[75] Winter DK, Davis SF. Organic foods. *Journal of Food Science*, 71, R117-R124, 2006.

[76] Quiroga AM, Berli FJ, Moreno D, Cavagnaro JB, Bottini R. Abscisic acid sprays significantly increase yield per plant in ineyyard-grown wine grapes (*Vitis vinifera* L.) cv. Cabernet Sauvignon through increases berry set with no negative effects on anthocyanin content and total polyphenol index of both juice and wine. *Journal of Plant Growth and Regulation* 28, 28-35, 2009.

[77] Campbell RJ, Marini RP. 1992. Light environment and time of harvest affect 'Delicious' apple fruit quality characteristics. *Journal of the American Society for Horticultural Science* 20, 551-557, 1992.

[78] Prange R. Deell JR. Preharvest factors affecting quality of berry crops. *HortScience,* 32, 824-830, 1997.

[79] Huang C, Yu B, Teng Y, Su J, Shu Q, Cheng Z, Zeng L. Effects of fruit bagging on coloring and related physiology, and qualities of red Chinese sand pears during fruit maturation *Scientia Horticulturae* 121, 149-158, 2009.

[80] Berli FJ, Fanzone M, Piccoli P, Bottini R. Solar UV-B and ABA are involved in phenol metabolism of *Vitis vinifera* L. increasing biosynthesis of berry skin polyphenols. *Journal of Agricultural and Food Chemistry* 59, 4874-4884, 2011.

[81] Logemann E., Tavernaro A, Schulz W, Somssich IE, Hahlbrock K. UV light selectively coinduces supply pathways from primary metabolism and flavonoid secondary product formation in parsley. *Proceedings of the National Academy of Sciences U.S.A.* 97, 1903-1907, 2000.

[82] Matus JTR, Loyola A, Vega A, Pena-Neira E, Bordeu P, Arce-Johnson, Alcalde JA. Post-veraison sunlight exposure induces MYB-mediated transcriptional regulation of anthocyanin and flavonol synthesis in berry skins of Vitis vinifera. *Journal of Experimental Botany* 60, 853-867, 2009.

[83] Butelli E, Titta L, Giorgio M, Mock HP, Matros A, Peterek S, Schijlen EG, Hall RD, Bovy AG, Luo J, Martin C. Enrichment of tomato fruit with health-promoting anthocyanins by expression of select transcription factors. *Nature Biotechnology* 26, 1301-1308, 2008.

[84] Espley RV, Hellens RP, Putterill J, Stevenson DE, Kutty-Amma S, Allan AC. Red colouration in apple fruit is due to the activity of the MYB transcription factor, MdMYB10. *The Plant Journal* 49, 414-427, 2007.

[85] Rogez H, Pompeu DR, Akwie SNT, Larondelle Y. Sigmoidal kinetics of anthocyanin accumulation during fruit ripening: A comparison between acai fruits (*Euterpe oleracea*) and other anthocyanin-rich fruits. *Journal of Food Composition and Analysis,* 24, 796-800, 2011.

[86] Ayala-Zavala JF, Wang SY, Wang CY, Gonzalez-Aguilar GA. Effect of storage temperatures on antioxidant capacity and aroma compounds in strawberry fruit. *Lebensmittel-Wissenschaft und-Technologie* 37, 687-695, 2004.

[87] Manganaris GA, Vicente AR, Crisosto CH, Labavitch JM. Effect of delayed storage and continuous ethylene exposure on flesh reddening of 'Royal Diamond' plums. *Journal of the Science of Food and Agriculture*, 88, 2180-2185, 2008.

[88] Reichel M, Carle R, Sruamsiri P, Neidhart S. Changes in flavonoids and nonphenolic pigments during on-tree maturation and postharvest pericarp browning of litchi (*Litchi*

chinensis Sonn.) as shown by HPLC-MS^n. *Journal of Agricultural and food Chemistry* 27, 3924-3939, 2011.

[89] Zhang ZQ, Pang XQ, Duan XW, Ji ZL, Jiang JM. Role of peroxidase in anthocyanin degradation in litchi fruit pericarp. *Food Chemistry* 90, 47-52, 2005.

[90] Somboonkaew N, Terry LA. Altered physiology and biochemistry of imported litchi fruit held under different vapour pressure deficits. *Journal of Agricultural and Food Chemistry* 58, 6209-6218, 2010.

[91] Ruenroengklin N, Sun J, Shi J, Xue SJ, Jiang YM. Role of endogenous and exogenous phenolics in litchi anthocyanin degradation caused by polyphenol oxidase. *Food Chemistry* 115, 1253-1256, 2009.

[92] Vicente AR, Martínez GA, Civello PM, Chaves AR. Quality of heat-treated strawberry fruit during refrigerated storage. *Postharvest Biology and Technology* 25, 59-71, 2002.

[93] Holcroft DM, Kader AA. Controlled atmosphere-induced changes in pH and organic acid metabolism may affect color of stored strawberry fruit. *Postharvest Biology and Technology* 17, 19-32, 1999.

[94] Duarte C, Guerra M, López Camelo DA, Yommi A. Quality changes of highbush blueberries bruit stored in CA with different CO_2 levels. *Journal of Food Science* 74, S154-S159, 2009.

[95] Siomos A, Gerasopoulos D, Tsouvaltzis P. Prestorage hot water treatments inhibit postharvest anthocyanin synthesis and retain overall quality of white asparagus spears. *Postharvest Biology and Technology* 38, 160-168, 2005.

[96] Civello, PM, Martínez, GA, Chaves, AR, Añón, MC. Heat treatments delay ripening and postharvest decay of strawberry fruit. *Journal of Agricultural and Food Chemistry* 45, 4589-4594, 1997.

[97] Civello, PM, Vicente, AR, Martínez, GA. Recent Advances in Alternative Postharvest Technologies to Control Fungal Diseases in Fruits & Vegetables: UV-C technology to control postharvest diseases of fruits and vegetables. Editors: Rosalba Troncoso-Rojas, Martín E. Tiznado-Hernández and Alberto González-León Kerala India, Transworld Research Network; 2006.

[98] Pan J, Vicente AR, Martínez GA, Chaves AR, Civello PM. Combined use of UV-C irradiation and heat treatment to improve postharvest life of strawberry fruit. *Journal of the Science of Food and Agriculture* 84, 1831-1838, 2004.

[99] Baka M, Mercier J, Corcuff F, Castaigne F, Arul J. Photochemical treatment to improve storability of fresh strawberries. *Journal of Food Science* 64, 1068-1072, 1999.

[100] Cantos E, García-Viguera C, de Pascual-Teresa S, Tomás-Barberán F.A. Effect of postharvest ultraviolet irradiation on resveratrol and other phenolics of cv. Napoleón table grapes. *Journal of Agricultural and Food Chemistry 48*, 4606-4612, 2000.

[101] Cantos E, Espín JC, Tomás-Barberán FA. Postharvest induction modeling method using UV irradiation pulses for obtaining resveratrol-enriched table grapes: A new "Functional" fruit? *Journal of Agricultural and Food Chemistry* 49, 5052-5058, 2001.

[102] Hagen SF, Borge GIA, Bengtsson GB, Bilger W, Berge A, Haffner K, Solhaug KA. Phenolic contents and other health and sensory related properties of apple fruit (*Malus domestica* Borkh., cv. Aroma): Effect of postharvest UV-B irradiation. *Postharvest Biology and Technology* 45, 1-10, 2007.

[103] Frohnmeyer H, Staiger Jenkins GI, Long JC, Wade HK, Shenton MR, Bibikova TN. UV and blue light signalling: pathways regulating chalcone synthase gene expression in *Arabidopsis*. *New Phytologist* 151, 121-131, 2001.

[104] Ulm R, Nagy F. Signalling and gene regulation in response to UV light. *Current Opinion in Plant Biology* 8, 477-482, 2005.

[105] Brown BA, Jenkins GI. UV-B signalling pathways with different fluence-rate response profiles are distinguished in mature Arabidopsis leaf tissue by requirement for UVR8, HY5, and HYH. *Plant Physiology* 146, 576-588, 2008.

[106] Yang S, Chen Y, Feng L, Yang E, Su X, Jiang Y. Effect of methyl jasmonate on pericarp browning of postharvest lychees. *Journal of Food Processing and Preservation* 35, 417-422, 2011.

[107] de la Peña Moreno, M, Monagas, GP, Blanch, BB, Ruiz del Castillo, ML. Enhancement of phenolic and aroma compounds in strawberry fruit through methyl jasmonate vapour treatment. *European Food Research and Technology* 230, 989–999, 2010.

[108] Wang K, Jin P, Shang J, Zheng Y. Effect of methyl jasmonate in combination with ethanol treatment on postharvest decay and antioxidant capacity in chinese bayberries. *Journal of Agricultural and Food Chemistry* 58, 9597-9604, 2010.

[109] Valero D, Díaz-Mula HM, Zapata PJ, Castillo S, Guillén F, Martínez-Romero D, Serrano M. Postharvest treatments with salicylic acid, acetylsalicylic acid or oxalic acid delayed ripening and enhanced bioactive compounds and antioxidant capacity in Sweet cherry. *Journal of Agricultural and Food Chemistry* 59, 5483-5489, 2011.

[110] Sayyaria M, Castillo S, Valero D, Huertas M, Díaz-Mula M, Serrano M. Acetyl salicylic acid alleviates chilling injury and maintains nutritive and bioactive compounds and antioxidant activity during postharvest storage of pomegranates. *Postharvest Biology and Technology* 60, 136-142, 2011.

[111] Cao SF, Hu ZC, Zheng YH, Lu BH. Effect of BTH on anthocyanin content and activities of related enzymes in strawberry after harvest. *Journal of Agricultural and Food Chemistry*, 58, 5801-5805, 2010.

[112] Botondi R, Lodola L, Mencarelli F. Postharvest ethylene treatment affects berry dehydration, polyphenol and anthocyanin content by increasing the activity of cell wall enzymes in Aleatico wine grape. *European Food Research and Technology* 232, 679-685, 2011.

[113] Bellincontro A, Fardelli A, De Santis D, Botondi R, Mencarelli F. Postharvest ethylene and 1-MCP treatments both affect phenols, anthocyanins, and aromatic quality of Aleatico grapes and wine. *Australian Journal of Grape and Wine Research* 12, 141-149, 2006.

[114] Giusti MM, Wrolstad RE. Acylated anthocyanins from edible sources and their applications in food systems. *Biochemical Engineering Journal* 14, 217-225, 2003.

[115] Mazza G., Brouillard R. The mechanism of co-pigmentation of anthocyanins in aqueous solutions. *Phytochemistry* 29, 1097-1102, 1990.

[116] Cavalcanti RN, Santos DT, Meireles MAA. Non-thermal stabilization mechanisms of anthocyanins in model and food systems-An overview. *Food Research International* 44, 499-509, 2011.

[117] Torskangerpoll T, Andersen AM. Colour stability of anthocyanins in aqueous solutions at various pH values. *Food Chemistry* 89, 427-440, 2005.

[118] Rein M. Copigmentat Co-pigmentation reactions and color stability of berry anthocyanins. Academic Dissertation, Faculty of Agriculture and Forestry, Department of Applied Chemistry and Microbiology. University of Helsinki. Helsinski Finland, 2005.

[119] Osawa Y. Copigmentat Co-pigmentation of anthocyanins. In P. Markakis (Ed.), *Anthocyanins as food colors*. New York: Academic Press; 1982.

[120] Elhabiri M, Figueiredo P, Toki K, Saito N, Brouillard R. Anthocyanin aluminium and –gallium complexes in aqueous solution. *Journal Chemical Society Perkin Transactions* 2, 355-362, 1997.

[121] Fleschhut J, Kratzer F, Rechkemmer G, Kulling S.E. Stability and biotransformation of various dietary anthocyanins *in vitro*. *European Journal of Nutrition* 45, 7-18, 2006.

[122] Bobbio FO, Mercadante, AZ. Food colorants: Anthocyanins in foods: Occurrence and physicochemical properties. Boca Raton, FL USA, CRC Press, 2008.

[123] Bakowska A, Kucharska AZ, Oszmianski J. The effects of heating, UV irradiation, and storage on stability of the anthocyanin–polyphenol copigment complex. *Food Chemistry* 81, 349-355, 2003.

[124] Ochoa MR, Kesseler AG, De Michelis A, Mugridge A, Chaves AR. Kinetics of colour change of raspberry, sweet (*Prunus avium*) and sour (*Prunus cerasus*) cherries preserves packed in glass containers: light and room temperature effects. *Journal of Food Engineering* 49, 55-62, 2001.

[125] Mazza G, Brouillard R. Recent developments in the stabilization of anthocyanins in food products. *Food Chemistry* 25, 207-225, 1987.

[126] Thakur, BR, Arya, SS. Studies on stability of blue grape anthocyanins. *International Journal of Food Science and Technology* 24, 321-326, 1989.

[127] García-Viguera C, Bridle P. Influence of structure on colour stability of anthocyanins and flavylium salts with ascorbic acid. *Food Chemistry* 64, 21-26, 1999.

[128] Iacobucci GA, Sweeny JG. The chemistry of anthocyanins, anthocyanidins, and related flavilium salts. *Tetrahedron Letters* 39, 3005-3012, 1983.

[129] Pacheco-Palencia LA, Hawken P, Talcott ST. Phytochemical, antioxidant and pigment stability of açaí (*Euterpe oleracea* Mart.) as affected by clarification, ascorbic acid fortification and storage. *Food Research International* 40, 620-628, 2007.

[130] Pacheco-Palencia LA, Duncan CE, Talcott ST. Phytochemical composition and thermal stability of two commercial acai species, *Euterpe oleracea* and *Euterpe precatoria*. *Food Chemistry* 115, 1199-1205, 2009.

[131] Pacheco-Palencia LA, Mertens-Talcott SU, Talcott ST. *In vitro* absorption and antiproliferative activities of monomeric and polymeric anthocyanin fractions from acai fruit (*Euterpe oleracea* Mart.). *Food Chemistry* 119, 1071-1078, 2010.

[132] Koca I, Karadeniz B. Antioxidant properties of blackberry and blueberry fruits grown in the Black Sea Region of Turkey. *Scientia Horticulture* 121, 447-450, 2009.

[133] Del Caro A, Piga A. Polyphenol composition of peel and pulp of two Italian flesh fig fruits cultivars (*Ficus carica* L.). *European Food Research and Technology* 226, 715-719, 2008.

[134] Duenas M, Perez-Alonso JJ, Santos-Buelga C, Escribano-Bailon T. Anthocyanin composition in fig (*Ficus carica* L.). *Journal of Food Composition and Analysis* 21, 107-115, 2008.

[135] Dimitrovska M, Bocevska M, Simitrovski, D, Murkovic, M. Anthocyanin composition of Vranec, Cabernet Sauvignon, Merlot and Pinot Noir grapes as indicator of their varietal differentiation. *European Food Research and Technology* 232, 691-600, 2011.

[136] Fischer UA, Carle R, Kammerer DR. Identification and quantification of phenolic compounds from pomegranate (*Punica granatum* L.) peel, mesocarp, aril and differently produced juices by HPLC-DAD-ESI-MS^n. *Food Chemistry* 127, 807-821, 2011.

[137] Buendia B, Gil MI, Tudela JA, Gady AL, Medina JJ, Soria C, López JM, Tomás-Barberán FA. HPLC-MS analysis of proanthocyanidin oligomers and other phenolics in 15 strawberry cultivars. *Journal of Agricultural and Food Chemistry* 58, 3916-3926, 2010.

[138] Goulas V, Manganaris GA. The effect of postharvest ripening on strawberry bioactive composition and antioxidant potential. *Journal of the Science of Food and Agriculture* 91, 1907-1914, 2011.

In: Anthocyanins: Structure, Biosynthesis and Health Benefits ISBN: 978-1-62257-329-5
Editor: Noboru Motohashi

Chapter 11

RESVERATROL AND HEALTH

Sang Yoon Choi*
Korea Food Research Institute, Baekhyeon-dong,
Bundang-gu, Seongnam-si, Kyunggi-do, Korea

ABSTRACT

Resveratrol (3,5,4'-trihydroxystilbene), which is a phytoalexin mainly found in grape skin. Resveratrol has been recognized for its cardiovascular protective effects, but its effects are widening based on many results. Despite its low bioavailability and content, resveratrol is directly related to the physiological activity of wine and plays a leading role as a substance for new drug development and in increasing economic value due to the increase in wine consumption worldwide.

Keywords: Resveratrol, wine, bioavailability, bioactivity, stilbene derivative

1. INTRODUCTION

Resveratrol (Figure 1), which is mainly found in grape skin, is a type of stilbene with three hydroxyl groups. Resveratrol is a representative non-flavonoid stilbene type of phytoalexin, which is synthesized from *p*-coumaroyl-CoA and malonyl-CoA by stilbene synthase [1,2]. Various plants synthesize phytoalexin to resist microbial invasion and stimulation by UV light.

In 1940, resveratrol was first isolated from the root of *Veratrum grandiflorum*. Resveratrol has now been identified in 72 kinds of plants and occurs in high concentrations in fresh grape skin [3]. Resveratrol has some beneficial physiological effects [4-6] and, particular interest is as an active component of the "French Paradox", because of a significantly low incidence of cardiovascular diseases is found among French people in spite of their relatively high consumption of meat as the clinical phenomenon. The French Paradox

* Corresponding author: sychoi@kfri.re.kr

is thought to be attributable to the action of wine which French people consume very frequently. Because, various polyphenols including resveratrol in wine had a cardiovascular protective effect by the preventing of the low-density lipoprotein (LDL) oxidation and by the inhibition against the scavenging free radicals that cause the oxidative damage [3,7].

Various effects of resveratrol including its anti-carcinogenic effect, lipid metabolism regulation effect, anti-inflammatory effect, neuroprotective effect, and its cardiovascular protective effect have been known [4,8,9]. In particular, the anti-cancer effect of resveratrol was presented in *Science* 1997 [10]; and the effect for antimetabolic disease by resveratrol was presented in *Cell* 2006 [11]; and the effect for antiobesity related disease was presented in *Nature* 2006 [12].

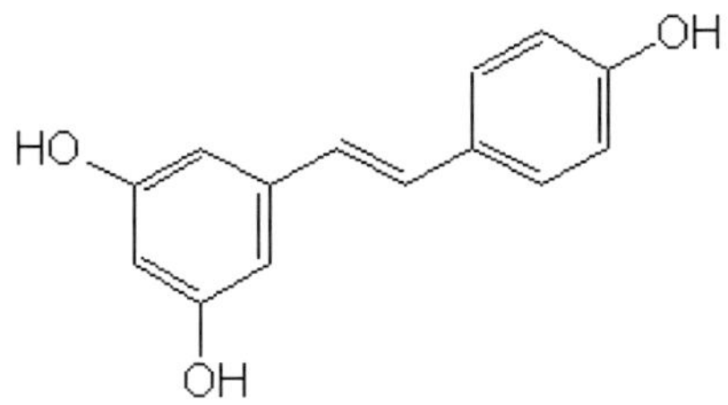

Figure 1. Chemical structure of resveratrol

2. Physiological Activity of Resveratrol

2.1. Cardiovascular Protective Effect

The cardiovascular protective effect of resveratrol has been verified in many *in vitro* and *in vivo* models, and many related studies have also been reported. The cardiovascular protective effect is related to antioxidant, anti-inflammatory, anti-proliferation, and effects on cell signaling regulation by resveratrol [3,7].

Resveratrol inhibited LDL oxidation and reduces free radicals in the blood to prevent oxidative damage to cardiac vessels [13]. The anti-inflammatory related activities, including inhibiting lipopolysaccharide-induced vascular cell adhesion molecule-1 (VCAM-1) and tissue factor expression as well as lipooxygenase activity related to the inflammatory mediator synthesis contribute to the effect of anti-cardiovascular disease by resveratrol [3]. Additionally, a hypothesis of cardiovascular protective action has been proposed in which the platelet aggregation inhibitory effect, anti-atherosclerotic property, ischemic injury protective effect, vascular cell fuction modulation and estrogen-like action of resveratrol [3, 14].

2.2. Anti-Cancer Effect

Resveratrol inhibited the proliferation and induces apoptosis in several cancer cell lines including colon cancer, lung cancer, ovarian cancer, prostate cancer, breast cancer, and hematological malignancy. Their mechanisms of the decreased nuclear facter κB (NF-κB) and Fas receptor (CD95), blocking the Akt and Wnt pathway, STAT3 and cyclooxygenase-2 (COX-2) inhibition were examined [15,16]. Additionally, resveratrol eliminated the carcinogenic toxicity by acting as an antioxidant and mutation inhibitory substance during carcinogenesis, and suppressed the cell differentiation by inhibiting cyclooxygenase-1 (COX-1) and hydrogen peroxidase activities through its anti-inflammatory action [4,17-19]. In fact, resveratrol inhibited the precancerous damage in carcinogen-treated mammary gland tissues and inhibited the tumorigenesis in a mouse skin cancer model [4].

2.3. Neuroprotective Effects

Based on the antioxidant activity of resveratrol, the protective effect of resveratrol for the nerve cells has been studied. The oxidative stress and reactive oxygen species (ROS) production were the major causes of age-related diseases such as brain ageing, stroke, dementia, and Parkinson's disease [20]. The antioxidant activity of resveratrol could reduce the reactive oxygen species (ROS) [21]. Resveratrol inhibited the apoptosis caused by the oxidative damage to PC12 cells and effectively inhibited the nerve cell death in region I of hippocampus proper [CA1] due to oxygen and glucose deprivation (ODG) in the brains of rats [22]. The protective effects on the various nerve cells and brain diseases by resveratrol were studied [23, 24].

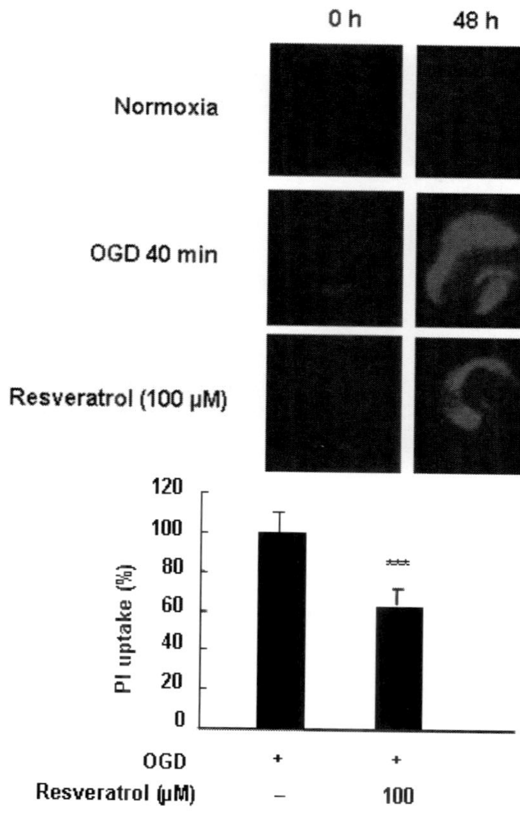

Figure 2. Neuroprotective effect of resveratrol on the oxygen and glucose deprivation (ODG)-exposed hippocampal region [22].

2.4. Lipid Metabolism Regulation Effect

Resveratrol increased the insulin sensitivity and decreases the insulin-like growth factor-1 concentrations, while increasing AMP-activated protein kinase, peroxisome proliferator-activated receptor-gamma co-activator 1 alpha activity, and the number of mitochondria. Resveratrol also reduced a cholesterol ester and triglyceride production [12,25]. In animal models, resveratrol increased the survival rate of mice with high-calorie diet-induced obesity [12].

2.5. Other Physiological Activities

2.5.1. Platelet Aggregation Inhibitory Effect

Resveratrol inhibited the platelet aggregation induced by thrombin, ADP, arachidonic acid, and collagen. Furthermore, resveratrol inhibited the free radical formation in the platelets activated by thrombin [4].

2.5.2. Tyrosinase Inhibitory Effect

Resveratrol inhibited tyrosinase activity, which has an important role in browning in plants and melanin production in animal skin. Resveratrol also caused a decrease in food quality and skin pigmentation diseases [26,27].

3. Resveratrol Content in Different Red Wines

Although resveratrol exists as both *cis* and *trans* isomers, it is found mainly in the *trans* form in grapes. The *trans*-resveratrol content in red wine varies depending on the origin, species, infection, and variety type of grape and the method of manufacturing the wine. According to a report by Fremont, the resveratrol content among ten variety red wines from different origins ranged from 0.02 to 13.4 mg/L [4].

4. Bioavailability

Resveratrol was quickly absorbed and transported to various organs when ingested and was then metabolized into glucuronide and sulfate conjugates [28,29]. Because of this fast and extensive metabolic activity, the bioavailability of resveratrol is very low, with only 2% being detectable in the blood after oral administration [28]. The blood concentration reached a maximum at 30 minutes after ingestion and then quickly disappeared, and a small amount is detected again after 6 hours due to the hydrolysis of conjugated resveratrol metabolites in the small intestine. This fast metabolism of resveratrol in the body has been observed after intravenous administration. Therefore, it is expected that continuous wine intake for a certain period of time will show the effects of resveratrol.

Studies on Derivatives

Various studies on resveratrol derivatives have been synthetized to overcome the resveratrol deficiencies such as low bioavailability and stability, and to increased its activities. New resveratrol derivatives were 100 times on the level of cytochrome P450 inhibition [30], 150 times on the level of tyrosinase inhibition [27], and twice on the cell protective effect of resveratrol itself [22]. From these results, these new resveratrol derivatives could be introduced as leading compounds for developing new drugs [31].

References

[2] Schroder J, Schroder G. Stilbene and chalcone synthases: Related enzymes with key functions in plant-specific pathways. *Z Naturforsch* 45, 1-8, 1990.

[3] Becker JV, Armstrong GO, van der Merwe MJ, Lambrechts MG, Vivier MA, Pretorius IS. Metabolic engineering of Saccharomyces cerevisiae for the synthesis of the wine-related antioxidant resveratrol. *FEMS Yeast Res* 4(1), 79-85, 2003.

[4] Hao H, He L. Mechanisms of cardiovascular protection by resveratrol. *J Med Food* 7(3), 290-298, 2004.

[5] Fremont L. Biological effects of resveratrol. *Life Sci* 66(8), 663-673, 2000.

[6] Nassiri-Asl M, Hosseinzadeh H. Review of the pharmacological effects of Vitis vinifera(Grape) and its bioactive compounds. *Phytother Res* 23(9), 1197-204, 2009.

[7] Pirola L, Frojdo S. Resveratrol: one molecule, many targets. IUBMB *Life* 60(5), 323-332, 2008.

[8] Das S, Das DK. Resveratrol: a therapeutic promise for cardiovascular diseases. *Recent Pat Cardiovasc Drug Discov* 2(2), 133-8, 2007.

[9] Kim YE, Lim SY, Rhee SH, Park KY, Lee WH, Choi YH. Induction of Cdk inhibitor p21 and inhibition of cyclooxygenase-2 by resveratrol in human lung carcinoma A549 cells. *Journal of Life Science* 14(5), 800-808, 2004.

[10] Min HY, Park EJ, Lee SK, Cho YJ. Effects of grape extracts on free radical scavenging activity and inhibition of pro-inflammatory mediator production in mouse macrophage cells. *Korean J Food Sci Technol* 35(1), 132-137, 2003.

[11] Jang M, Cai L, Udeani G, Slowing K, Thomas C, Beecher C, Fong H, Farnsworth N, Kinghorn D, Mehta R, Moon R, Pezzuto J. Cancer chemopreventive activity of resveratrol, a natural product derived from grape. *Science* 275, 218-220, 1997.

[12] Lagouge M, Argmann C, Gerhart-Hines Z, Meziane H, Lerin C, Daussin F, Messadeq N, Milne J, Lambert P, Elliott P, Geny B, Lambert M, Puigserver P, Auwerx J. Resveratrol improves mitochondrial function and protects against metabolic disease by activating SIRT1 and PGC-1alpha. *Cell* 127(6), 1109-1122, 2006.

[13] Baur JA, Pearson KJ, Price NL, Jamieson HA, Lerin C, Kalra A, Prabhu VV, Allard JS, Lopez-Lluch G, Lewis K, Pistell PJ, Poosala S, Becker KG, Boss O, Gwinn D, Wang M, Ramaswamy S, Fishbein KW, Spencer RG, Lakatta EG, Le Couteur D, Shaw RJ, Navas P, Puigserver P, Ingram DK, de Cabo R, Sinclair DA. Resveratrol improves health and survival of mice on a high-calorie diet. *Nature* 444(7117), 337-342, 2006.

[14] Bradamante S, Barenghi L, Villa A. Cardiovascular protective effects of resveratrol. *Discov Today* 15(17-18), 757-765, 2010.

[15] Bradamante S, Barenghi L, Villa A. Cardiovascular protective effects of resveratrol. *Cardiovasc Drug Rev* 22(3), 169-88, 2004.

[16] Subbaramaiah K, Chung WJ, Michaluart P, Telang N, Tanabe T, Inoue H, Jang M, Pezzuto JM, Dannenberg AJ. Resveratrol inhibits cyclooxygenase-2 transcription and activity in phorbol ester-treated human mammary epithelial cells. *J Biol Chem* 21, 273(34), 21875-21882, 1998.

[17] Shukla Y, Singh R. Resveratrol and cellular mechanisms of cancer prevention. *Ann N Y Acad Sci* 1215, 1-8, 2011.

[18] Lastra C, Villegas I. Resveratrol as an antioxidant and pro-oxidant agent: mechanisms and clinical implications. *Biochemical Society Transactions* 15(5), 1156-1160, 2007.

[19] Jang M, Cai L, Udeani GO, Slowing KV, Thomas CF, Beecher CW, Fong HH, Farnsworth NR, Kinghorn AD, Mehta RG, Moon RC, Pezzuto JM. Cancer chemopreventive activity of resveratrol, a natural product derived from grapes. *Science* 275, 218–220, 1997.

[20] Bhat KP, Pezzuto JM. Cancer chemopreventive activity of resveratrol. *Ann N Y Acad Sci.* 957, 210-229, 2002.

[21] Sun A., Simonyi A, Sun G. The French paradox and beyond: Neuroprotective effects of polyphenols. *Free Radical Biology & Medicine* 32(4), 314-318, 2002.

[22] Simão F, Matté A, Matté C, Soares FM, Wyse AT, Netto CA, Salbego CG. Resveratrol prevents oxidative stress and inhibition of Na(+)K(+)-ATPase activity induced by transient global cerebral ischemia in rats. *J Nutr Biochem.* 22(10), 921-8, 2011.

[23] Choi SY, Kim S, Son D, Lee P, Lee J, Lee S, Kim DS, Park Y, Kim SY. Protective effect of (4-methoxybenzylidene)-(3-methoxynophenyl) amine against neuronal cell death induced by oxygen and glucose deprivation in rat organotypic hippocampal slice culture. *Biol Pharm Bull* 30(1), 189-192, 2007.

[24] Pallàs M, Casadesús G, Smith MA, Coto-Montes A, Pelegri C, Vilaplana J, Camins A. Resveratrol and neurodegenerative diseases: activation of SIRT1 as the potential pathway towards neuroprotection. *Curr Neurovasc Res* 6(1), 70-81, 2009.

[25] Quincozes-Santos A, Gottfried C. Resveratrol modulates astroglial functions: neuroprotective hypothesis. Ann N Y Acad Sci. 1215, 72-8, 2011.

[26] Cho IJ, Ahn JY, Kim S, Choi MS, Ha TY. Resveratrol attenuates the expression of HMG-CoA reductase mRNA in hamsters. *Biochem Biophys Res Commun* 29, 367(1), 190-194, 2008.

[27] Kim Y, Yun J, Lee C, Lee H, Min K, Kim Y. Oxyresveratrol and hydroxystilbene compounds. *J Biol Chem* 277(18), 16340-16344, 2002.

[28] Shin NH, Ryu SY, Choi EJ, Kang SH, Chang IM, Min KR, Kim Y. Oxyresveratrol as the potent inhibitor on dopa oxidase activity of mushroom tyrosinase. *Biochem Biophys Res Commun* 24, 243(3), 801-803, 1998.

[29] Wenzel E, Somoza V. Metabolism and bioavailability of *trans*-resveratrol. *Mol Nutr Food Res* 49, 472-481, 2005.

[30] Asensi M, Medina I, Ortega A, Carretero J, Baño MC, Obrador E, Estrela JM. Inhibition of cancer growth by resveratrol is related to its low bioavailability. *Free Radic Biol Med* 1, 33(3), 387-398, 2002.

[31] Kim S, Ko H, Park JE, Jung S, Lee SK, Chun YJ. Design, synthesis, and discovery of novel trans-stilbene analogues as potent and selective human cytochrome P450 1B1 inhibitors. *J Med Chem* 3, 45(1), 160-164, 2002.

[32] Fulda S. Resveratrol and derivatives for the prevention and treatment of cancer. *Drug Cardiovasc Drug Rev.* 22(3), 169-88, 2004.

In: Anthocyanins: Structure, Biosynthesis and Health Benefits
Editor: Noboru Motohashi
ISBN: 978-1-62257-329-5

Chapter 12

COMPARATIVE HUMAN *IN VITRO* AND *IN VIVO* BIOAVAILABILITY INVESTIGATION OF BILBERRY ANTHOCYANINS IN DIFFERENT COMPLEX LIGANDS WITH DIFFERENT COPIGMENTATION STATUS

Thomas Eidenberger**, *Manuel Selg and Sigrid Fuerst
Upper Austrian University of Applied Sciences, Faculty of Natural and Environmental Sciences, Wels, Austria

ABSTRACT

The colorful anthocyanins are well recognized members of the bioflavonoid phytochemicals. Anthocyanins have gained much attention as the food ingredients with health-promoting functions for recent years. Bilberries have been particularly known as one of the richest sources of anthocyanins.

The physiological effects of anthocyanins in humans are dependent on the absorption after ingestion. Clinical studies have demonstrated that the bioavailability of anthocyanins is very low and highly variable because of their instability in physiological absorption conditions. The targets for the development of new anthocyanin products are improved absorption and reduced absorption variability. One way to achieve these targets is an increase of the stability of anthocyanins under physiological conditions.

Amongst others, copigmentation of anthocyanins is a natural occurring mechanism to stabilize anthocyanins. For this reason, anthocyanin extracted from bilberry fruits have been investigated for their copigmentation effects in solution.

Bilberry Anthocyanin Extracts (BAEs) exerting strong copigmentation effects were tested by both, an *in vitro* CaCo-2 cell culture system and an *in vivo* human clinical trial. It could be shown convincingly that copigmentation effects of BAEs were correlated with an increased absorption as well as with reduced absorption variability.

* Corresponding author: Thomas Eidenberger, E-mail address: thomas.eidenberger@fh-wels.at.

Keyword: BAE 2 (MyrtArgos), *vaccinium myrtillus* anothocyanins (VMA), copigmentation, bioavailability, CaCo-2 cell, randomized human clinical trial (RTC)

Abbreviations

Dp-Gal : delphinidin 3-*O*-galactoside; Dp-Glu:delphinidin 3-*O*-glucoside; Dp-Ara:delphinidin 3-*O*-arabinoside;

Cy-Gal : cyanidin 3-*O*-galactoside;Cy-Glu:cyanidin 3-*O*-glucoside; Cy-Ara:cyanidin 3-*O*-arabinoside;

Pt-Gal : petunidin 3-*O*-galactoside; Pt-Glu: petunidin 3-*O*-glucoside; Pt-Ara: petunidin 3-*O*-arabinoside;

Pe-Gal: peonidin 3-*O*-galactoside; Pe-Glu: peonidin 3-*O*-glucoside; Pe-Ara: peonidin 3-*O*-arabinoside;

Mv-Gal: malvidin 3-*O*-galactoside; Mv-Glu: malvidin 3-*O*-glucoside; Mv-Ara: malvidin 3-*O*-arabinoside.

1. Introduction

Anthocyanins belong to the flavonoid group of polyphenolic phytochemicals, generally occurring in fruits, vegetables, nuts, plant oils, cocoa and cereals. Flavonoids are comprised of a huge group (> 10.000) of diverse aromatic plant compounds with multiple substitution patterns. Among the flavonoid family, anthocyanidins are of special interest because their intake from the daily human diet is estimated to be more than 100 mg/day [1-3]. However, the bioavailability of anthocyanins in humans has been regarded very low and in studies investigating the urinary excretion the bioavailability was actually ranging only from 0.018% to 0.37% of the ingested does [4].

2. General

Based on increasing consumer awareness for natural instead of synthetic food ingredients, anthocyanins have been introduced into the food industries as coloring food ingredients and natural food colorants. More recently anthocyanins were recognized as functional food ingredients due to a number of physiological effects attributed to a daily ingestion of anthocyanins [5-8].

The increasing awareness of anthocyanins has promoted research into the stability and coloring properties, the structural elucidation and the distribution of anthocyanins in functional foods. Anthocyanin extracts prepared from edible berry fruits have gained much attention in many functional foods applications [9-13]. Among berry fruits, bilberries, *Vaccinium Myrtillus* L., (Photo 1) have been used with special interest for health-promoting foods because of their high natural contents of different types of anthocyanins [14-16]. *Vaccinium myrtillus* anthocyanins (VMA) consist of at least 15 anthocyanins which are composed of 5 anthocyanidin skeletons, each occurring as 3-O-glycoside with glucose (Glu), galactose (Gal) or arabinose (Ara) (Figure 1) [17, 18].

2.1. *In Vivo* Studies with Berry Anthocyanins

Pharmacokinetic studies in humans and animals demonstrated that edible berry anthocyanins were absorbed intact into the blood. However, the bioavailability is poor most likely because of the high instability of anthocyanins under physiological absorption conditions. Anthocyanins are quickly degradaded before absorption. Therefore, the low bioavailability of anthocyanins can be improved simply by improvement of absorption [19-24].

Poor bioavailability of nutrients is of no major concern unless the absorption is highly variable. Several published studies strongly indicated that the absorption of anthocyanins is highly variable. High variability of poorly absorbed compounds causes variable physiological effects and finally makes prediction for a dose-depending cause-and-effect relationship difficult, if not impossible.

2.2. Stability of Anthocyanins Depending on pH

Development of novel functional food products rich in anthocyanins must overcome the instability of anthocyanins under different physiological conditions. The stability of anthocyanins in solution and the color appearance are strongly correlated [25-31]. Depending on the pH value, the color of anthocyanins changes substantially (Figure 2). The changes observed depend on a structural change of the anthocyanidin skeleton (Figure 3). The change substantially influences the maximum absorption wavelength (λ_{max}) and the extinction value (E) at λ_{max} (Figure 4). In the acidic region, the color is stable over time indicating stability of the corresponding anthocyanidin structure. However, at physiologically relevant weakly acidic and neutral pH, the color of anthocyanins is unstable over time indicating degradation the corresponding anthocyanin structures. Determination of color stability over time using visual spectrum or color measurement (CIE-Lab) could be used to estimate the stability of anthocyanins [32-35]. To obtain quantitative information on the degradation of anthocyanins, high performance liquid chromatography (HPLC) analyses are used [36, 37].

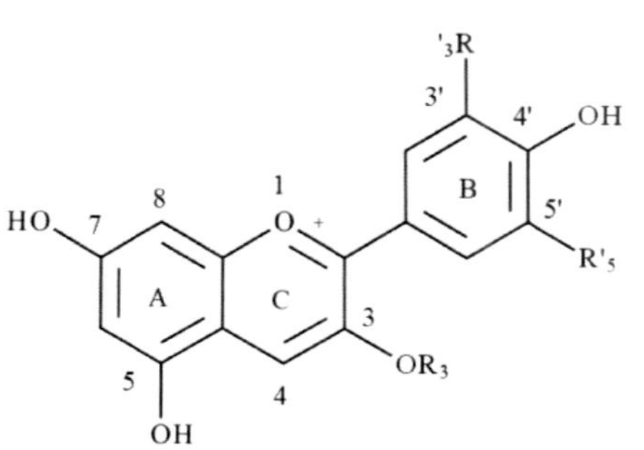

Compound	R'3	R'5
cyanidin (Cy)	OH	H
delphinidin (Dp)	OH	OH
petunidin (Pt)	OH	OCH_3
peonidin (Pe)	OCH_3	H
malvidin (Pt)	OCH_3	OCH_3

Figure 1. Structures of 5 anthocyanidins occurring in bilberry.

Photo 1. Bilberry fruits in harvest time. Photographed by Hannu Huttsu, Aug 2006.

In their natural environment, anthocyanins appear to be stable under conditions resembling the physiological absorption conditions. Several, naturally occurring mechanisms for anthocyanin stabilization in plants have been identified [38]. In addition to the stabilization of anthocyanin structures, those mechanisms cause usually significant changes in the color appearances which explains the enormous color variation of anthocyanins in nature. **Color changes could be explained by both, a bathochromic** (λ_{max} drifts to higher wavelengths) **and a hyperchromic** (ϵ **at** λ_{max} increases) shift.

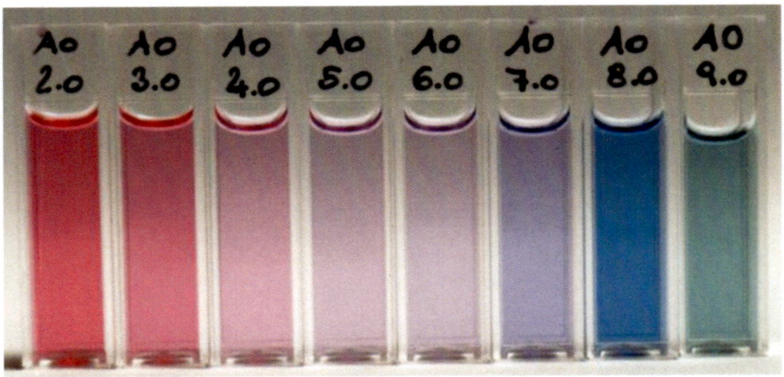

Figure 2. pH-dependent shift of the color in aqueous solutions of VMA extracts (BAEs) (McIlvaine buffer system, numbers correspond to pHs).

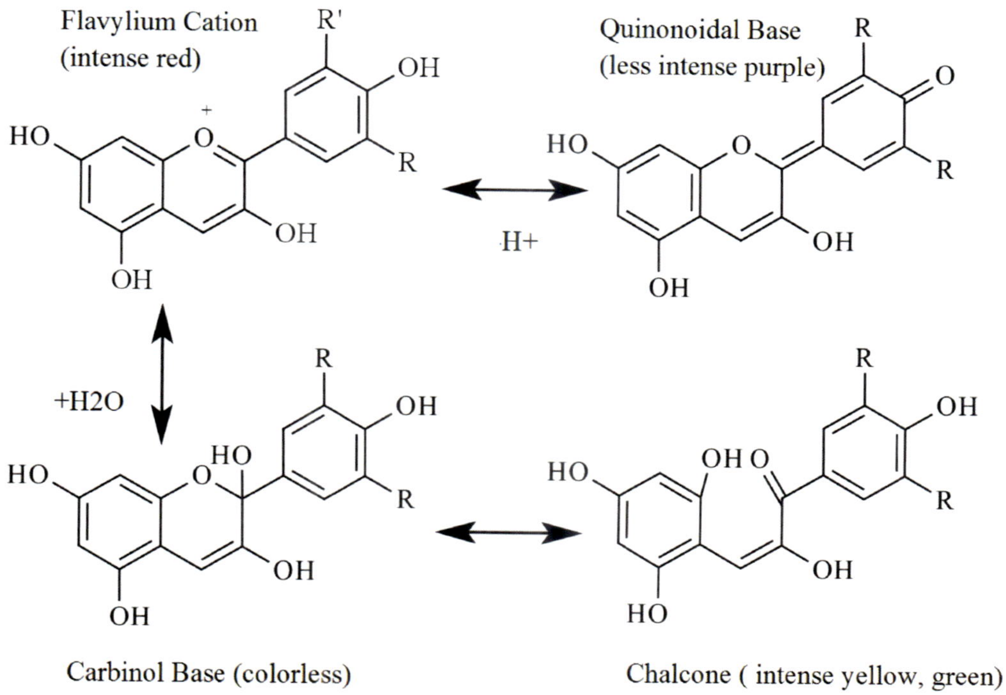

Figure 3. pH-dependent change of the anthocyanidin skeleton structure.

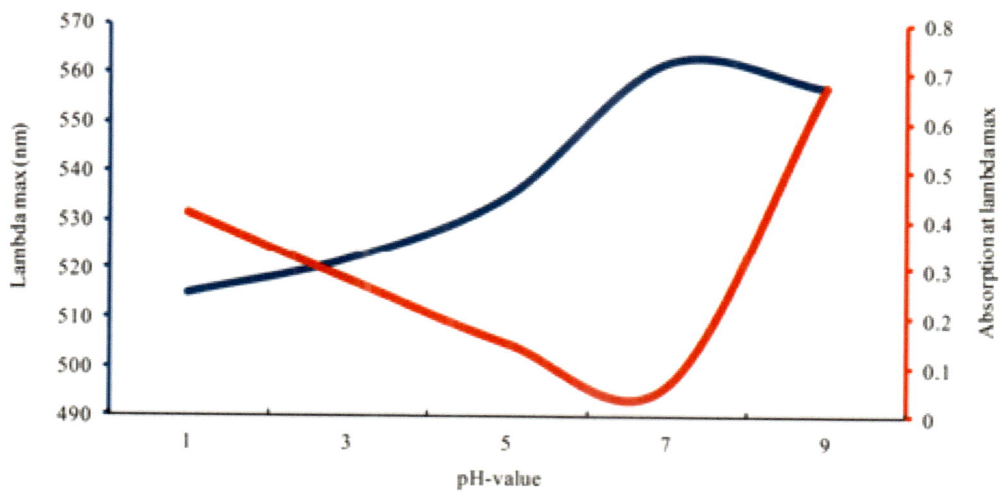

Figure 4. pH-**dependent shift of λ_{max} [blue line] and Є at λ_{max}** [red line] (10^{-3} M aqueous solution Cy-Glu in McIlvaine buffer system).

2.3. Formation of Anthocyanin-Complex with Metal Ions

Anthocyanins can form stable complexes in presence of metal ions such as Al^{3+} (to a lesser extent with Mg^{2+}, Fe^{2+} and other metal ions) as shown in Figure 5 [39].

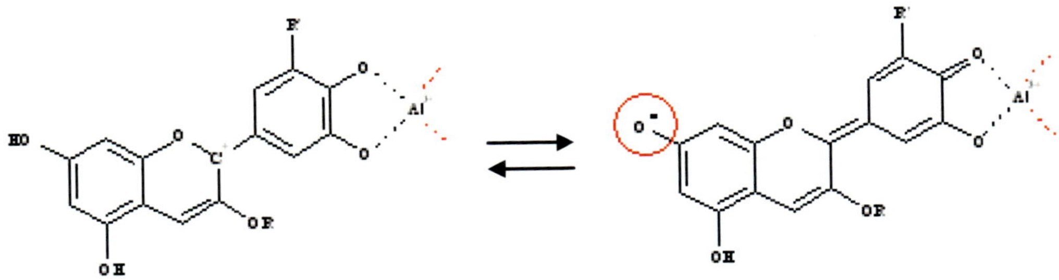

Figure 5. Complex of anthocyanin with Al^{3+}. Left: pH around 3-4; Right: pH around 4-6 (red parts indicate anchor point for other anthocyanin molecules).

Anthocyanins with acyl groups bound to the sugars may form complexes with other flavonoids around a central metal ion. Such complexes are known as metalloanthocyanins possessing interesting properties in terms of color, stability and solubility [40]. Examples are the metalloanthocyanins occuring in *Centaurea cyanus* L. (protocyanin) and of *Commelina communis* L. (commelinin) [41, 42].

2.4 Anthocyanin – Copigmentation

Copigmentation comprises of several mechanisms for anthocyanin stabilization (Figure 6) and is an important phenomenon strongly contributing to the stability and color variation of anthocyanins. Also, copigmentation effects can be observed as batho- and hyperchromic shifts in solutions. The magnitude of both batho- and hyperchromic shifts depend on the copigmentation mechanism, the content of the copigmentation formation, the ratio of anthocyanins to copigments, and the pH of the solution. Although there is no strict correlation, larger batho- and hyperchromic shifts indicate higher copigmentation strength. Copigmentation effects are also described for the anthocyanins occurring in bilberries [43-47].

Current production techniques of commercial anthocyanin extracts used in functional food ingredients are focused on optimum anthocyanin extraction yield and neglect to investigate whether copigmentation effects are retained in the final products. Innovative products with a strong copigmentation effect should lead to higher anthocyanin stability under physiological conditions and provide a possibility to increase the absorption of anthocyanins.

For this reason, it was attempted to develop novel anthocyanin extracts from bilberries with a special focus on copigmentation effects observed under physiological conditions. These candidate extracts were tested using both *in vitro* and *in vivo* to confirm whether copigmentation could increase the absorption or reduce variability of the absorption of anthocyanins.

This chapter presents the results of both *in vitro* and *in vivo* experiments to investigate the absorption of VMA from extracts with different copigmentation strengths.

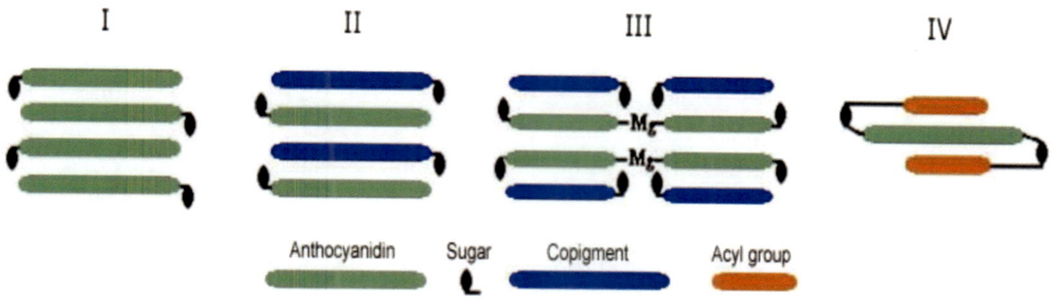

Figure 6. Overview of copigmentation mechanisms.
I : Intermolecular self-association of anthocyanins.
II : Intermolecular association of anthocyanins and copigments (flavone or flavonol).
III: Intermolecular association of anthocyanins and copigments augmented by metal ions.
IV: Intramolecular association of acylated anthocyanins (acyl group is the copigment).

3. MATERIALS

Solvents (HPLC-grade), standard chemicals including *p*-coumaric acid (p.a.-grade or better) and stationary phase material (Sephadex LH20, Amberlite XAD-7 and poly(vinylpolypyrrolidone 110 µm mesh) were obtained from Sigma-Aldrich Inc. (St. Louis, MO, USA) or VWR International, LLC (Radnor, USA). Anthocyanin reference compounds were obtained from Extrasynthese (Lyon, France). Bilberries (crop 2006) were harvested from Savukoski area Finland (Photo 1). BAE 1 (Standard Bilberry extract) and BAE 2 (MyrtArgos) were kindly supplied by Omnica Co., Ltd. in Tokyo, Japan.

4. METHODS

4.1. Copigmentation Test

Single anthocyanins, VMA extracts and copigments were dissolved in McIlvaine citrate-phosphate buffer (pH 4.5 or 7.5). Copigmentation of single anthocyanins with copigments was determined at a molar ratio of 1:20. BAEs were dissolved directly in buffer and cleared by filtration. For comparative purposes, tentative copigments present in BAEs were removed by purification over poly(vinylpolypyrrolidone) at the conditions described below.

Visible spectrum analyses and color measurements were performed with a Hunter Lab ColorQuest XE. Incubations were performed at 37 °C for up to 360 min with samples taken every 10-15 minutes. For HPLC conditions see 4.4. All tested samples were adjusted for the same content of anthocyanins.

4.2. Preparation of Bilberry Extracts

An almost pure anthocyanin concentrate lacking copigments (PBAC) was prepared as a control following previously published techniques [48-51]. Briefly, 3 kg bilberries collected in Finland and provided by Omnica Co., Ltd., were crushed and extracted 3 times with ethanol/formic acid=9/1 (v/v). The combined extracts (9 L) were evaporated by vacuum distillation. The dry residue was dissolved in 100 mL deionized water/formic acid=99.95/0.05 (v/v). Aliquots of 5 mL were loaded on a semi-preparative column (Amberlite XAD-7, 40 x 3.5 mL bed volume) equilibrated with water/formic acid=99.95/0.05 (v/v). The column was washed with deionized water/formic acid=99.95/0.05 (v/v), the anthocyanin fraction was then eluted with ethanol/formic acid=9/1 (v/v). The combined anthocyanin fractions were evaporated by vacuum distillation. After dissolution in 50 mL water/formic acid=99.95:0.05 (v/v) aliquots of 5 mL were loaded onto on a semi-preparative column (Sephadex LH-20, 40 x 3.5 mL bed volume) equilibrated with deionized water/formic acid=99.95/0.05 (v/v). After washing of the column with deionized water/formic acid=99.95/0.05 (v/v), elution of anthocyanins was accomplished by a stepwise gradient from 5 to 70 % ethanol in deionized water/formic acid=99.95/0.05 (v/v). The combined anthocyanin fractions were evaporated by vacuum distillation. After dissolution in 50 mL deionized water/trifluoroacetic acid=99/1 (v/v) aliquots of 5 mL were loaded on a semi-preparative column (poly(vinylpolypyrrolidone) 110 µm mesh, 40 x 3.5 mL bed volume) equilibrated with deionized water/trifluoroacetic acid=99/1 (v/v). After washing of the column with deionized water/trifluoroacetic acid=99/1 (v/v), elution of anthocyanins was accomplished with ethanol/trifluoroacetic acid=95/5 (v/v). The combined fractions were evaporated by vacuum distillation and dissolved in 10 mL trifluoroacetic acid followed by drop-wise addition to 1 Liter ice-cold water free diethylether for crystallization of anthocyanins. The overall extraction yield was 7.2 g PBAC.

In cooperation with Omnica Co., Ltd. (Tokyo, Japan) BAEs were produced and characterized for copigmentation effects. For series I of BAEs, bilberries were grinded and extracted in an ethanol/water/phosphoric acid solution. After concentration by vacuum distillation, the anthocyanins were further concentrated by preparative resin absorption. Afterwards they were fractionated and eluted in ethanol/water mixtures. The anthocyanin containing fractions were concentrated by vacuum distillation and finally spray dried. Variations included the ethanol/water ratio in the extraction solvent and the addition of antioxidant (*L-cysteine*) to prevent oxidation processes during manufacturing.

For series II of BAEs, bilberries were squeezed to obtain juice. The press residue from juice preparation was grinded and extracted in an ethanol/water/phosphoric acid solution. The juice was further processed as described above. Further on, the juice and the extract were combined in varying ratio and processed as described above. Again, variations included the ethanol/water ratio in the extraction solvent and addition of antioxidants.

For comparative purposes, potential copigments were removed from the BAEs produced by purification over poly(vinylpolypyrrolidone) chromatography as described above.

4.3. CaCo-2 Cell Test System

The CaCo-2 cell culture test system was chosen to estimate the rate and extent of absorption *in vitro* [52-54]. CaCo-2 cells were cultured in Dulbeccos's Modified Eagle

Medium containing 20% fetal bovine serum, 1.2 % non-essential amino acids, 0.83 mM L-glutamine, 1.2 % penicillin-streptomycin and 0.1 % mercaptoethanol in an atmosphere of 5% CO_2 and 95% air at 37°C. Cells were grown in 75 cm^2 culture-flasks (T75) and sub-cultured after one week. Prior to experiments, cells were seeded in wells at a density of $3x10^5$ cells/well (standard 6-well cell culture plate) and grown in an atmosphere of 5% CO_2 and 95% air at 37°C until confluence was reached. For the experiments, BAEs were dissolved in incubation medium (adjusted to 1 mg anthocyanins/ml medium) and transferred into the wells. The cells were incubated for 30, 60 and 90 minutes at 37 °C. At each time point, the cells from 5 wells were removed for further processing (removal of medium, washing of cells with buffer, harvest of cells, and disruption of cells in acidified methanol, centrifugation and storage of the clear supernatant until analysis by HPLC/UV. Each experiment was performed in three replicates providing 15 samples/experiment/time-point. The individual anthocyanin concentrations observed in each well were summed up. Results given are mean values of the summarized anthocyanin concentration observed per well and time point (Figure 13, Table 3).

4.4. HPLC Analysis

Separation of anthocyanins was achieved on a Knauer, Hypersil ODS (250×4.6 mm) column with a flow rate of 1.5 mL/min using a linear gradient from 100 % deionized water/formic acid =90/10 (v/v) to 30 % methanol/acetonitrile/formic acid/water=20/20/10/50 (v/v) in 40 minutes. The column was kept at 45 °C. Detection was performed with UV/VIS (520 nm) coupled to positive mode ESI/MS [55, 56]. Quantification was performed against external standardization (Cy-Glu, Dp-Glu, Mv-glu, Pe-Glu) using single ion monitoring selective for anthocyanins and anthocyanidins occurring in bilberries. The qualitative elucidation of anthocyanin structures was performed using total ion current settings.

During the copigmentation and the *in vitro* experiments the samples collected were submitted to HPLC without further treatment. Samples obtained during the clinical study were subjected to solid phase extraction before analysis [57, 58].

4.5. Pharmacokinetic Study in Healthy Human

For pharmacokinetic investigations in humans, a randomized crossover design with multiple treatments in 12 subjects of either gender was chosen (Figure 12). The study design followed GCP guidelines, the Declaration of Helsinki, and fulfilled the guideline for bioequivalence testing [59-61]. The design of the study was approved by the Responsible Ethics Committee.

Primary aim of the study was to investigate the pharmacokinetic profile of VMA after single and multiple intakes of two selected BAEs (BAE 1 or BAE 2) in order to compare the rate and extent of absorption of the anthocyanins. During each period, the participants received once daily treatments for 7 consecutive days with 185 mg anthocyanins from either BAE 1 or BAE 2. Between the 2 periods a wash-out period of 7 days was introduced. During each period blood samples were drawn for pharmacokinetic assessments on the corresponding treatment days 1 (single dose) and 7 (repeated dose) immediately before and 0.5, 1.5, 2.5 and 4 hours after administration (Figure 13). Blood samples drawn were immediately cooled to

4 °C and centrifuged at 4 °C to obtain plasma. Plasma samples were acidified with formic acid (1800 µL blood + 200 µL formic acid), vortexed for 5 minutes and centrifuged again at 13.000 rpm for 5 minutes. The supernatant was collected, transferred into an appropriately labeled vial and stored at -80 °C until analysis. From the plasma concentrations observed, the pharmacokinetic parameters C_{max} (maximum concentration) (Table 4, Figure 13) and $AUC_{0\text{-}inf}$ (area under the concentration time curve extrapolated from 0 to infinity) (Tables 6, 7) were calculated using non-compartmental methods. Bioequivalence assessment was performed according to published methods [62, 63]. Bioequivalence assessment consisted of the calculation of the intra-individual ratios of C_{max} and $AUC_{0\text{-}inf}$ values BAE 2/BAE 1, the calculation of the corresponding mean ratio (comparative bioavailability) and the confidence interval (C.I.). If the mean ratio, together with the confidence interval, lies within the margins of 80-125 % the two products tested are bioequivalent; otherwise the products are not bioequivalent (Tables 4,5,6,7).

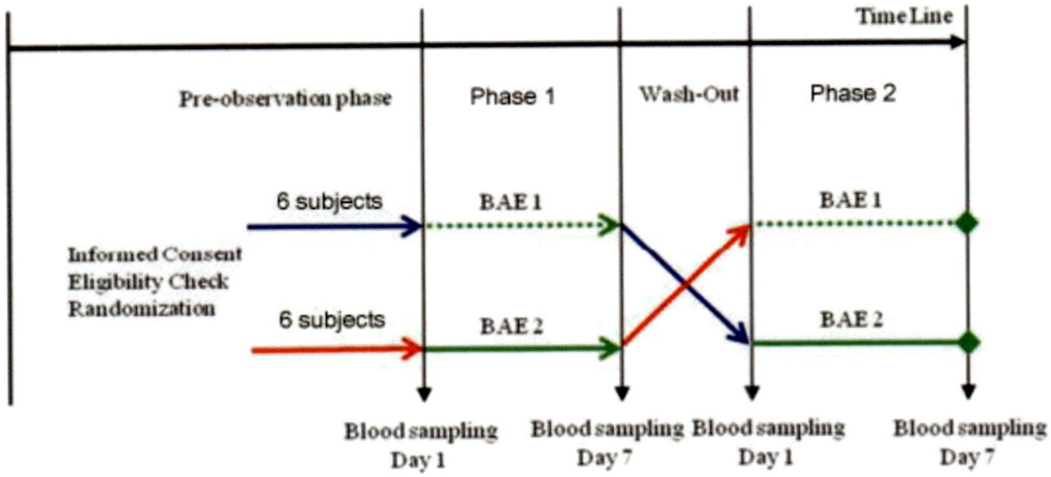

Figure 12. Flowchart of clinical study interventions.
Total (n=12) average age 33 y; male (n=3) average age 34 y; female (n=9) average age 38 y.

5. Results

5.1. Effects of Copigmentation

Exemplifying for the intermolecular association of anthocyanins and copigments, the batho- and hyperchromic shift obtained for copigmentation of Cy-Glu with *p*-coumaric acid at pH 4.5 are shown in Figure 7. In Figure 8, the stabilization observed for copigmentation of Cy-Glu with coumaric acid at pH=7.5 and 37 °C is presented. The results obtained indicate that copigmented anthocyanins are substantially more stable under the physiologically relevant conditions.

5.2. Development of Copigmented Bilberry Anthocyanin Extracts (BAEs)

The primary target was to elucidate the influence of different manufacturing processes on copigmentation effects of BAEs.

A reference PBAC completely devoid of copigments was prepared. The anthocyanin content in PBAC was determined to be >85 %. All 15 anthocyanins occurring in bilberry were retrieved by HPLC in their approximate relative distribution. This indicated that the applied purification technique yielded an apparently pure anthocyanin fraction with the naturally occurring anthocyanin pattern. The visible spectrum and color characteristics of PBAC obtained were interpreted as typical for VMA without copigmentation effects.

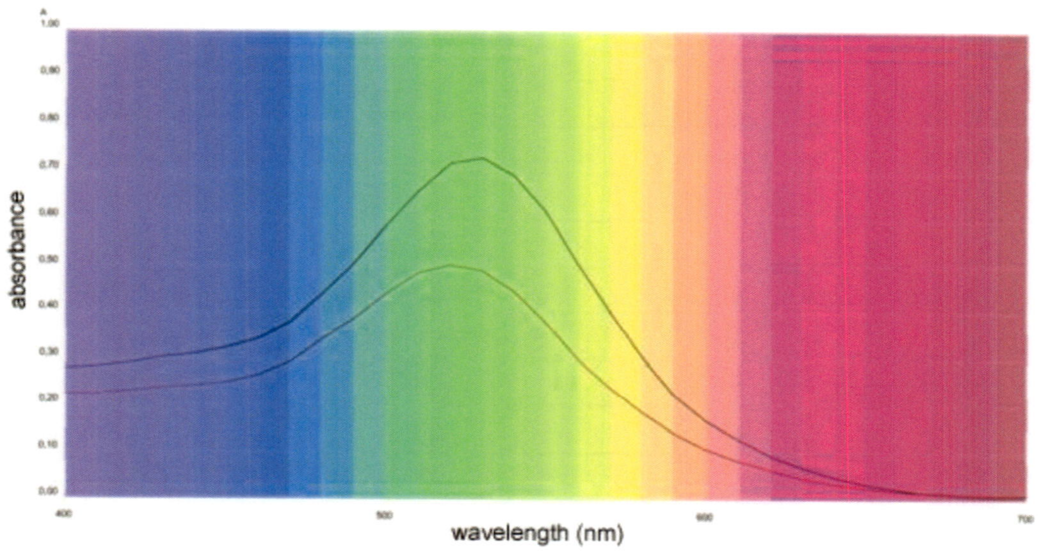

Figure 7. Visible absorption spectrum of pure (lower trace: red) and Cy-Glu copigmented with coumaric acid (upper trace: black) at pH 4.5 (bathochromic shift: +15 nm; hyperchromic shift: +50 %).

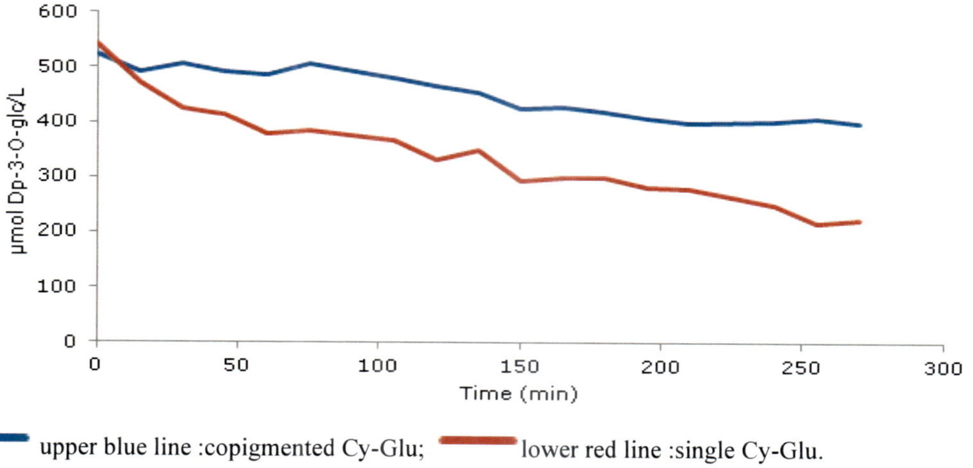

upper blue line :copigmented Cy-Glu; lower red line :single Cy-Glu.

Figure 8. Degradation of Cy-Glu copigmented *p*-coumaric acid at pH 7.5 and 37 °C.

The BAEs prepared were compared with PBAC for copigmentation effects. Based on the results obtained, two extracts denominated as BAE 1 (low copigmentation effect) and BAE 2 (high copigmentation effect, MyrtArgos) were selected for further experimental works. The visible spectrum and color characteristics as well as the stability (aqueous buffer pH=7.5, 37 °C, up to 6 hours) of BAE 1 and BAE 2 are shown in Tables 1, 2, and Figures 9, 10. BAE 1 was obtained from series I with 80 % ethanol in the extraction solvent and antioxidants included. The anthocyanin content was 34.5 %. BAE 2 was obtained from series II with supplementary antioxidants and combining the whole press residue extract (extracted with 80 % ethanol) with the juice. The anthocyanin content was 38.3 %.

The results strongly indicate that BAE 1 and BAE 2 show spectroscopic properties of copigmentation. That copigmentation reactions are the reason for the spectroscopic changes could be derived from the comparison of BAEs prior to and after removal of copigments. In a general agreement, the magnitude of the copigmentation effect was taken as an estimate of the copigmentation strength. The results of color measurement indicated that the color stability correlated with the copigmentation strength. The results of the stability revealed that the stability of anthocyanins under physiological conditions correlates with the copigmentation strength observed. The removal of copigments diminished the stabilization effect.

Table 1. Summary of visible spectra characteristics of three VMA extracts (aqueous buffer pH 4.5) PBAC, BAE 1 and BAE 2

Sample characteristic	Bathochromic shift (difference in λ_{max}, nm)	Hyperchromic shift (difference in E_{max}, %)
PBAC	Reference=520 nm	Reference=100 %
BAE 1	+10	123
BAE 1 (after removal of copigments)	+2	104
BAE 2	+25	167
BAE 2 (after removal of copigments)	+5	107

Table 2. Color characteristics (CIE-lab) of three VMA extracts in aqueous solution (pH=4.5) obtained immediately after dissolution, and after 24 hours at room temperature

Sample characteristic	L*	a*	b*	dL*	da*	db*	dE*
PBAC	83.37	11.87	-2.65	-12.42	11.92	-2.53	12.03
PBAC after 24 hours	89.67	5.28	0.23	-6.12	5.33	0.35	5.16
BAE 1	77.90	12.40	-0.02	-17.89	12.45	0.10	12.27
BAE 1 after 24 hours	81.79	10.22	-0.07	-14.00	10.27	0.05	10.09
BAE 2	83.47	10.87	1.30	-12.33	10.93	1.42	10.82
BAE 2 after 24 hours	82.56	12.55	-0.55	-13.24	12.60	-0.43	12.43

L*:lightness; a*:color strength between purple-red and blue-green; b*:color strength between yellow and blue; dL*, da*, db*: L*, a*, b* corrected for standard; dE*:total color difference.

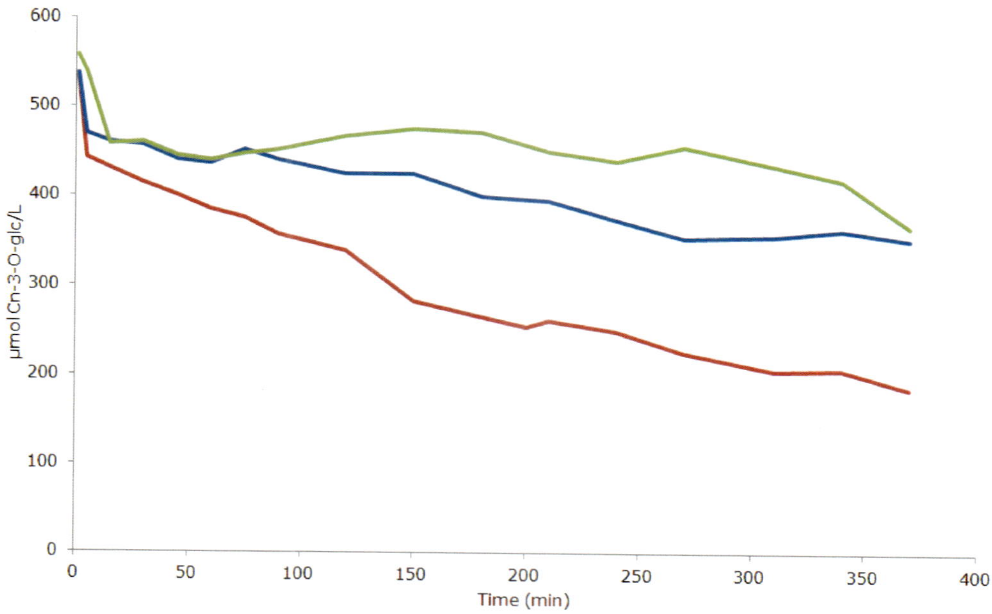

lower red line:PBAC; medium blue line:BAE 1; upper green line:BAE 2.

Figure 9. Comparative stability of 3 VMA extracts (aqueous buffer pH=7.5, 37 °C, HPLC/UV with external standardization).

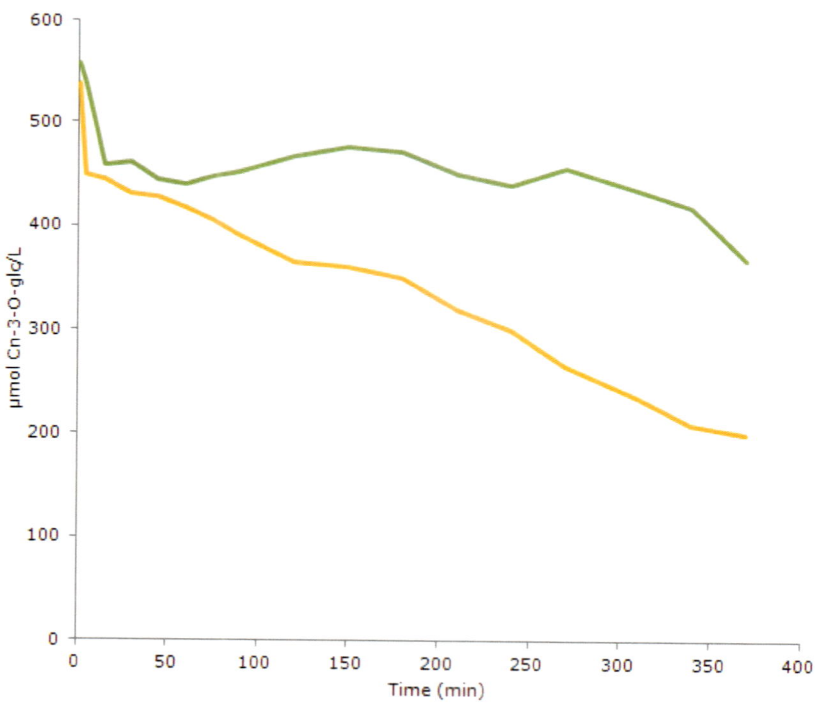

upper green line: BAE 2; lower yellow line: BAE 2 after removal of copigments.

Figure 10. Stability of VMA extracts in relation to copigmentation (aqueous buffer pH=7.5, 37 °C, HPLC/UV).

5.3. *In Vitro* Investigations with CaCo-2 Cell

The absorption of anthocyanins from PBAC, BAE 1 and BAE 2 were investigated *in vitro* using the CaCo-2 cells culture system. Figure 11 shows the amount of anthocyanins recovered in the cells at the indicated time points, whereas Table 3 contains the amount of 15 anthocyanins determined after 60 minutes of incubation. These results strongly indicate that the absorption of anthocyanins into CaCo-2 Cell correlates with the copigmentation strength. An estimate of the increase could be derived from the concentrations of 15 anthocyanins observed after 60 minutes of incubation. Strong copigmentation doubled the amount of anthocyanins absorbed when compared to PBAC lacking copigmentation. It should be noted that the amount of all 15 anthocyanins was increased suggesting that copigmentation could protect all anthocyanins present in bilberries.

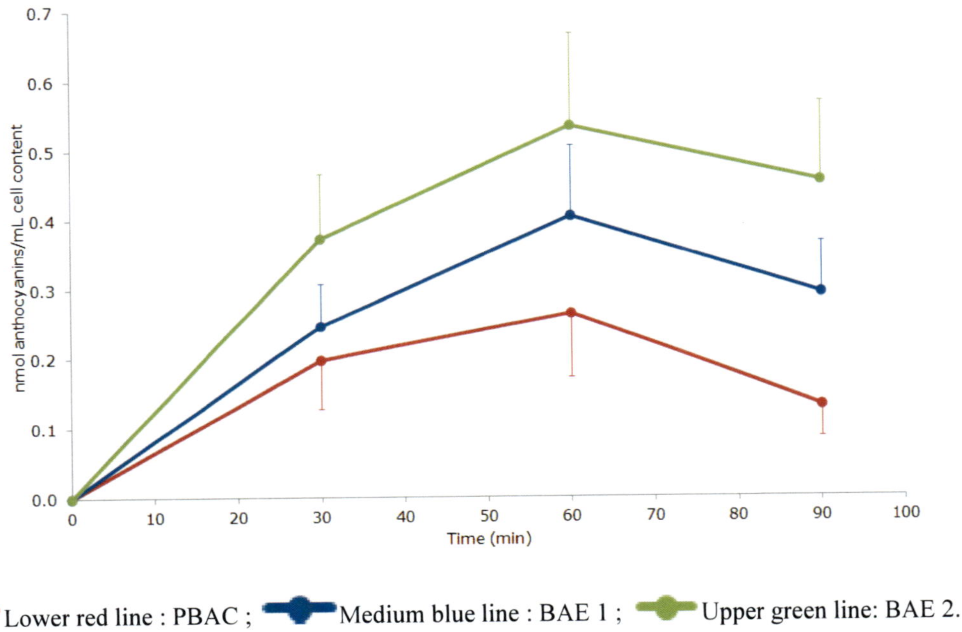

Figure 11. Mean values of anthocyanins ± s.d. (n=15) absorbed into CaCo-2 cells from 3 bilberry extracts.

5.4. Pharmacokinetic Study in Humans

The absorption of anthocyanins from BAE 1 and BAE 2 was investigated *in vivo* in a randomized, cross-over pharmacokinetic study in 12 subjects with single and repeated oral administration.

Figure 13 shows the mean values of anthocyanidin-related structures observed in plasma after single and repeated administration of either BAE 1 or BAE 2. The values shown refer to the sum of all individual structures bearing the corresponding anthocyanidin. Table 4 gives the maximum plasma concentrations observed for 15 anthocyanins after treatment with BAE 1 or BAE 2. Table 5 shows the scattering of C_{max} values observed. The results of the bioequivalence assessment are shown in Tables 6 and 7.

Table 3. Mean values of 15 anthocyanins (nmol/mL cell content) absorbed into CaCo-2 cell from three bilberry extracts after 60 minutes incubation (n=15) at 37C

Compound	PBAC	BAE 1	BAE 2
Dp-Gal	0.0303	0.0450	0.0601
Dp-Glc	0.0306	0.0443	0.0629
Dp-Ara	0.0268	0.0367	0.0460
Cy-Gal	0.0365	0.0557	0.0669
Cy-Glu	0.0287	0.0454	0.0620
Cy-Ara	0.0110	0.0132	0.0211
Pt-Gal	0.0284	0.0319	0.0511
Pt-Glu	0.0174	0.0270	0.0318
Pt-Ara	n.a.	0.0124	0.0169
Pe-Gal	n.a.	0.0114	0.0130
Pe-Glu	0.0166	0.0240	0.0258
Pe-Ara	n.a.	n.a	n.a
Mv-Gal	0.0170	0.0251	0.0240
Mv-Glc	0.0201	0.0329	0.0358
Mv-Ara	n.a.	n.a	0.0194
Sum	0.264	0.405	0.537
(%)	(100 %)	(153 %)†	(203 %)†

n.a.:not applicable (below limit of detection); †:relative to sum of PBAC.

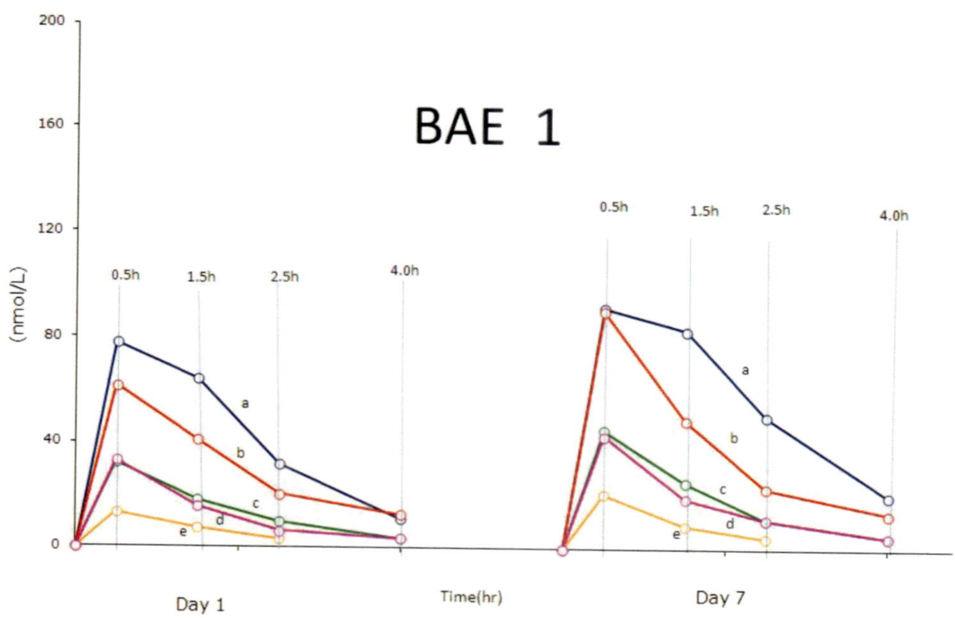

Figure 13. (A) Mean concentrations of anthocyanidin-relates compounds (nmol/L) after single (day 1) and repeated (day 7) oral administrations of BAE 1 (n=12). Each plasma level/time curve covers 0-4 hours (at 0.5h/1.5h/2.5h/4.0h, respectively) after treatment; blue line a: cyanidin; red line b: delphinidin; green line c: petunidin; orange line d: peonidin; magenta line e: malvidin.

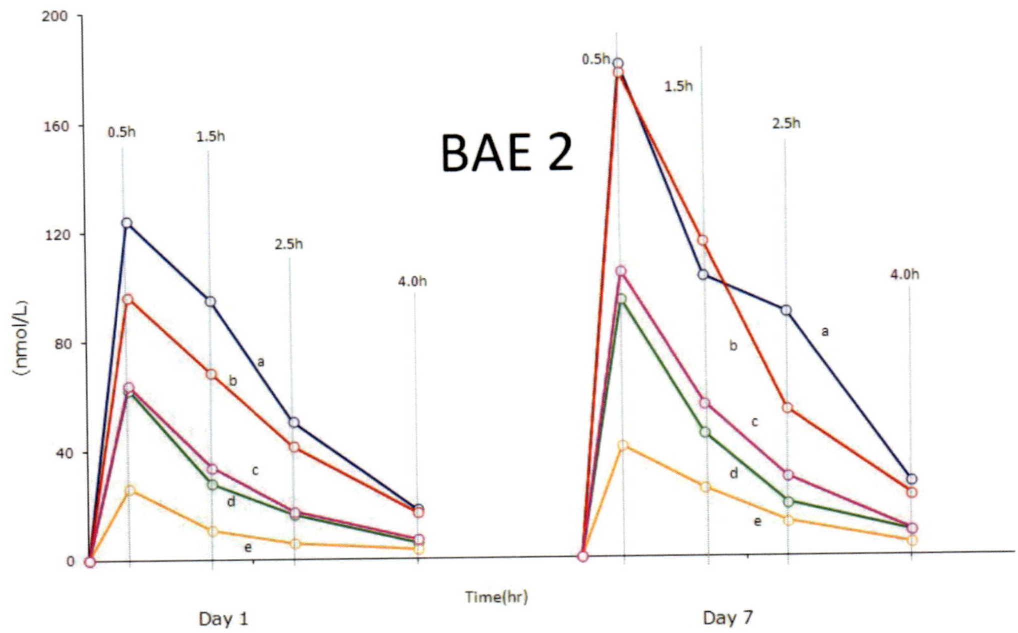

Figure 13. (B) Mean concentrations of anthocyanidin-relates compounds (nmol/L) after single (day 1) and repeated (day 7) oral administrations of BAE 2 (n=12). Each plasma level/time curve covers 0-4 hours (at 0.5h/1.5h/2.5h/4.0h, respectively) after treatment; blue line a: cyanidin; red line b: delphinidin; green line c: petunidin; orange line d: peonidin; magenta line e: malvidin.

Table 4. Mean maximum concentrations (nmol/L) of VMA after single and repeated administration of BAE 1 and BAE 2 (n=12, respectively)

Compound	C_{max}, single administration		C_{max}, repeated administration	
	BAE 1	BAE 2	BAE 1	BAE 2
Cy-Glu	21.4	42.5	31.0	64.8
Cy-Glu	11.6	17.1	16.5	33.9
Cy-Ara	8.35	15.5	12.9	26.3
Dp-Glu	11.6	23.4	19.6	37.0
Dp-Gal	9.03	16.3	13.1	30.3
Dp-Ara	9.20	16.8	14.3	25.7
Pt-Glc	11.9	20.0	16.7	35.3
Pt-Gal	4.18	7.10	5.64	12.7
Pt-Ara	2.96	7.74	4.33	12.8
Pe-Glc	6.26	13.8	10.4	23.8
Pe-Gal	1.99	4.75	3.02	8.86
Pe-Ara	2.54	4.39	2.77	8.31
Mv-Glc	12.2	24.9	17.0	44.0
Mv-Gal	4.06	8.32	6.29	12.8
Mv-Ara	3.46	7.56	5.18	12.5

Table 5. Scattering of C_{max} values (Sum of anthocyanidin-related structures) after single (day 1) and repeated (day 7) treatments with BAE 1 and BAE 2, coefficient of variation (%) (n=12, respectively)

Compound	Single treatment		Repeated treatment	
	BAE 1	BAE 2	BAE 1	BAE 2
Cyanidin-relates	26.6	21.2	24.0	20.5
Delphinidin-relates	29.4	23.3	27.8	24.9
Petunidin-relates	30.2	26.1	29.8	25.4
Peonidin-relates	27.8	24.6	29.5	25.0
Malvidin-relates	24.3	19.7	23.9	20.1

The results strongly indicate that the mean anthocyanin plasma levels were considerably higher after treatment with BAE 2 when compared to BAE 1. Likewise the values for C_{max} and $AUC_{0\text{-}inf}$, indicative for the rate and extent of absorption, of BAE 2 were higher when compared to that of BAE 1. The effects observed are more significant after repeated treatments when compared to single treatment. Interestingly, the individual values for C_{max} after treatment with BAE 1 indicated higher variability when compared to BAE 2 (Table 7). Whether this points to reduced variability in the absorption due to higher copigmentation strength cannot be concluded at this stage. However, the finding could indicate that copigmented anthocyanins were stabilized irrespectively of the individual physiological conditions at the absorption site whereas the stability of anthocyanins devoid of copigmentation effects was more dependent on individual conditions.

The results of the bioequivalence assessment demonstrated a significantly increased relative bioavailability (supra-bioavailability) of BAE 2 when compared to BAE 1 after single (day 1) and repeated treatment (day 7).

Table 6. Results for the bioequivalence test of pharmacokinetic parameters after single treatments with BAE 1 and BAE 2

Compound	Comparative bioavailability BAE 2/BAE 1 (%)		Confidence interval BAE 2/BAE 1 (%)	
	C_{max}	$AUC_{0\text{-}inf}$	C_{max}	$AUC_{0\text{-}inf}$
Cyanidin-relates	162.0[1)]	161.1[1)]	145.5-180.5[1)]	150.6-172.2[1)]
Delphinidin-relates	158.4[1)]	126.4[1)]	142.9-175.6[1)]	89.2-178.2[1)]
Petunidin-relates	182.0[1)]	n.a.	160.9-205.8[1)]	n.a.
Peonidin-relates	215.1[1)]	n.a.	185.0-250.2[1)]	n.a.
Malvidin-relates	197.6[1)]	n.a.	173.1-225.5[1)]	n.a.

1) statistically significant better bioavailability of BAE 2; n.a. : not applicable (AUC could not be determined).

Table 7. Results for the bioequivalence test of pharmacokinetic parameters d after repeated (day 7) treatment with BAE 1 and BAE 2

Compound	Comparative bioavailability BAE 2/BAE 1 (%)		Confidence interval BAE 2/BAE 1 (%)	
	C_{max}	$AUC_{0\text{-}inf}$	C_{max}	$AUC_{0\text{-}inf}$
Cyanidin-relates	196.3[1)]	187.1[1)]	181.3-212.7[1)]	163.4-202.2[1)]
Delphinidin-relates	188.5[1)]	206.4[1)]	163.3-217.7[1)]	165.3-257.8[1)]
Petunidin-relates	201.2[1)]	n.a.	177.6-228.0[1)]	n.a.
Peonidin-relates	202.6[1)]	n.a.	183.9-223.2[1)]	n.a.
Malvidin-relates	237.9[1)]	n.a.	208.8-271.1[1)]	n.a.

1) statistically significant better bioavailability of BAE 2; n.a.: not applicable (AUC could not be determined).

6. DISCUSSION

VMA have gained much attention as functional food ingredients. However, the bioavailability in humans is poor because VMA are instable under physiological conditions, especially under the pH conditions at the absorption site. Moreover, published studies indicate that absorption of anthocyanins is highly variable.

The current knowledge of anthocyanin copigmentation effects could provide a way to investigate whether copigmentation can overcome the instability of anthocyanins under physiological conditions leading to better absorption and reduced variability of absorption.

The investigations at hand showed that the conditions of the manufacturing process for bilberry anthocyanin extracts influence the copigmentation strength of the resulting product. As described in the experimental setups, the direct extraction of bilberries yields a low copigmentation effects only. Strong copigmentation effects were only achieved by obtaining a berry juice, extraction of the press residue and combining both, the juice and the press residue extract. Moreover, a positive effect of antioxidants on the resulting copigmentation strength was observed. The investigations showed that different manufacturing processes can yield extracts with substantially higher copigmentation strength regardless of their content of anthocyanins. To provide conclusive results, further investigations into the detailed phytochemical composition of differently prepared VMA extracts following up published procedures should be performed [64-66]. Comparison of the phytochemical composition in the extracts with special emphasis laid on copigment structures and the copigmentation strength should provide the conclusive results for the phenomena observed.

In this study, the stability of VMA from different bilberry extracts under physiological conditions correlated with the copigmentation strength. Moreover it could be demonstrated that the removal of the copigment fraction diminished these effects. It is therefore reasonable to conclude that the copigmentation phenomena observed in the natural environment can be retained in VMA extracts if a special manufacturing process is chosen.

Following the hypothesis, it was expected that the stabilization of anthocyanins under physiological conditions could provide an opportunity for improved absorption. Two BAEs, selected on the basis of their copigmentation strengths (low for BAE 1 and high for BAE 2, respectively), were further submitted to *in vitro* and *in vivo* absorption studies.

The *in vitro* tests for the absorption of anthocyanins into CaCo-2 cells provided strong evidence that the copigmentation strength observed in the BAEs tested correlated with improved absorption of anthocyanins The detailed mechanism of absorption of anthocyanins is unknown; hence, the improved absorption observed cannot be explained conclusively yet. The increased stability of anthocyanins resulting in higher concentrations of intact anthocyanins over time might be just one explanation. It might also be speculated that copigmented anthocyanins are better absorbed than single anthocyanins or that copigmented anthocyanins follow a different absorption route than single anthocyanins. In both cases, this might be due to changes in the physiochemical properties in response to copigmentation. Moreover it cannot be completely excluded that the correlation of copigmentation strength and improved absorption pretends a true, yet unknown, mechanism for improving absorption.

The *in vivo* test investigating the pharmacokinetics in humans confirmed the *in vitro* test results observed. Based on the C_{max} and $AUC_{0\text{-}inf}$ values calculated from the anthocyanin plasma concentrations, a substantially improved absorption correlating with the copigmentation strength can be derived. Bioequivalence assessment demonstrated that VMA from an extract with strong copigmentation effects are significantly better bioavailable than from an extract with low copigmentation effects.

The results observed in our *in vitro* test system and in our *in vivo* pharmacokinetic study in humans could support the copigmentation hypothesis discussed above. Based on the C_{max} and $AUC_{0\text{-}inf}$ values calculated from the anthocyanin plasma concentrations observed in the human *in vivo* study, a substantially improved absorption correlating with the copigmentation strength could be derived. Bioequivalence assessment demonstrated that VMA from an extract with strong copigmentation effects were significantly higher bioavailable than that of an extract with low copigmentation effects.

Based on the scattering of the C_{max} values, also a trend to the reduced variability of the absorption could also be observed for the bilberry extract with a strong copigmentation effect. Although statistically not significant, it can be speculated that increased stability based on copigmentation effects at the physiological site of absorption might lead to and maintain a more consistent anthocyanin absorption. Consistency of absorption points to reduced variability in plasma concentrations and would make a dose-effect relationship more predictable.

CONCLUSION

Taken together, the extent of anthocyanins absorbed correlates *in vitro* and *in vivo* with copigmentation effects observed. Bilberry anthocyanin extracts with a strong copigmentation effect -observed as changes in the visible spectrum and increased stability at physiologically relevant conditions- are statistically better absorbed in humans. Further investigations will focus on the detailed composition of bilberry anthocyanin extracts which should provide insights into the exact mechanisms of copigmentation.

REFERENCES

[1] Hertog MG, Hollman PC, Katan MB, Kromhout D. Intake of potentially anticarcinogenic flavonoids and their determinants in adults in Netherlands. *Nutr Cancer,* 20, 21–29, (1993).

[2] Prior RL. Absorption and metabolism of anthocyanins: potential health effects in phytochemicals – mechanisms of action. In *Phytochemicals: mechanism of action.* Meskin MS, Bidlack WR, Davies AJ, Lewis DS, Randolph RK. Eds. CRC Press, Boca Boton, 1-19 ,(2004).

[3] Kühnau J. The flavonoids. A class of semi-essential food components: their role in human nutrition. *World Rev Nutr Diet* 24, 117-191, (1976).

[4] Galvano F, La Fauci L, Vitaglione P, Fogliano V, Vanella L, Felgines C. Bioavailability, antioxidant and biological properties of the natural free-radical scavengers cyanidin and related glycosides. *Ann Ist Super Sanita,* 43(4), 382-393, 2007.

[5] Delgado-Vargas F, Jimenez AR, Paredes-Lopez O. Natural pigments: Carotenoids, anthocyanins, and betalains-characteristics, biosynthesis, processing, and stability. *Crit Rev Food Sci Nutr* 40, 173-289, (2000).

[6] Tsuda T, Shiga K, Ohshima K, Kawakishi S, Osawa T. Inhibition of lipid peroxidation and the active oxygen radical scavenging effect of anthocyanin pigments isolated from Phaseolus vulgaris L. *Biochem Pharmacol,* 52, 1033-1039 ,(1996).

[7] Wang H, Cao G, Prior RL. The oxygen radical absorbing capacity of anthocyanins. *J Agric Food Chem,* 45, 304-309, (1997).

[8] Russo A, Acquaviva R, Campisi A, et al. Bioflavonoids as antiradicals, antioxidants and DNA cleavage protectors. *Cell Biol Toxicol,* 16, 91-98, (2000).

[9] Kähkönen MP, Heinämäki J, Ollilainen V, Heinonen M. Berry anthocyanins: isolation, identification and antioxidant activities. *J Sci Food Agric,* 83, 1403-1411, (2003).

[10] Prior RL, Cao G, Martin A, et al. Antioxidant capacity as influenced by total phenolic and anthocyanin content, maturity, and variety of *Vaccinium* species. *J Agric Food Chem,* 46, 2686-2693, (1998).

[11] Hong V, Wrolstad RE. Characterization of anthocyanin-containing colorants and fruit juices by HPLC/photodiode array detection. *J Agric Food Chem,* 38, 698-708, (1990).

[12] Kong JM, Chia LS, Goh NK, Chia TF, Brouillard R. Analysis and biological activities of anthocyanins. *Phytochem,* 64, 923-933, (2003).

[13] Beattie J, Crozier A, Duthie GG. Potential Health Benefits of Berries. *Current Nutr Food Sci,* 1, 71-86, (2005).

[14] Canter PH, Ernst E. Anthocyanosides of *Vaccinium myrtillus* (bilberry) for night vision-a systematic review of placebo-controlled trials. *Surv Ophthalmol,* 49, 38-50, (2004).

[15] Baj A, Bombardelli E, Gabetta B, Martinelli EM. Qualitative and Quantitative evaluation of *Vaccinium Myrtillus* anthocyanins by high-resolution gas chromatography and high-performance liquid chromatography. *J Chrom,* 279, 365-372, (1983).

[16] Muth ER, Laurent JM, Jasper P. The effect of bilberry nutritional supplementation on night visual acuity and contrast sensitivity. *Altern Med Rev,* 5, 164-173, (2000).

[17] Morazzoni P, Bombardelli E. *Vaccinium myrtillus L. Fitoterapia,* 67, 3-29, (1996).

[18] Canter PH, Ernst E. Anthocyanosides of *Vaccinium myrtillus* (bilberry) for night vision-a systematic review of placebo-controlled trials. *Surv Ophthalmol,* 49, 38-50, (2004).

[19] McKhie MG, Ainge GD, Barnett LE, Cooney JM, Jensen DJ. Anthocyanin Glycosides from Berry Fruit Are Absorbed and Excreted Unmetabolized by Both Humans and Rats. *J. Agric. Food Chem,* 51, 4539-4548, (2003).

[20] Kay CD, Mazza G, Holub BJ, Wang J. Anthocyanin metabolites in human urine and serum. *Br J Nutr,* 91, 933–942, (2004).

[21] Felgines C, Texier O, Besson C, Fraisse D, Lamaison JL, Remesy C. Blackberry Anthocyanins Are Slightly Bioavailable in Rats. *J Nutr,* 132, 1249-1253, (2002).

[22] Talavera S, Felgines C, Texier O, Besson C, Lamaison JL, Remesy C. Anthocyanins Are Efficiently Absorbed from the Stomach in Anesthetized Rats. *J Nutr,* 133, 4178-4182, (2003).

[23] Cao G, Muccitelli HU, Sánchez-Moreno C, Prior RL. Anthocyanins are absorbed in glycated forms in elderly women: a pharmacokinetic study. *Am J Clin Nutr,* 73, 920-926, (2001).

[24] McGhie TK, Walton MC. The bioavailability and absorption of anthocyanins: Towards a better understanding. *Mol Nutr Food Res,* 51, 702-713, (2007).

[25] Fossen T, Cabrita L, Andersen OE. Colour and stability of pure anthocyanins influenced by pH including the alkaline region. *Food Chem,* 63, 435-440, (1998).

[26] Cabrita L, Fossen T, Andersen OE. Colour and stability of the six common anthocyanidin 3-glucosides in aqueous solutions. *Food Chem,* 68, 101-107, (2000).

[27] Sarni P, Fulcrand H, Souillol V, Souquet JM, Cheynier V. Mechanisms of Anthocyanin Degradation in Grape Must-Like Model Solutions. *J Sci Food Agric,* 69, 385-391, (1995).

[28] Brouillard R. Chemical structure of anthocyanins. In: *Anthocyanins as Food Colors. Pericles Markakis* (ed.), Academic Press Inc, New York, 1-38, (1982).

[29] Garcia-Viguera C, Zafrilla B, Artes F, Romero F, Abellan P, Tomas-Barberan F. Color and anthocyanin stability of red rasperrry Jam. *J Sci Food Agric,* 78, 565-573, (1998).

[30] Torskangerpoll K, Andersen OM. Colour stability of anthocyanins in aqueous solutions at various pH values. *Food Chem,* 89, 427-440, (2005).

[31] Wrolstad RE, Dursta RW, Leeb J. Tracking color and pigment changes in anthocyanin products. *Trends Food Sci Technol,* 16, 423-428, (2005).

[32] Gonnet JF. Colour effects of co-pigmentation of anthocyanins revisited-1. A calorimetric definition using the CIELAB scale. *Food Chem,* 63, 409-415, (1998).

[33] Gonnet JF. Colour effects of co-pigmentation revisited—2. A colorimetric look at the solutions of cyanin co-pigmented by rutin using the CIELAB scale. *Food Chem,* 66, 387-394, (1999).

[34] Gonnet JF. Colour effects of co-pigmentation of anthocyanin revisited—3. A further description using CIELAB differences and assessment of matched colours using the CMC model. *Food Chem,* 75, 473-485, (2001).

[35] Wallace TC, Giusti MM. Determination of Color, Pigment, and Phenolic Stability in Yogurt Systems Colored with Nonacylated Anthocyanins from Berberis boliviana L. as Compared to Other Natural/Synthetic Colorants. *J Food Sci,* 73, C241-C248, (2008).

[36] Prior RL, Lazarus SA, Cao G, Muccitelli H, Hammerstone JF. Identification of procyanidins and anthocyanins in blueberries and cranberries (*Vaccinium* spp.) using high-performance liquid chromatography/mass spectrometry. *J Agric Food Chem,* 49, 1270-1276, (2001).

[37] Santos-Buelga C, Bravo-Haro S, Rivas-Gonzalo JC. Interactions between catechin and malvidin-3-monoglucoside in model solutions. *Z Lebensm Unters Forsch,* 201, 269-274, (1995).

[38] Castañeda-Ovando A, Pacheco-Hernández M, Páez-Hernández E, Rodríguez JA, Galán-Vida CA. Chemical studies of anthocyanins: A review. *Food Chem,* 113, 859-871, (2009).

[39] Elhabiri M, Figueiredo P, Toki K, Saito N, Brouillard R. Anthocyanin–aluminium and –gallium complexes in aqueous solution. *J Chem Soc, Perkin Trans,* 2, 355-362, (1997).

[40] Yoshida K, Mori M, Kondo T. Blue flower color development by anthocyanins: from chemical structure to cell physiology. *Nat Prod Rep,* 26, 884-915, (2009).

[41] Takeda K, Osakabe A, Saito S, Furuyama D, Tomita A, Kojima Y, Yamadera M, Sakuta M. Components of protocyanin, a blue pigment from the blue flowers of Centaurea cyanus. *Phytochem,* 66, 1607-1613, (2005).

[42] Kondo T, Uedat M, Isobe M Gotoi T. A New Molecular Mechanism of Blue Color Development with Protocyanin, a Supramolecular Pigment from Cornflower, Centaurea cyanus. *Tetrahedron Letters,* 39, 8307-8310, (1998).

[43] Asen S, Stewart RN, Norris KH. Copigmentation of anthocyanins in plant tissues and its effect on color. *Phytochem,* 11, 1139-1144, (1972).

[44] Malien-Aubert C, Dangles O, Amiot MJ. Color Stability of Commercial Anthocyanin-Based Extracts in Relation to the Phenolic Composition. Protective Effects by Intra- and Intermolecular Copigmentation. *J Agric Food Chem,* 49, 170-176, (2001).

[45] Willstatter R, Zollinger EH. Anthocyans. XVI. Anthocyans of the grape and of the bilberry. II. *J Chem Soc, Abs,* 112, 47-48, (1916).

[46] Boulton R. The Copigmentation of Anthocyanins and Its Role in the Color of Red Wine: A Critical Review. *Am J Enol Vitic,* 52, 67-87, (2001).

[47] Mazzaracchio P, Pifferi P, Kindt M, Munyaneza A, Barbirol G. Interactions between anthocyanins and organic food molecules in model systems. *Int J Food Sci Technol,* 39, 53-59, (2004).

[48] Teraharaa N, Konczak-Islam I, Nakatani M, Yamakawa O, Goda Y, Honda T Anthocyanins in callus induced from purple storage root of Ipomoea batatas L. *Phytochem,* 54, 919-922, (2000).

[49] Degenhardt A, Knapp H, Winterhalter P. Separation and purification of anthocyanins by high-speed countercurrent chromatography and screening for antioxidant activity. *J Agric Food Chem,* 48, 338-343, (2000).

[50] Fossen T, Rayyan , Andersen OM. Dimeric anthocyanins from strawberry (Fragaria ananassa) consisting of pelargonidin 3-glucoside covalently linked to four flavan-3-ols. *Phytochem,* 65, 1421-1428, (2004)

[51] Andersen, OM. Semipreparative isolation and structure determination of pelargonidin 3-O-a-l-rhamnopyranosyl-(1!2)-b-dglucopyranoside and other anthocyanins from the tree Dacrycarpus dacrydioides. *Acta Chem Scand,* 42,462-468, (1988).

[52] Weiguang Y, Akoh CC, Fischer J, Krewer G. Absorption of Anthocyanins from Blueberry Extracts by Caco-2 Human Intestinal Cell Monolayers. *J Agric Food Chem,* 54, 5651-5658, (2006).

[53] Delie F, Rubas W. A human colonic cell line sharing similarities with enterocytes as a model to examine oral absorption: Advantages and limitations of the caco-2 model, *Crit Rev Ther Drug Carrier Syst,* 14, 221-286, (1997).

[54] Yamashita S, Tanaka Y, Endoh Y, Taki Y, Sakane T, Nadai T, Sezaki H. Analysis of Drug Permeation Across Caco-2 Monolayer: Implication for Prediction of In *Vivo Drug Absorption, Pharmacol Res,* 14, 486-491, (1997).

[55] Cooke DN, Thomasset S, Boocock DJ, et al. Development of Analyses by High-Performance Liquid Chromatography and Liquid Chromatography/Tandem Mass Spectrometry of Bilberry (*Vaccinium myrtillus*) Anthocyanins in Human Plasma and Urine. *J Agric Food Chem,* 54, 7009-7013, (2006).

[56] Woodward G, Kroon P, Cassidy A, Kay C. Anthocyanin Stability and Recovery: Implications for the Analysis of Clinical and Experimental Samples. *J Agric Food Chem,* 57, 5271-5278, (2009).

[57] Cao G, Prior RL. Anthocyanins Are Detected in Human Plasma after Oral Administration of an Elderberry Extract. *Clin Chem,* 45, 574-576, (1999).

[58] Cao G, Muccitelli HU, Sánchez-Moreno C, Prior RL. Anthocyanins are absorbed in glycated forms in elderly women: a pharmacokinetic study. *Am J Clin Nutr,* 73, 920-926, (2001).

[59] European Medicines Agency. *Note for Guidance on Good Clinical Practice* CPMP/ICH/135/95, (2002).

[60] *World Medical Association Declaration of Helsinki.* Ethical Principles for Medical Research Involving Human Subjects, Adopted by the 18th WMA General Assembly, Helsinki, Finland, June 1964, last amended at the 59th WMA General Assembly, Seoul, Korea, (2008).

[61] *European Medicines Agency.* Note for Guidance on the Investigation of Bioavailability and Bioequivalence, CPMP/EWP/QWP/1401/98 Rev. 1, (2010).

[62] Schuirmann DJ, A Comparison of the two One-sided Test procedure and the Power Approach for Assessing Equivalence of Average Bioavailability, *J Pharmacokin Biopharm,* 15, 657-680, (1987).

[63] Sauter R, Steinijans VW, Diletti E, Böhm A, Schulz HU, presentation of results from bioequivalence studies, *Int. J Clin Pharmacol Ther Toxicol,*30, S7-S30, (1992).

[64] Kähkönen MP, Hopia AI, Heinonen M. Berry Phenolics and Their Antioxidant Activity. *J Agric Food Chem,* 49, 4076-82, (2001).

[65] Määttä-Riihinen KR, Kähkönen MP, Törrönen AR, Heinonen M. Catechins and Procyanidins in Berries of *Vaccinium* Species and Their Antioxidant Activity. *J Agric Food Chem,* 53, 8485-8491, (2005)

[66] Riihinen K, Jaakola L, Kärenlampi S, Hohtola A. Organ-specific distribution of phenolic compounds in bilberry (*Vaccinium myrtillus*) and 'northblue' blueberry (*Vaccinium corymbosum* x *V. angustifolium*). *Food Chem,* 110, 156-160, (2008).

In: Anthocyanins: Structure, Biosynthesis and Health Benefits ISBN: 978-1-62257-329-5
Editor: Noboru Motohashi

Chapter 13

ANTHOCYANIN PROFILE AND THEIR BIOLOGICAL PROPERTIES IN POPULAR MEDITERRANEAN RED WINES

Gian Carlo Tenore*

Department of Pharmaceutical and Toxicological Chemistry,
Faculty of Pharmacy, University of Naples "Federico II", Naples, Italy

ABSTRACT

The anthocyanin composition, their antioxidant properties and their contribution to antioxidant activities of selected red wines from Morocco were evaluated. The results obtained revealed that the wine samples were characterized by a higher anthocyanin concentration and a higher antioxidant capacity than the same variety of wines with a different geographical origin and other more consumed red wines; the highest quantity was found in Cabernet Sauvignon, followed by Merlot, and last by Syrah wine. All of the wines tested showed very similar anthocyanin patterns: individual compound contents resulted generally higher in comparison to conventional wines. In addition, the contribution given by each class of polyphenolic compounds to the antioxidant activities of wines was estimated. Results revealed a good linear correlation (R) between antioxidant capacity determined by both assays, DPPH (antiradical potential test) and FRAP (reducing capacity test), and total phenols (R ranging from 0.9675 to 0.9989), anthocyanins (R ranging from 0.9349 to 0.9982) and flavonols (R ranging from 0.8599 to 0.9729) of all of the wines tested. Most anthocyanins revealed very significant correlation coefficients (R mean value 0.8413). As regards the antimicrobial properties, each fraction exhibited activity against a broad spectrum of microorganisms, especially food-borne pathogens, revealing not only a moderate to high natural preserving capacity, but also potentially beneficial influence on human health.

---

* Tel/Fax: +39081678610 – e-mail: giancarlo.tenore@unina.it.

INTRODUCTION

Over the last decade health effects of wine consumption have been studied in depth. Owing to their beneficial influence on health, wines are the subject of increasing numbers of investigations. Special attention has been focused on protection against cancer and cardiovascular diseases which may derive from a moderate intake of wine [1]. Generally, it is established that an oxidation process is involved in the initial development steps of these diseases. Indeed, reactive oxygen species (ROS), naturally formed during normal metabolism, can damage biological structures such as proteins, lipids or DNA. Human metabolism counts on an antioxidant defensive system involving enzymes and proteins to prevent these effects. However, the defences can be overwhelmed under certain circumstances so that harmful effects occur. It is accepted that the intake of antioxidant substances reinforces the defences against ROS. Therefore, the role of antioxidant nutrients, such as vitamins (A, E and C) or selenium, and other diet-derived antioxidants, such as polyphenols, is attracting considerable attention nowadays from health, food and nutrition groups.

Particularly, attention has been focused on the nonalcoholic fractions of wine as an important source of polyphenols, which are capable of inhibiting the oxidation process involved in the initial development steps of these diseases [2]. Red wine contains up to 3000 mg/L of phenolic compounds, which can be grouped into hydroxybenzoic acids, hydroxycinnamic acids, flavanols, flavonols, anthocyanins, stilbenes, and tannins. Each of the winemaking stages involves various chemical changes, which ultimately modify the phenolic composition and antioxidant activities of the wine [3].

The health-protective properties of wines are attributed to their antioxidant activities, i.e., their capability to eliminate free radicals. Consequently, numerous papers have been focused on the determination of the antioxidant capacity of wines [4] as well as on their polyphenolic content, probably responsible for the antioxidant action. In addition, polyphenols play an essential role in sensory properties since they contribute to colour, flavour and bitterness of wines [5].

Several studies indicated a significant correlation between total polyphenolic content and the antimicrobial activity of wine, especially its effectiveness against some gastrointestinal pathogens such as *E. coli, Salmonella, Listeria, Shigella,* and *Helicobacter pylori*. Thus, it has been suggested that the antibacterial properties of red and white wines may be strictly correlated to their ability to stimulate digestion [6-10]. It is well known that the antimicrobial activity of intact wine is largely due to its low pH and alcohol content and that the antimicrobial role of phenolic constituents strictly depends on the physical-chemical properties of the medium in which they are contained [11]. However, despite the large body of evidence for the antimicrobial effectiveness of intact wine and of various isolated phenolic compounds [12, 13], direct comparative studies on antimicrobial activity of different components of wines under the same experimental conditions are scarce.

Morocco has been a leading wine producer for several years and its bold red and white grapes have become popular mainly among the French and Americans. Three quarters of the output is red, twenty percent rose and the rest white. The city of Meknès concentrates more than 60% of the Moroccan wine production. Its sandy, clay-lime soil, combined with a hot, humid climate, is very suitable for a good maturity of vineyards. The Moroccan wines more popular abroad include different varieties of the bold red wines: Bonassia Cabernet

Sauvignon (rich and smooth, hints of black currants with floral finish), Halana Merlot (hints of raspberry, jasmine and spice) and Halana Syrah (hints of plum, raspberry, licorice and cinnamon). Despite the increasing participation of these Moroccan wines in the market, there is a scarcity of data concerning their composition and biological properties.

POLYPHENOLIC COMPOSITION AND ANTIOXIDANT CAPACITY

The results obtained for the wine polyphenolic composition [14] (Figure 1) were generally higher than those reported elsewhere for the same variety of wines from France [3], South America [15], Spain [16] and Australia [17]. Among the different Moroccan viticulture regions, the city of Meknès is characterized by a hot, humid climate, with very high levels of exposure to sunlight, which is known to exert a marked influence on the polyphenolic content of wines. Thus, it has been shown that sun-exposed grapes can contain up to 10 times more total phenolics than grapes cultivated in the shade [18].

For all of the antioxidant classes of compounds the highest quantity was found in Cabernet Sauvignon, followed by Merlot, and last by Syrah, reflecting a trend similar to earlier studies [3]. As reported in literature [19], Merlot skin contains higher anthocyanin contents than Cabernet Sauvignon and Syrah grapes. However, not all are extractable during conventional winemaking or liquid-alcoholic extraction methods [20]. Studies on different grape skins [21] suggested that the low anthocyanin extractability from Merlot skin may be caused by its higher level of skin cell wall material (firmer and thicker skin) and higher contents of polysaccharides and lignin.

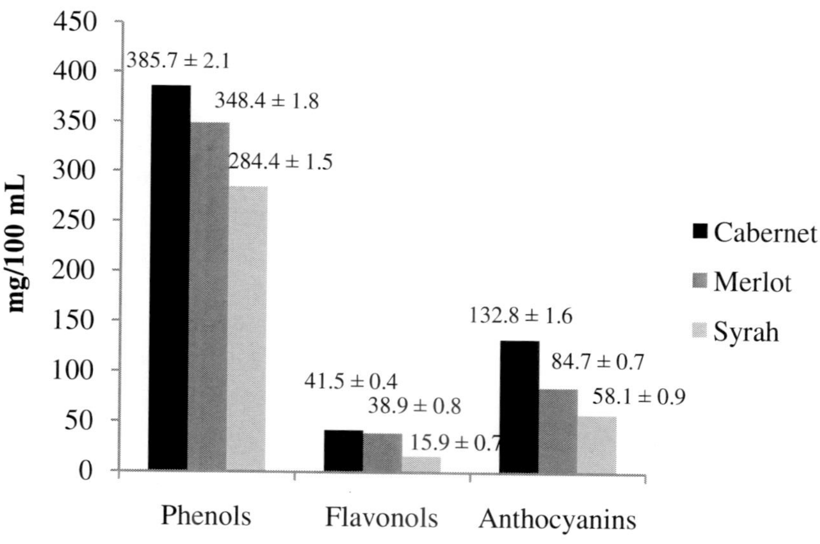

Figure 1. Polyphenolic contents in Morocco wines. Phenol contents are expressed as mg GAE/100 mL wine ± SD; Flavonol contents are expressed as mg QE/100 mL wine ± SD; Anthocyanin contents are expressed as mg ME/100 mL wine ± SD.

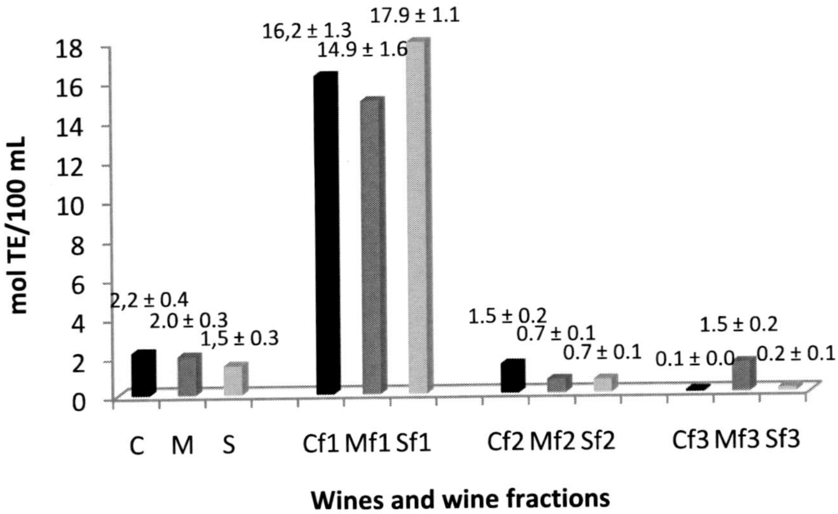

Figure 2. Reducing activity of Moroccan red wines and their fractions determined by FRAP test. Values are expressed ± SD; C=Cabernet, M=Merlot, S=Syrah; f1=fraction 1, f2=fraction 2, f3=fraction 3.

The content of flavonols in the selected wines in the study of Tenore [14] was higher than that reported in previous works [22]. The risk of an overestimation, which is possible at 360 nm for the presence of other phenolic compounds (e.g., cinnamic acids, anthocyanins) that also have some absorbance, was avoided in this study by using a more specific spectral method of flavonol estimation. This method, based on a complex formation with aluminum chloride [23], is rather specific to flavonols, because the aluminum complexation requires a 4-keto group and at least one neighboring (3- or 5-) hydroxyl group, which are common features of flavonols. Similar features are also present in flavones and flavanones, but these compounds are not common in red wines.

Owing to the complex reactivity of phytochemicals, the antioxidant activities of food and food extracts cannot be evaluated by only a single method, but at least two test systems have been recommended for the determination of antioxidant activity to establish authenticity [24]. For this reason, the antioxidant activity of wines was determined by two spectrophotometric methods, DPPH and FRAP tests, and expressed as trolox equivalents (TEs). The reduction of DPPH absorption is indicative of the capacity of the samples to scavenge free radicals, while the FRAP method is used to determine the capacity of reductants in a sample.

Antioxidant activity at the steady state resulted higher in FRAP test (ranging from 1513.9 to 2247.4 μmol TE/100 mL) than in DPPH test (ranging from 326.4 to 330.6 μmol TE/100 mL) for all of the wine samples [14]. The probable reason for the lower DPPH values of wines tested could be the presence of compounds not reactive towards the DPPH free radical due to steric hindrance. Generally, DPPH assay has a relatively small linear reaction range of only 2-3 fold, and small molecules that have better access to the radical site have higher apparent antioxidant activity with this test [25]. Cabernet Sauvignon showed the highest antioxidant capacity in FRAP test (2247.4 ± 0.4 μmol/100 mL), followed by Merlot (2003.4 ± 0.3 μmol/100 mL) and last by Syrah (1513.9 ± 0.3 μmol/100 mL). Interestingly, differently from FRAP test, in DPPH assay samples reached almost identical results at the end of the reaction kinetic, revealing comparable antiradical activities for all of the wines tested.

Table 1. LC/MS data of tentatively identified anthocyanins in selected Moroccan wines and their quantitative analysis using DAD at 520 nm

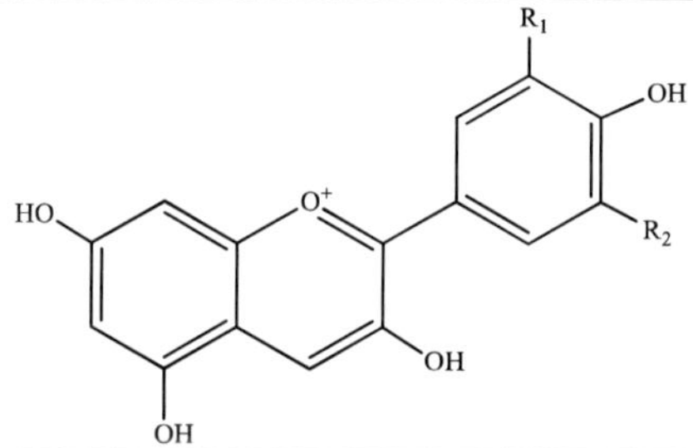

Name	R_1	R_2
Cyanidin	OH	H
Peonidin	OCH_3	H
Delphinidin	OH	OH
Petunidin	OCH_3	OH
Malvidin	OCH_3	OCH_3

Peak	Compound	mg ME/100 mL of wine*			Retention time** (min)	*m/z* $[M+H]^+$	MS^{2a}	MS^{3a}	MS^{4a}
		C	M	S					
1	Delphinidin-3-*O*-glucoside	19.53±1.1	21.20±1.3	5.02±0.1	5.14±0.2	465	303	229, 257, *303*	229, 257
2	Cyanidin-3-*O*-glucoside	13.88±1.2	9.55±1.1	3.00±0.1	6.62±0.3	449	287	213, 231, 241, 259, *287*	213, 231, 241, 259, *287*
3	Petunidin-3-*O*-glucoside	16.89±1.4	14.69±1.0	7.71±0.3	8.12±0.4	479	317	257, 274, *302*, 317	218, 228, 246, *274*
4	Peonidin-3-*O*-glucoside	9.18±1.0	10.44±0.9	4.16±0.4	10.07±0.5	463	301	286	230, 258, 268
5	Malvidin-3-*O*-glucoside	65.29±1.9	44.79±1.7	26.55±1.2	11.50±0.6	493	331	179, 242, 270, 287, 299, *315*/316	213, 257, 285, 287, 313, 315
6	Delphinidin-3-*O*-acetylglucoside	14.40±1.5	13.36±0.3	7.24±0.7	15.44±0.7	507	303	229, 257, *303*	229, 257, *303*
7	Cyanidin-3-*O*-acetylglucoside	9.61±0.9	6.27±0.4	4.19±0.4	16.86±0.7	491	287	213, 231, 259, *287*	213, 231, 259, *287*
8	Malvidin-3-(6-*O*-coumaroyl)glucoside (*cis* isomer)	4.32±0.8	3.94±0.4	2.47±0.3	22.35±1.02	639	331	179, 242, 270, 287, 299, *315*/316, 331	257, 285, 287, 313, 315
9	Malvidin-(6-*O*-caffeoyl)glucoside	22.60±1.2	11.56±0.6	11.25±1.0	24.00±1.09	655	331	179, 242, 270, 287, 299, *315*/316, 331	257, 285, 287, 313, 315
10	Peonidin-3-(6-*O*-coumaroyl)glucoside (*trans* isomer)	4.32±1.0	3.11±0.2	3.08±0.6	32.15±0.7	609	301	286	230, 258, 268
11	Malvidin-3-(6-*O*-coumaroyl)glucoside (*trans* isomer)	6.10±0.8	5.01±0.1	4.07±0.6	32.73±0.6	639	331	179, 242, 270, 287, *299*, 315/316, 331	225, 253, *281*, 299

C: Cabernet Sauvignon; M: Merlot; S: Syrah.

*ME: Malvidin-3-*O*-glucoside Equivalents, expressed as value ± SD.

**Expressed as mean value ± SD.

[a] Base peak (100%) is underlined.

Table 2. LC/MS data of tentatively identified flavonols in selected Moroccan wines and their quantitative analysis using DAD at 353 nm

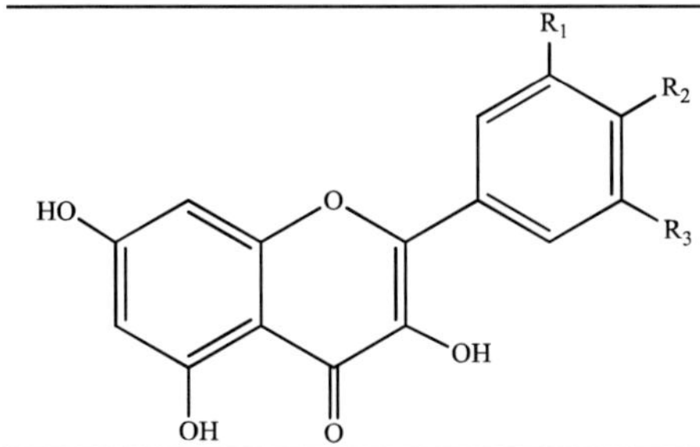

Name	R_1	R_2	R_3
Kaempferol	H	OH	H
Quercetin	OH	OH	H
Isorhamnetin	OCH_3	OH	H
Myricetin	OH	OH	OH
Laricitrin	OCH_3	OH	OH
Syringetin	OCH_3	OH	OCH_3

Peak	Compound	mg QE/100 mL of wine*			Retention time** (min)	m/z $[M-H]^-$	MS^{2a}	MS^{3a}	MS^{4a}
		C	M	S					
1	Myricetin-3-*O*-glucoside	1.98±0.5	0.75±0.04	2.62±0.4	9.32±0.3	479	*316*/317	242, 270/*271*, 287	171, 199, 227
2	Quercetin-3-*O*-glucuronide	2.48±0.2	2.18±0.3	1.70±0.3	13.01±0.4	477	301	151, *179*,193, 257, 273	151
3	Quercetin-3-*O*-glucoside	1.00±0.07	0.46±0.08	1.21±0.08	13.55±0.4	463	301	151, *179*,193, 257, 273	151
4	Laricitrin-3-*O*-galactoside	0.74±0.05	0.47±0.04	1.00±0.07	14.96±0.5	493	330, *331*	151, 179,193, *316*, 317	151, 164, *179*, 219, 244, 270/271,287/288
5	Kaempferol-3-*O*-glucoside	0.49±0.02	0.18±0.01	0.21±0.02	15.68±0.4	447	255, *284*/285, 327, 401, 419, 429	227, 239, *255*/256	212, 227
6	Laricitrin-3-*O*-rhamnose-7-*O*-trihydroxycinnamic acid	1.79±0.1	1.49±0.05	0.95±0.04	16.86±0.3	655	303, 314, *329*, 347, 475, 501, 509	314	299
7	Kaempferol-3-*O*-caffeoylate	0.46±0.02	0.34±0.02	0.21±0.02	17.77±0.4	447	*284*/285	227, 239, *255*/256	212, 227
8	Isorhamnetin-3-*O*-glucoside	0.52±0.03	0.34±0.01	0.80±0.05	19.35±0.4	477	271, 285, *314*/315, 357	243, 257, 271, *285*/286, 299/300	241/270
9	Syringetin-3-*O*-galactoside	1.08±0.05	0.66±0.03	1.07±0.06	19.95±0.4	507	*344*/345, 387, 479, 489	330	

C: Cabernet Sauvignon; M: Merlot; S: Syrah.

*QE: Quercetin-3-*O*-glucoside Equivalents, expressed as value ± SD.

**Expressed as mean value ± SD.

[a] Base peak (100%) is underlined.

Table 3. Antimicrobial activity of Moroccan red wines and their fractions[a]

Micro-organism	C	M	S	Cf1	Mf1	Sf1	Cf2	Mf2	Sf2	Cf3	Mf3	Sf3	CTAX	PEN	TET	AMB	ECN
Gram (+) bacteria																	
B. cereus	256 ± 0.1	64 ± 0.0	64 ± 0.1	32 ± 0.1	16 ± 0.0	16 ± 0.1	64 ± 0.1	32 ± 0.1	32 ± 0.3	256 ± 0.1	32 ± 0.2	32 ± 0.1	R	7.5 ± 0.0	R	NT	NT
S. aureus	128 ± 0.2	64 ± 0.2	32 ± 0.1	16 ± 0.2	32 ± 0.2	8 ± 0.1	32 ± 0.4	16 ± 0.4	8 ± 0.1	256 ± 0.4	32 ± 0.1	16 ± 0.0	2.0 ± 0.4	0.03 ± 0.0	2.0 ± 0.3	NT	NT
E. faecalis	128 ± 0.1	64 ± 0.2	64 ± 0.1	32 ± 0.1	16 ± 0.2	16 ± 0.1	64 ± 0.2	32 ± 0.3	32 ± 0.2	128 ± 0.2	32 ± 0.0	32 ± 0.1	R	8.0 ± 0.0	2.0 ± 0.6	NT	NT
L. mono-cytogenes	256 ± 0.4	256 ± 0.1	256 ± 0.2	64 ± 0.4	64 ± 0.1	32 ± 0.2	128 ± 0.3	128 ± 0.4	64 ± 0.4	256 ± 0.1	128 ± 0.3	128 ± 0.4	16 ± 0.0	R	R	NT	NT
Gram (-) bacteria																	
E. coli	256 ± 0.3	256 ± 0.2	256 ± 0.0	64 ± 0.3	64 ± 0.2	32 ± 0.0	128 ± 0.4	128 ± 0.2	64 ± 0.3	256 ± 0.1	128 ± 0.1	128 ± 0.1	32 ± 0.1	64 ± 0.4	32 ± 0.2	NT	NT
P. mirabilis	128 ± 0.0	128 ± 0.4	128 ± 0.3	32 ± 0.0	64 ± 0.4	16 ± 0.3	64 ± 0.1	64 ± 0.1	32 ± 0.2	256 ± 0.0	64 ± 0.3	64 ± 0.0	0.03 ± 0.0	4.0 ± 0.0	32 ± 0.1	NT	NT
P. vulgaris	128 ± 0.1	128 ± 0.2	128 ± 0.2	32 ± 0.1	64 ± 0.2	32 ± 0.2	64 ± 0.2	128 ± 0.1	32 ± 0.1	256 ± 0.1	64 ± 0.1	64 ± 0.0	2.0 ± 0.1	4.0 ± 0.3	R	NT	NT
P. aerugi-nosa	256 ± 0.1	128 ± 0.0	128 ± 0.4	32 ± 0.1	64 ± 0.0	32 ± 0.4	128 ± 0.3	128 ± 0.0	32 ± 0.3	128 ± 0.0	64 ± 0.4	128 ± 0.2	16 ± 0.0	R	32 ± 0.1	NT	NT
S. typhi	128 ± 0.0	256 ± 0.0	256 ± 0.1	32 ± 0.0	32 ± 0.0	64 ± 0.1	64 ± 0.0	128 ± 0.1	32 ± 0.0	256 ± 0.4	128 ± 0.2	128 ± 0.3	0.5 ± 0.1	4.0 ± 0.0	1.0 ± 0.3	NT	NT
E. cloaceae	128 ± 0.0	128 ± 0.4	64 ± 0.1	32 ± 0.0	64 ± 0.4	32 ± 0.1	64 ± 0.1	64 ± 0.3	32 ± 0.1	256 ± 0.2	32 ± 0.3	32 ± 0.0	R	4.0 ± 0.0	R	NT	NT
E. aerogenes	128 ± 0.4	128 ± 0.4	64 ± 0.3	64 ± 0.4	32 ± 0.4	16 ± 0.3	32 ± 0.1	64 ± 0.4	32 ± 0.0	256 ± 0.3	32 ± 0.0	32 ± 0.4	R	4.0 ± 0.0	R	NT	NT
Y. entero-colitica	256 ± 0.3	128 ± 0.3	128 ± 0.2	64 ± 0.3	64 ± 0.3	64 ± 0.2	64 ± 0.0	128 ± 0.1	64 ± 0.2	128 ± 0.2	64 ± 0.1	64 ± 0.3	0.1 ± 0.0	18 ± 0.6	8.0 ± 0.0	NT	NT
K. pneu-moniae	256 ± 0.3	128 ± 0.2	128 ± 0.4	64 ± 0.3	64 ± 0.2	64 ± 0.4	128 ± 0.2	128 ± 0.0	32 ± 0.4	256 ± 0.4	64 ± 0.1	128 ± 0.1	0.1 ± 0.0	R	16 ± 0.1	NT	NT

Table 3. (Continued)

Micro-organism	C	M	S	Cf1	Mf1	Sf1	Cf2	Mf2	Sf2	Cf3	Mf3	Sf3	CTAX	PEN	TET	AMB	ECN
Yeasts																	
C. albicans 10231	128 ± 0.2	128 ± 0.3	128 ± 0.4	32 ± 0.2	64 ± 0.3	32 ± 0.4	64 ± 0.4	128 ± 0.1	64 ± 0.1	256 ± 0.2	64 ± 0.4	128 ± 0.1	NT	NT	NT	1 ± 0.0	NT
C. albicans 90028	128 ± 0.0	128 ± 0.1	256 ± 0.1	32 ± 0.0	64 ± 0.1	64 ± 0.1	64 ± 0.1	128 ± 0.1	64 ± 0.1	256 ± 0.1	64 ± 0.2	128 ± 0.4	NT	NT	NT	1 ± 0.1	NT
Moulds																	
F. oxysporum	256 ± 0.3	256 ± 0.2	256 ± 0.2	64 ± 0.3	64 ± 0.2	64 ± 0.2	64 ± 0.0	128 ± 0.3	64 ± 0.0	256 ± 0.0	128 ± 0.3	128 ± 0.0	NT	NT	NT	NT	4 ± 0.0
C. herbarum	256 ± 0.4	128 ± 0.0	256 ± 0.0	64 ± 0.4	64 ± 0.0	64 ± 0.0	64 ± 0.1	128 ± 0.2	64 ± 0.1	256 ± 0.3	64 ± 0.4	256 ± 0.4	NT	NT	NT	NT	4 ± 0.1
B. cinerea	128 ± 0.1	128 ± 0.0	128 ± 0.1	32 ± 0.1	32 ± 0.0	32 ± 0.1	64 ± 0.0	64 ± 0.4	64 ± 0.3	128 ± 0.2	128 ± 0.1	128 ± 0.2	NT	NT	NT	NT	4 ± 0.0
A. flavus	256 ± 0.1	128 ± 0.0	256 ± 0.1	64 ± 0.1	64 ± 0.0	64 ± 0.1	64 ± 0.1	128 ± 0.1	64 ± 0.4	256 ± 0.4	64 ± 0.0	128 ± 0.1	NT	NT	NT	NT	3 ± 0.1

[a] Values are expressed as minimum inhibitory concentration (MIC, μg/mL) and represent the average of three determinations ± SD.

C: Cabernet Sauvignon; M: Merlot; S: Syrah.

f1: fraction 1; f2: fraction 2; f3: fraction 3.

CTAX: cefotaxime; PEN: penicillin; TET: tetracycline; AMB: amphotericin B; ECN: econasol.

NT: not tested; R: resistant.

In addition, the contribution given by each class of polyphenolic compounds to the antioxidant activities of wines was estimated. Tenore [14] revealed a good linear correlation (*R*) between antioxidant capacity determined by both assays, DPPH and FRAP, and total phenols (*R* ranging from 0.9675 to 0.9989), anthocyanins (*R* ranging from 0.9349 to 0.9982) and flavonols (*R* ranging from 0.8599 to 0.9729) of all of the wines tested.

ANTHOCYANIN AND FLAVONOL PROFILES

Cabernet Sauvignon, Merlot and Syrah red wine samples showed very similar anthocyanin and flavonol patterns [14] (Table 1). Single anthocyanin contents resulted generally higher than those reported elsewhere for the same variety of wines from other countries [3, 15, 16, 17]. As reported in literature on *Vitis vinifera* wines [26], the compound malvidin-3-*O*-glucoside was found as the main monomeric anthocyanin in all of the wines tested (ranging from 26.55±1.2 to 65.29±1.9 mg ME/100 mL). Differently from these studies, this anthocyanin was followed by malvidin-(6-*O*-caffeoyl)-glucoside in Cabernet Sauvignon and in Syrah, by delphinidin-3-*O*-glucoside in Merlot (Table 1). The extractability of the anthocyanins from grapes during fermentation depends on their variety, geographical origin, vintage and wine-making techniques. The anthocyanin composition or profile has been used in several studies to fingerprint grape varieties. It was found that the anthocyanin profile of Spanish red wines could be used to distinguish grape cultivar, wine origin, and also the wine-making method [27].

All of the wine samples showed a higher concentration of flavonols [14] in comparison to conventional wines [28], as reported in Table 2. It has long been known that the increased biosynthesis of polyphenols, especially flavonols, is greatly influenced by sunlight exposure and temperature, so it is expected that the wines made from grapes which are grown in warmer, sunnier areas, have a higher level of flavonols. Tenore [14] confirmed that the main flavonol glycosides in red wines are derivatives of quercetin and myricetin [29], namely quercetin-3-*O*-glucuronide (ranging from 1.70±0.3 to 2.48±0.2 mg QE/100 mL) and myricetin-3-*O*-glucoside (ranging from 0.75±0.04 to 2.62±0.4 mg QE/100 mL) (Table 2). In contrast with literature [29], low levels of kaempferol-3-*O*-glucoside (mean value 0.29 mg QE/100 mL), kaempferol-3-*O*-caffeoylate (mean value 0.34 mg QE/100 mL) and quercetin-3-*O*-glucoside (mean value 0.89 mg QE/100 mL) were found. The results suggested that the rates of hydrolysis of the flavonol glycosides in wine were different according to the type of flavonol aglycone and also with respect to the nature of the glycoside moiety [30]. The 3-*O*-glucosides of kaempferol, quercetin and isorhamnetin look to be easily hydrolyzed. As quercetin-3-*O*-glucoside is one of the main flavonol glycosides found in grapes, it is usual to observe an important peak of free quercetin in wines due to its hydrolysis, while the high remaining amounts of quercetin-3-*O*-glucuronide in wines suggested its resistance to acid hydrolysis [30]. Similarly, the wines tested revealed an evident peak of free quercetin probably due to hydrolysis of its 3-*O*-glucoside. Generally, wine samples have shown an important peak corresponding to free myricetin [30], but, on the contrary, the wines object of this study did not show that peak and revealed interesting amounts of myricetin-3-*O*-glucoside; thus, this flavonol seems to be not as easily hydrolyzed as quercetin-3-*O*-

glucoside. Finally, the amounts of the 3-glycosides of both laricitrin and syringetin resulted much higher than those of their corresponding free aglycones.

The correlation of the individual polyphenolic contents with the antioxidant properties was also measured. Many authors have reported on the contribution given by total polyphenolic classes to the antioxidant activities of wines, but very little has been written about the influence exerted by single components. In agreement with literature [30], just few flavonols seemed to largely influence both assays, DPPH and FRAP, mainly quercetin-3-*O*-glucuronide (*R* ranging from 0.9720 to 0.9983), laricitrin-3-*O*-rhamnose-7-*O*-trihydroxycinnamic acid (*R* ranging from 0.9643 to 0.9996), kaempferol-3-*O*-caffeoylate (*R* ranging from 0.9859 to 0.9919) and kaempferol-3-*O*-glucoside (*R* ranging from 0.6892 to 0.8743), while the other compounds gave quite low correlations (*R* mean value 0.3970). On the contrary, most anthocyanins revealed very significant correlation coefficients (*R* mean value 0.8413). Interestingly, not all of these compounds showed positive results in both tests. Particularly, delphinidin-3-*O*-glucoside demonstrated a higher reducing capacity (R vs $AA_{FRAP} = 0.9133$; R vs $AA_{DPPH} = 0.7504$) and the same was observed for peonidin-3-*O*-glucoside (R vs $AA_{FRAP} = 0.8696$; R vs $AA_{DPPH} = 0.6830$). On the contrary, malvidin-(6-*O*-caffeoyl)-glucoside showed to be more powerful radical scavenger (R vs $AA_{FRAP} = 0.7663$; R vs $AA_{DPPH} = 0.9230$) and the same was observed for peonidin-3-(6-*O*-coumaroyl)-glucoside (*trans* isomer) (R vs $AA_{FRAP} = 0.7645$; R vs $AA_{DPPH} = 0.9219$).

Wine Sample Fractionation

Wines were extracted using a previously described fractionation scheme [31]. An aliquot of each wine sample (100 mL) had its pH adjusted to 2.0 and was then extracted with 100 mL of ethyl acetate (three times). The aqueous phase was concentrated under reduced pressure with ethanol addition (at 50 °C) until dryness, producing Fraction 1. The remaining organic phases were combined and concentrated under reduced pressure to dryness, re-dissolved with water (100 mL), and extracted with ethyl acetate (100 mL) three times. The combined organic phase was concentrated under reduced pressure (at 50 °C) until dryness to produce Fraction 2; the remaining aqueous phase had its pH adjusted to 7.0 and was then extracted with ethyl acetate (100 mL) three times. The resulting organic phases were combined and concentrated under reduced pressure (at 50 °C) to dryness, producing Fraction 3, and the aqueous phase was discarded. This fractionation scheme concentrates wine anthocyanins in the aqueous Fraction 1; procyanidins, catechins, epicatechins, flavonoids and quercetin glycosides in Fraction 2; and phenolic acids and quercetin glucuronides in Fraction 3. Stock solutions (pH 4.5) of the wine fractions were prepared in absolute ethanol at a concentration of 1 mg/mL.

Antioxidant Capacity of Wine Extracts

All of the wine fractions revealed a higher reducing capacity (ranging from 0.1 to 17.9 mol TE/100 mL) [31] (Figure 2) than radical scavenging potential (ranging from 0.06 to 4.1 mol TE/100 mL) (Figure 3). Polyphenolic compounds are reported to have a high radical scavenging capacity, particularly those showing *O*-dihydroxy structures which confer great stability to the radical form and participate in the electron delocalization [32]. Thus, this result was of great interest because it highlighted that polyphenols occurring in the wine samples were preferentially involved in electron-transfer reactions rather than hydrogen atom-transfer mechanisms.

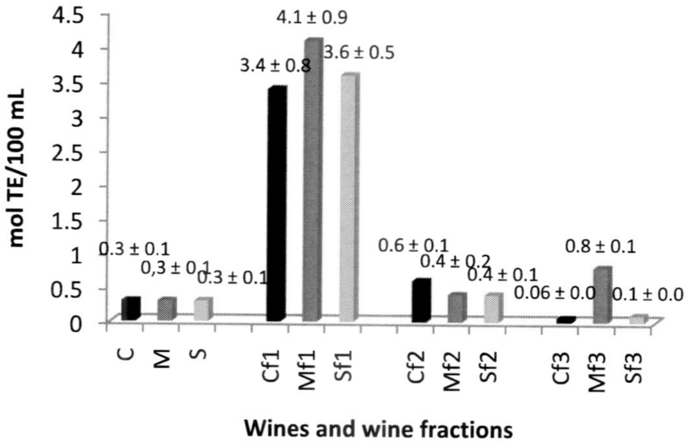

Figure 3. Radical scavenging activity of Moroccan red wines and their fractions determined by DPPH test. Values are expressed ± SD; C=Cabernet, M=Merlot, S=Syrah; f1=fraction 1, f2=fraction 2, f3=fraction 3.

Fractions 1 exhibited the highest antioxidant activity in both FRAP and DPPH assays, although showing a polyphenolic content in some cases comparable to or even lower than fractions 2 and 3. Anthocyanins have been proven very efficient antioxidants in a number of systems, both red grape extracts and various red wines, and anthocyanin contents have shown a good correlation with antioxidant activities of red grape extracts and various red wines [33-36]. Nevertheless, the difference could be attributed to the different kinds of polyphenolics occurring in the samples. Evidence indicated that phytochemical biological properties may be deeply affected by even small structural changes [37]. Structural diversity, infact, is supposed to account for a different reactivity of polyphenols towards antioxidant assays and the same may occur for the polyphenolic classes tested. As regards what measured within the same kind of polyphenolic fractions, Tenore [31] suggested that the different relative percentage of the fraction constituents may deeply influence the fraction biological properties owing to synergisms or antagonisms so that between two fractions characterized by a different composition the one with the highest content may exhibit the lowest biological property value [38-40] (Figure 2).

In addition, the contribution given by polyphenolic contents of wine fractions to wine fraction antioxidant activities was evaluated (Table 2). Tenore [31] revealed a good linear correlation (*R*) between antioxidant capacity determined by both assays, FRAP and DPPH, and total polyphenols (*R* ranging from 0.7676 to 0.9955 and from 0.7509 to 0.9974, respectively), confirming the important role as antioxidants played by phenols in the wine samples tested. Moreover, a strong correlation (*R* ranging from 0.9863 to 0.9995) was obtained between the extract antioxidant values from DPPH and FRAP assays.

Antimicrobial Activity of Wine Extracts

The in vitro antimicrobial activity of wine extracts was evaluated by determining the minimum inhibitory concentration (MIC) [41] against two yeasts, four moulds and 13 bacteria species selected as representative of the classes of Gram (+) and Gram (-) and known to be food-borne pathogens, to cause respiratory, gastrointestinal, skin and urinary disorders. According to MIC values reported in Table 3, wine fractions exerted a broad antimicrobial spectrum by inhibiting the growth of all of the human pathogenic and/or food spoilage bacteria, moulds and yeasts tested [31]. On average, wine fractions 1, the one containing monomeric anthocyanins, revealed a higher antimicrobial activity than the other extracts. It is widely reported about the important antimicrobial role played by organic acids in intact wine, immediately followed by phenolic acids and then by non-antocyanin flavonoids [11, 42]. Despite the great number of papers devoted to the different biological effects exerted by anthocyanins only few studies are focused on the antimicrobial activity of these compounds; particularly, the attention have been focused on the anthocyanin profiles of different fruits (mainly berries) and their activity against different microbes, mainly Gram (+) pathogens which usually are more susceptible to the anthocyanin action than Gram (-) ones [43]. Actually, phenolic acids occurring in fractions 3 are present in ionized form at the physiological pH value (7.4) and are too polar compounds to penetrate the semipermeable bacterial membrane and react with the cytoplasm or cellular proteins [44]. This is the same reason for which the lipidic wall of Gram (-) pathogens represents a great barrier for most polyphenols hence a scarce inhibition is achieved. Thus, this result was of great interest because it revealed an important antimicrobial role of monomeric anthocyanins in the wines tested. There is no doubt that the antimicrobial efficacy of the phenolic compounds in wine may be deeply affected by other components of the inherent environment present in wine and that separated polyphenolic compounds from wine extracts, often divided into different fractions according to molecular weight and chemical class, are not likely to demonstrate antimicrobial activity as if they were in the physical-chemical environment of intact wine [11]. In addition, wine fractions were tested on several microbial strains among which some bacteria and all of the moulds and yeasts were taken into account for the first in this work.

Conclusion

The present study is the first effort to the elucidation of biological and toxicological properties of the main Moroccan red wines. A specific aim was to evaluate the contents of total polyphenolic classes and individual compounds and their contribution to wine antioxidant activities. The results obtained revealed that the wine samples were characterized by a higher phytochemical concentration and a higher antioxidant capacity than the same variety of wines with a different geographical origin and other more consumed red wines. A specific aim was to evaluate the antioxidant and antimicrobial properties of extracts from selected Moroccan red wines. Wine fractions showed a higher reducing capacity than radical scavenging activity and good correlations between polyphenolic contents and fraction antioxidant activities measured by both assays (FRAP and DPPH) were revealed. A broad antimicrobial spectrum was exerted by all of the wine fractions, as indicated by the appreciable values of their minimum inhibitory concentration (MIC) measured against both bacterial and fungal strains tested. Particularly, our results pointed out the attention on the leading role in the wine samples tested of the anthocyanin fraction in terms of antioxidant and antimicrobial properties. Thus, these compounds may be reasonably regarded as the main responsible for an appreciable self-preserving capacity and potentially healthy effects of the wines object of this study.

REFERENCES

[1] Urquiaga, I., & Leighton, F. (2005). Wine and health: evidence and mechanisms. *World Review of Nutrition and Dietetics*, 95, 122-139.

[2] Renaud S, Guèguen R, Siest G, Salamon R. 1999. Wine, beer, and mortality in middle-aged men from eastern France. Archives of *Internal Medicine* 159(16):1865−1870.

[3] Landrault N, Poucheret P, Ravel P, Gasc F, Cros G, Teissedre PL. **2001.** Antioxidant capacities and phenolics levels of French wines from different varieties and vintages. *Journal of Agricultural and Food Chemistry* 49(7):3341−3348.

[4] Fernandéz-Pachón, M. S., Villaño, D., Troncoso, A. M., & García-Parrilla , M. C. (2006) Determination of the phenolic composition of sherry and table white wines by liquid chromatography and their relation with antioxidant activity. *Analytica Chimica Acta,* **563,** 101-108.

[5] Rodríguez-Bernaldo de Quirós, A., Lage-Yusty, M. A. & López-Hernández, J. (2009). HPLC-analysis of polyphenolic compounds in Spanish white wines and determination of their antioxidant activity by radical scavenging assay. *Food Research International,* 42, 1018–1022.

[6] Weisse ME, Eberly B, Person DA. 1995. Wine as a digestive aid: comparative antimicrobial effects of bismuth salicylate and red and white wine. *British Medical Journal* 311(7021):1657–1660.

[7] Bellido JB, González F, Arrendó A, Galiano JV, Safont L. 1996. Brote de infección alimentaria por Salmonella enteritidis Posible efecto protector de las bebidas alcohólicas. *Medicina Clinica* (Barcelona) 107(3):641–644.

[8] Fernández J, Gomes F, Couto JA, Hogg T. 2007. The antimicrobial effect of wine on Listeria innocua in a model stomach system. *Food Control* 18(12):1477–1483.

[9] Just JR, Daeshchel MA. 2003. Antimicrobial effects of wine on Escherichia coli O157:H7 and Salmonella typhimurium in a model stomach system. *Journal of Food Science* 68(1):285–290.

[10] Marimon JM, Bujanda L, Gutiérrez-Stampa MA, Cosme A, Arenas JI. 1998. In vitro bactericidal effect of wine against Helicobacter pylori. *American Journal of Gastroenterology* 93(8):1392.

[11] Boban N, Tonkic M, Budimir D, Modun D, Sutlovic D, Punda-Polic, V, Boban, M. 2010. Antimicrobial Effects of Wine: Separating the Role of Polyphenols, pH, Ethanol, and Other Wine Components. *Journal of Food Science* 75(5):322−326.

[12] Waite JG, Daeschel MA. 2007. Contribution of wine components to inactivation of food-borne pathogens. *Journal of Food Science* 72(7):286−291.

[13] Cushnie TP, Lamb AJ. 2005. Antimicrobial activity of flavonoids. *Intenational Journal of Antimicrobial Agents* 26(5):343–356.

[14] Tenore, G. C., Troisi J., Di Fiore R., Manfra M., Novellino E. (2011). Nutraceutical value and toxicological profile of selected red wines from Morocco. *Food Chemistry,* 129(3), 792-798.

[15] Minussi, R. C., Rossi, M., Bologna, L., Cordi, L., Rotilio, D., Pastore, G. M., & Duran, N. (**2003**). Phenolic compounds and total antioxidant potential of commercial wines. *Food Chemistry,* 82, 409-416.

[16] Monagas, M., Gomez-Cordoves, C., & Bartolome, B. (2006). Evolution of the phenolic content of red wines from Vitis vinifera L. during ageing in bottle. *Food Chemistry,* 95, 405-412.

[17] Ginjom, I. R., D'Arcy, B. R., Caffin, N. A., & Gidley, M. J. (2010). Phenolic Contents and Antioxidant Activities of Major Australian Red Wines throughout the Winemaking Process. *Journal of Agricultural and Food Chemistry,* 58, 10133-10142.

[18] Spayd, S. E., Tarara, J. M., Mee, D L., & Ferguson, J. C. (2002). Separation of sunlight and temperature effects on the composition of Vitis vinifera cv. Merlot berries. *American Journal Enology and Viticulture,* 53, 171-182.

[19] Arnous, A., & Meyer, A. S. (2008). Comparison of methods for compositional characterization of grape (Vitis vinifera L.) and apple (Malus domestica) skins. *Food and Bioproduct Processing,* 86, 79-86.

[20] Jensen, J. S., Demiray, S., Egebo, M., & Meyer, A. S. (2008). Prediction of wine color attributes from the phenolic profiles of red grapes (Vitis vinifera). *Journal of Agricultural and Food Chemistry,* 56, 1105-1115.

[21] Arnous, A., & Meyer, A. S. (2010). Discriminated release of phenolic substances from red wine grape skins (Vitis vinifera L.) by multicomponent enzymes treatment. *Biochemical Engineering Journal,* 49, 68-77.

[22] Cliff, M. A., King, M. C., & Schlosser, J. (2007). Anthocyanin, phenolic composition, colour measurement and sensory analysis of BC commercial red wines. *Food Research International,* 40, 92-100.

[23] Zhishen, J., Mengcheng, T., & Jianming, W. (1999). The determination of flavonoid contents in mulberry and their scavenging effects on superoxide radicals. *Food Chemistry,* 64, 555-559.

[24] Schlesier, K., Harwat, M., Bohm, V., & Bitsch, R. (2002). Assessment of antioxidant activity by using different in vitro methods. *Free Radical Research,* 36, 177-187.

[25] Delgado-Andrade, C., Rufián-Henares, J. A., & Morales, F. J. (2005). Assessing the antioxidant activity of melanoidins from coffee brews by different antioxidant methods. *Journal of Agricultural and Food Chemistry,* 53, 7832-7836.

[26] Monagas, M., Nunez, V., Bartolome, B., & Gomez-Cordoves, C. (2003). Anthocyanin-derived pigments in Graciano, Tempranillo, and Cabernet Sauvignon wines produced in Spain. *American Journal of Enology and Viticulture,* **54**, 163–169.

[27] Gonzalez-San Jose, M. L., Santa-Maria, G., & Diez, C. (1990). Anthocyanins as parameters for differentiating wines by grape variety, wine-growing region, and wine-making methods. *Journal of Food Composition and Analysis,* **3**, 54-66.

[28] Rastija V., Srečnik G., & Medić-Šarić, M. (2009). Polyphenolic composition of Croatian wines with different geographical origins. *Food Chemistry,* 115, 54-60.

[29] Gawel, R. (1998). Red wine astringency: A review. *Australian Journal of Grape and Wine Research,* 4, 74-95.

[30] Castillo-Muñoz, N., Gómez-Alonso, S., García-Romero E., & Hermosín-Gutiérrez, I. (2007). Flavonol Profiles of Vitis vinifera Red Grapes and Their Single-Cultivar Wines. *Journal of Agricultural and Food Chemistry,* 55, 992-1002.

[31] Tenore, G. C., Basile A., Novellino E.. (2011). Antioxidant and antimicrobial properties of polyphenolic fractions from selected Moroccan red wines. *Journal of Food Science,* 76(9), C1342-C1348.

[32] Francisco M, Moreno DA, Cartea ME, Ferreres F, García-Viguera C, Velasco P. 2009. Simultaneous identification of glucosinolates and phenolic compounds in a representative collection of vegetable Brassica rapa. *Journal of Chromatography,* A 1216(38):6611−6619.

[33] Tung YT, Cheng KC, Ho ST, Chen YL, Wu TL. Hung KC, Wu JH. 2011. Comparison and characterization of the antioxidant potential of 3 wild grapes-Vitis thunbergii, V. flexuosa, and V. kelungeusis. *Journal of Food Science* 76(5):701−706.

[34] Frankel EN, Bosanek CA, Meyer AS, Silliman K, Kirk LL. 1998. Commercial grape juices inhibit the in vitro oxidation of human low-density lipoproteins. *Journal of Agricultural and Food Chemistry* 46(3):834−838.

[35] Burns J, Gardner PT, O'Neil J, Crawford S, Morecroft I, McPhail DB, Lister C, Matthews D, MacLean MR, Lean ME, Duthie GG, Crozier A. 2000. Relationship among antioxidant activity, vasodilatation capacity, and phenolic content of red wines. *Journal of Agricultural and Food Chemistry* 48(2):220−230.

[36] Sato M, Ramarathnam N, Suzuki Y, Ohkubo T, Takeuchi M, Ochi H. 1996. Varietal differences in the phenolic content and superoxide radical scavenging potential of wines from different sources. *Journal of Agricultural and Food Chemistry* 44(1):37−41.

[37] Rodriguez J, Yanez J, Vicente V, Alcaraz M, Benavente-Garcia O, Castillo J, Lorente J, Lozano JA. 2002. Effects of several flavonoids on the growth of B16F10 and SK-MEL-1 melanoma cell lines: relationship between structure and activity. *Melanoma Research* 12(2):99–107.

[38] Lee KW, Li J, Kim YW, Chung KW, Lee, YJ, Oh HB. 2011. Synergetic and antagonistic role of natural antioxidant in the autoxidation of soybean oil. *Journal* of Industrial and Engineering Chemistry 17(3):537–542.

[39] Hidalgo M, Sánchez-Moreno C, de Pascual-Teresa S. 2010. Flavonoid–flavonoid interaction and its effect on their antioxidant activity. *Food Chemistry* 121(3):691–696.

[40] Freeman BL, Eggett DL, Parker TL. 2010. Synergistic and antagonistic interactions of phenolic compounds found in navel oranges. *Journal of Food Science* 75(6):570–576.

[41] Barry A. 1976. *The Antimicrobic Susceptibility Test: Principles and Practices.* Philadelphia: Lea and Febiger.

[42] Friedman M, Henika PR, Levin CE, Mandrell RE. 2006. Antimicrobial wine formulations active against the foodborne pathogens Escherichia coli O157: H7 and Salmonella enterica. *Journal of Food Science* 71(7):245−251.

[43] Cisowska A, Wojnicz D, Hendrich AB. 2011. Anthocyanins as antimicrobial agents of natural plant origin. *Natural Product Communication* 6(1):149−156.

[44] Corrales M, Han JH, Tauscher B. 2009. Antimicrobial properties of grape seed extracts and their effectiveness after incorporation into pea starch films. International *Journal of Food Science and Technology* 44(2):425–433.

In: Anthocyanins: Structure, Biosynthesis and Health Benefits ISBN: 978-1-62257-329-5
Editor: Noboru Motohashi

Chapter 14

THE HEALTH EFFECTS OF TAROS (*COLOCASIA ESCULANTA* SHOTT) BASED ON THEIR EVIDENCES

Noboru Motohashi*
Meiji Pharmaceutical University, Noshio,
Kiyose-shi, Tokyo, Japan

ABSTRACT

Taros have been primarily used in their diverse culinary regions as major cereal crops in tropical and subtropical areas also including temperate region such as Japan and Korea. In these areas, taros also have been used as traditional medicines in preventions and treatments for hypertension, constipation, diuretic and other age-related diseases. By now, human has ingested their dietary anthocyanins with different colors in fruit, corm and vegetable to unknowingly. These having functionality could play a beneficial role in retarding or reversing the course of chronic degenerative diseases such as hypertension, diabetes and cardiac failure. Therefore, the purpose of this chapter is to describe the anthocyanins and their functionality in taros.

Keywords: Taro tuber, *Colocasia esculanta* Shott, flavonoids, anthocyanins

ABBREVIATIONS

Orientin (luteolin 8-*C*-glucoside, **1**)
Isoorientin (luteolin 6-*C*-glucoside, **2**)
Isovitexin (apigenin 6-*C*-glucoside, **3**)
Vicenin-2 (**4**)
Orientin 7-*O*-glucoside (**5**)

* noborumotohashi@jcom.home.ne.jp

Isovitexin 4' -*O*-glucoside (**6**)
Vitexin X" -*O*-glucoside (**7**)
Luteolin 7-*O*-glucoside (**8**)
Apigenin (**9**)
Luteolin (**10**)
Cyanidine -3-*O*-glucoside (**11**)
Cyanidin 3-*O*-rhamnosylglucoside (cyanidin 3-*O*-rutinoside, **12**)
Vitamin C (**13**)
Pelargonidin 3-*O*-glucoside (**14**)
Ferulic acid (**15**)
Rosmarinic acid (**16**)
Caffeic acid (**17**)
Cyanidin 3-*O*-rhamnoside (**18**)
Protocatechuic acid (**19**)

1. Introduction

Scientific name of taro is *Colocasia esculanta* Shott. Taro is perennial plant belonging to Arales, Araceae and Colocasia (Photo 1).

Taro of *Colocasia esculanta* Shott. might be native to India and surrounding areas such as Pakistan, southern China, Bangladesh and Nepal [1].

The roots of taro (*Colocasia esculenta*) is important as one of three major nutrients especially in tropical and subtropical regions. The rich starches and dietary fibers in the taro tubers as prebiotics might regulate the gastrointestinal tract to the health conditions. Therefore, the taro tubers enhance to increase their beneficial probiotics such as *Lactococcus lactis* and *Lactobacilli* of lactic acid-producing bacteria in gastrointestinal tract, and prevent or improve the diverse digestive system disease such as diarrhes, gastroenteritis, irritable bowel syndrome, and inflammatory bowel disease such as Crohn's disease and ulcerative colitis, cancer, depressed immune function, and inadequate lactase digestion [2, 3].

2. Phytochemicals and Their Functionality of Taro

2.1. Starches in Taro Tubers

The tubers of taro contains abundantly one of the major starchy root corns in fresh taro tubers. Then, the tubers of taro are a good resource of starches (Photo 2) [1].

Photo 1. Taro. at suburb of Accra, Ghana in west Africa. Photographed by Noboru Motohashi. 10/1/2005 Sat <Ghana 4. IMGP8870 (taro) 10/1/2005.JPG>.

Photo 2. Taro tubers. Photographed by Noboru Motohashi. 1/26/2012 Thu <494. IMGP7984 (taro tuber) 1/26/2012 Thu>.

2.2. Proteins in Taro Tubers

2.2.1. Proteins in Taro Tubers

The tubers of taro contains around 1.6% protein in fresh taro tubers. Then, the tubers of taro are a good resource of proteins [1].

The content of protein (g/100 g edible dry part) of taro varieties such as *Colocasia* of Araceae were 1 to 4.5g and the content (g/100 g edible dry part) of protein for one cultivar taro was the highest value protein 11.7g [4].

The contents of protein (g/100 g edible dry part) were 2 g for *Colocasia* of Araceae and *Xanthosen*, 0.8 g for *Cyrtosperma* of Araceae and 0.6 g for *Alocasia* of Araceae [5, 6].

In 2011, a group of Division of Horticulture and Forestry, Central Agricultural Research Institute, Port Blair, Andaman and Nicobar Islands, India measured their protein content (g/100 g dried edible part) of ten traditionally culinary dried vegetables in Andaman and Nicobar Islands of India. As a result, the crude protein content (g/100 g dried edible part) of *Colocasia esculanta* (taro of Araceae) was the highest value 6.75% and *Enhydra flactuans* Lour. (oxiladaceae) was the minumum value 2.63%. Then, the tubers of taro are also a good resource of proteins [7].

2.2.2. A Globulin Tarin of Globulins in Taro Tubers

By a dictionary in 2008,, the tubers of taro contain globulins accounting for 80% of the total tuber proteins, belonging to two unrelated globulin families. The total amino acids in the tubers are in the range of 1,380-2,397 mg/100 g. The lysine concentration was relatively low. [8, 9, 10].

In 1995, a group of EMBRAPA/National Center of Genetic Resource and Biotechnology, SAIN Parque Rural, Brasília, DF, Brazil was isolated a globulin tarin of main corm protein family from the tubers of taro, using a cDNA (complementary DNA) derived from a highly abundant corm-specific mRNA (messenger RNA) by a tar1 gene encoding globulin. [10].

Plasma protein contains albumin, fibrinogen, prothrombin of a protein in blood plasma that is the inactive precursor of thrombin, and the gamma (γ) globulins of a plasma protein containing the immunoglobulins that are responsible for immune responses because of a fraction of blood serum containing the most antibodies. Plasma protein constitute about 6% to 7% of the blood plasma in the body. Plasma protein help maintain water balance that affects osmotic pressure of the pressure exerted by a solution necessary to prevent osmosis into that solution when it is separated from the pure solvent by a semipermeable membrane, increase blood viscosity, and help maintain blood pressure. All the plasma protein except the gamma (γ) globulins are synthesized in the livey [11].

2.3. Flavonoids and Anthocyanins

2.3.1. Flavonoids and Anthocyanins in Taro Leaves

In 1999, a group of Tsukuba Botanical Garden, National Science Museum identified eight flavonoids such as orientin (**1**), isoorientin (**2**), isovitexin (**3**), vicenin-2 (**4**), orientin 7-*O*-glucoside (**5**), isovitexin 4' -*O*-glucoside (**6**), vitexin X" -*O*-glucoside (**7**) and luteolin 7-*O*-glucoside (**8**) from the leaves of taro, *Colocasia esculenta* (L.) Schott (Araceae), by using ultraviolet (UV) spectra, characterization of hydrolysates, and paper chromatography (PC) and HPLC comparisons with authentic specimens for the research of the useful substances which contained in the unused vegetable parts (Figure 1 and Figure 1 (cont. 1)) [12].

In 2008, the leaves of taro contain flavones such as apigenin (**9**), luteolin (10) and cyanidine -3-*O*-glucoside (11) by Khare's dictionary (Figure 2) [8].

2.3.2. Flavonoids and Anthocyanins in Taro Stalk (Zuiki)

In 2007, a group of Faculty of Education, Kanazawa University, Kakuma-machi, Kanazawa 920-1192, Japan identified cyanidin 3-*O*-rhamnosylglucoside (cyanidin 3-*O*-rutinoside, 12) by high performance liquid chromatography (HPLC) from 80% ethanol extract of a lyophilized taro stalk.

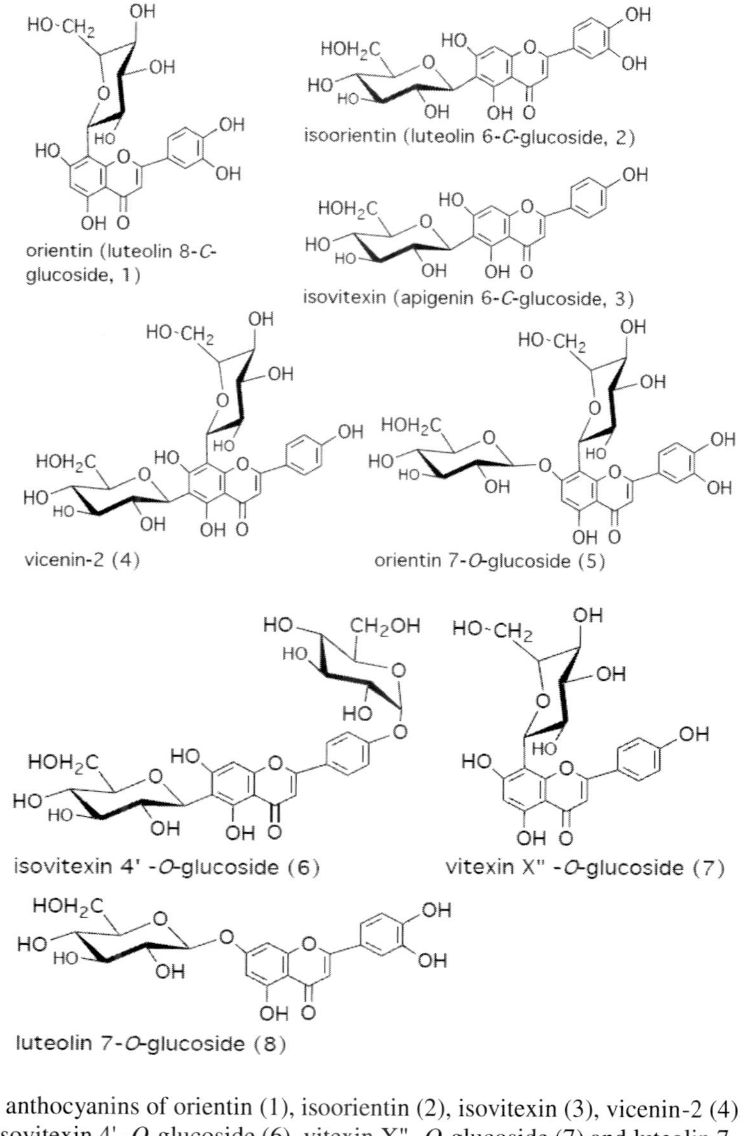

Figure 1. Eight anthocyanins of orientin (1), isoorientin (2), isovitexin (3), vicenin-2 (4), orientin 7-*O*-glucoside (5), isovitexin 4' -*O*-glucoside (6), vitexin X" -*O*-glucoside (7) and luteolin 7-*O*-glucoside (8) from leaves of taro.

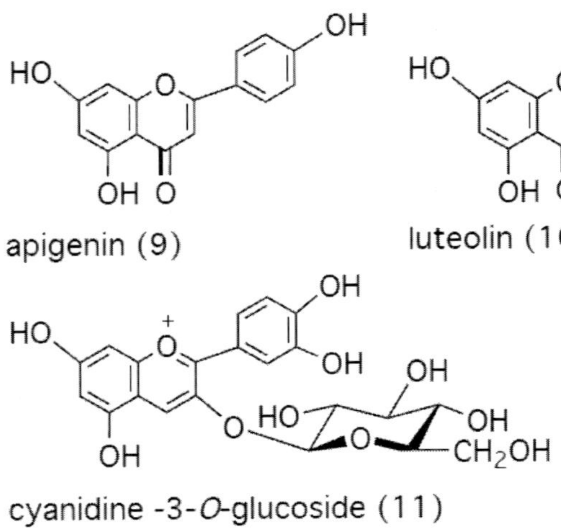

Figure 2. Apigenin (9), luteolin (10) and cyanidine -3-*O*-glucoside (11) from leaves of taro.

At the same time, vitamin C (13) also was measured from both raw and boiled taro stalks.

Cyanidin 3-*O*-rhamnosylglucoside (cyanidin 3-*O*-rutinoside, 12) and vitamin C (13) were effective on their radical scavenging activity using 1,1-diphenyl-2-picrylhydrazyl (DPPH) radical method (Figure 3) [13].

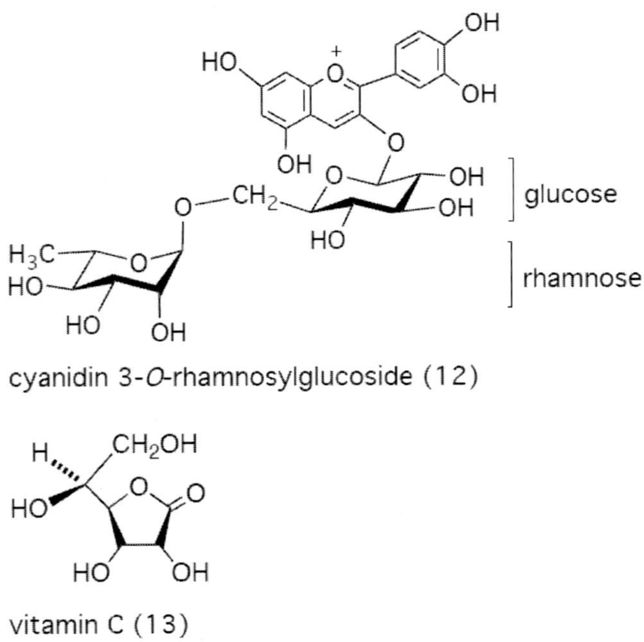

Figure 3. Cyanidin 3-*O*-rhamnosylglucoside (cyanidin 3-*O*-rutinoside, 12) and vitamin C (13) isolated from stalks of taro.

2.3.3. Flavonoids and Anthocyanins in Tubers

In 1997, a group of Lymphoma Research Unit, Tenovus Laboratory, General Hospital, Southampton, UK identified cyanidine-3-*O*-glucoside (11), cyanidin 3-*O*-rhamnosylglucoside (cyanidin 3-*O*-rutinoside, 12) and pelargonidin 3-*O*-glucoside (14) from 50% methanol extract of tato corm.

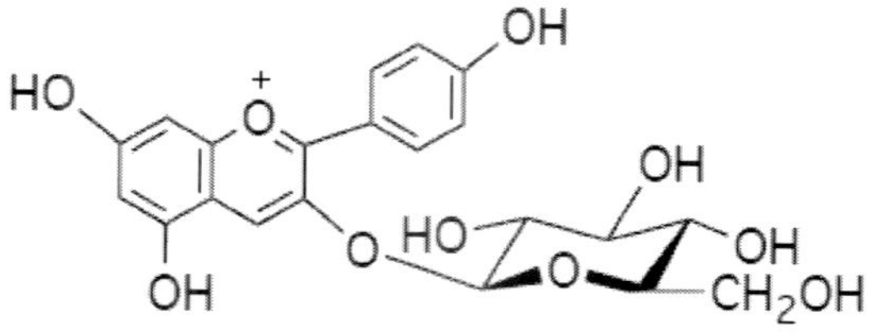

pelargonidin 3-*O*-glucoside (14)

Figure 4. Three anthocyanins of cyanidine-3-*O*-glucoside (11), cyanidin 3-*O*-rhamnosylglucoside (cyanidin 3-*O*-rutinoside, 12) and pelargonidin 3-*O*-glucoside (14) isolated from corm tubers of taro

Their level (mg/100 mg anthocyanin) of these anthocyanins was the highest 16.0% for the skin of the corm, and almost equal 3.29 % for both corm and petiole. Anthocyanogens of the anthocyanin precursor are also present in the plant (Figure 4) [14].

The intermolecular copigmentation reactions are significantly responsible for the manifold color expression on diverse fruits, berries and their fruit products. In 2002, based on this fact, a group of Division of Food Chemistry, Department of Applied Chemistry and Microbiology, University of Helsinki, Helsinki, Finland studied their anthocyanin color change and stability by effect of intermolecular copigmentation, using the copigmentation reaction of anthocyanins with phenolic acids such as a phytopolyphenol ferulic acid (15), rosmarinic acid (16) and a phenol caffeic acid (17) on their fruit storage of 6 months. Generally, it is known that when anthocyanins had an intermolecular copigmentation reaction with these plant phenolics, the larger bathochromic shift and stronger hyperchromic effect could be observed. The color intensity of pelargonidin 3-*O*-glucoside (14) could significantly represent the increased color intensity by the copigmentation with ferulic acid (15) and caffeic acid (17) during the 6 month storages (Figure 5) [15].

In 2003, a group of Department of Chemistry, The University of Auckland, Private Bag 92019, Auckland, New Zealand have described to contain the anthocyanins such as cyanidine-3-*O*-glucoside (11), pelargonidin 3-*O*-glucoside (14) and cyanidin 3-*O*-rhamnoside (18) in the corms of taro, which have antioxidant and anti-inflammatory properties (Figure 6) [16].

In 2011, a group of Division of Horticulture and Forestry, Central Agricultural Research Institute, Port Blair, Andaman and Nicobar Islands, India studied their phytochemicals on ten traditional vegetables such as *Colocasia asiatica*, *Colocasia esculanta* (taro of Araceae), *Eryngium foetidum* (shadon of Apiaceae), *Amaranthus viridis L* (slender amaranth or green amaranth of Ameranthaceae), *Portulaca oleracea L* (common purslane of Portulacaceae), *Alternanthera philoxeroides* (Mart.) Griseb. (alligatorweed of Amaranthaceae), *Enhydra flactuans* Lour. (oxiladaceae), *Hibiscus sabdariffa* L. (jelly okra of Malvaceae), *Sauropus androgynus* (L.) Merr. (sweet leaf bush of Euphorbiceae) and *Ipomoea aquatica* Forx. (water spinach of Convolvulaceae) in Andaman and Nicobar Islands of India.

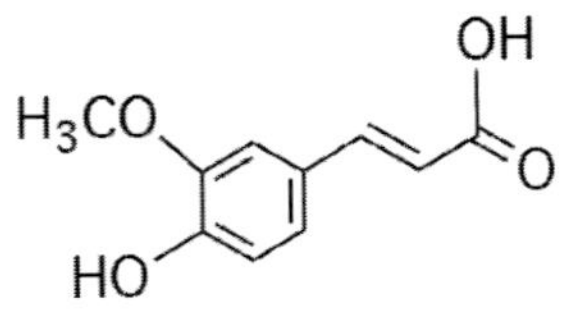

ferulic acid (15)

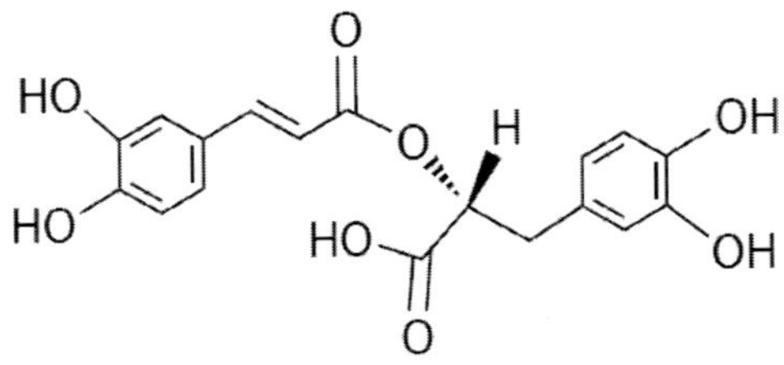

rosmarinic acid (16)

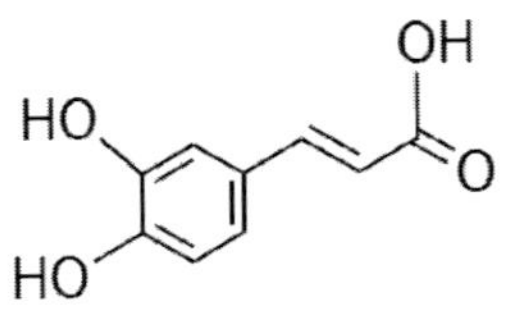

caffeic acid (17)

Figure 5. Three phenolics of ferulic acid (15), rosmarinic acid (16) and caffeic acid (17).

When their contents (mg/100 g edible part) of anthocyanins in ten these vegetables using cyanidine-3-*O*-glucoside (11) equivalent were compared, *Colocasia esculanta* (taro of Araceae) was the highest value 83.5, followed by 83.0 for *Sauropus androgynus* (L.) Merr. (sweet leaf bush of Euphorbiceae), 42.2 for *Eryngium foetidum* (shadon of Apiaceae), 41.7 for *Portulaca oleracea L* (common purslane of Portulacaceae), 41.7 for *Hibiscus sabdariffa* L. (jelly okra of Malvaceae), 37.6 for *Colocasia asiatica* (Araceae), 36.6 for *Ipomoea aquatica* Forx. (water spinach of Convolvulaceae), 33.4 for *Amaranthus viridis L* (slender amaranth or green amaranth of Ameranthaceae), 25.0 for *Alternanthera philoxeroides* (Mart.) Griseb. (alligatorweed of Amaranthaceae) and 12.5 for *Enhydra flactuans* Lour. (oxiladaceae). Their ranges of anthocyanin content (mg/100 g edible part) were 12.5 to 83.5. The content of *Colocasia esculanta* (taro of Araceae) was around 6.7 times when compared to content (12.5) for *Enhydra flactuans* Lour. (Oxiladaceae) and it means that *Colocasia esculanta* (taro of Araceae) might be a resouce of antioxidant anthocyanin-rich foods [7].

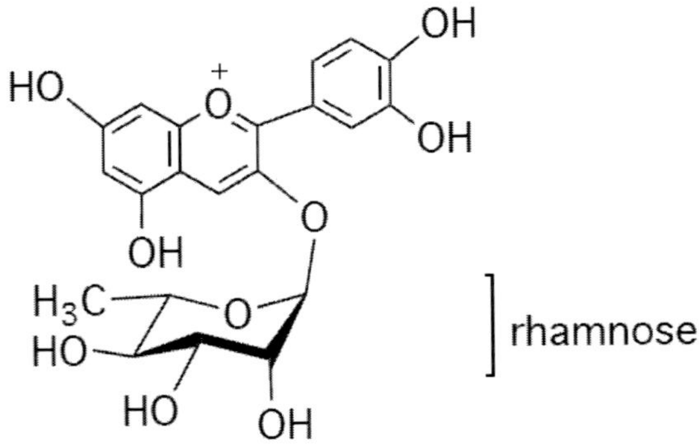

cyanidin 3- *O*-rhamnoside (18)

Figure 6. Cyanidine 3-*O*-glucoside (11), Pelargonidin 3-*O*-glucoside (14) and cyanidin 3-*O*-rhamnoside (18) isolated from corms of taro.

3 .Bioavailability of Anthocyanins

A water-soluble anthocyanin-related cyanidins (aglycones) and its glycosides (anthocyanines) could give a beneficial health effect to daily human health. Especially, the cyanidine glycosides might be helpful to their absorption and metabolism including detoxification in body, antioxidant activity, anticarcinogenic activity, vasoprotective, antiinflammatory, antiobesity and antidiabetes effects. Their metabolism by the intestinal microflora might be an important factor for controlling the bioavailability. Their degradation of cyanidine glycosides was studies by microflora *in vitro*.

It is known that the intact anthocyanidin glycosides were decreased extensively by the interactions in the gastrointestinal tract. Anthocyanidin glycosides were hydrolysed by the microflora within 20 min and 2 h of the incubation, depending on the different sugar moiety. Because the liberated aglycones had the high instability at neutral pH, the degradation products of primary phenolics were already detected after 20 min of incubation. Further metabolism of the phenolic acids was accompanied by demethylation. From these results, it suggests that because of their higher chemical and microbial stability, phenolic acids and anthocyanin metabolites might be mainly responsible for their antioxidant activities and other physiological effects *in vivo* [17].

In 2005, a group of VTT Biotechnology, 1500, 02044 VTT, Espoo, Finland studied their actual bioavailability of anthocyanins that if anthocyanin glycosides were deglycosylated, whether the resulting aglycones were degraded further to smaller phenolic compounds by colonic bacteria which characterise their metabolites.

Cyanidine-3-*O*-glucoside (11) and cyanidin 3-*O*-rutinoside (12) were fermented *in vitro* using human faecal microbiota as an inoculum. Their metabolites of two anthocyanins (11, 12) were analysed and characterised by high performance liquid chromatography (HPLC) - data automation system (DAS) (HPLC-DAS) and liquid chromatography-mass spectrometry (LC-MS). After 2 hr of their fernentation, protocatechuic acid (19) was detected as a major metabolite for two anthocyanins (11, 12) and the low molecular weight metabolites were accompaied with protocatechuic acid (19). This means that two these anthocyanins (11, 12) were converted by the fermentation of gut microflora. From these results, it suggests that their bacterial metabolism of anthocyanins in human gut might simultaneously involve their cleavage of glycosidic linkages and breakdown of the anthocyanidin heterocycle (Figure 7) [18].

In 2006, a group of Institute for Nutritional Physiology, Federal Research Centre for Nutrition and Food, Karlsruhe, Germany studied their stability and biotransformation of various dietary anthocyanins *in vitro*, for the understanding of their mechanism on biotransformation and the metabolism of anthocyanins.

Actually, their possible transformation pathways of anthocyanins by human faecal microflora and by rat liver microsomes as a source of cytochrome P450 enzymes, and a detoxification enzyme of glucuronyltransferases were studied.

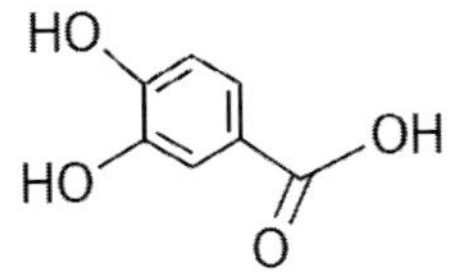

protocatechuic acid (19)

Figure 7. Protocatechuic acid (19) isolated after 2 hr fermentation of cyanidine -3-*O*-glucoside (11) and cyanidin 3-*O*-rutinoside (12).

Then, it is observed that glycosylated and acylated anthocyanins were rapidly degraded by the intestinal microflora after anaerobic incubation with a human faecal suspension. The corresponding phenolic acids derived from the B-ring of the anthocyanins skeleton were their major stable products after the degradation of their anthocyanins.

Interestingly, cytochrome P450 enzymes could not metabolize, neither hydroxylate nor demethylate the anthocyanins. However, rat liver microsomes could glucuronidate anthocyanins to several glucuronidated products.

From these results, their biotransformation of the anthocyanins might mostly be happened in their gut microflora. Then, because anthocyanins might immediately might be degrated in their gut microflora, their bioavailability of anthocyanins mght be too low in pharmacokinetic viewpoint. Therefore, it suggests that their actual bioactivity of *in vivo* anthocyanins also might depend on the bioactivity of phenolic acids of the major stable degradation products of anthocyanins [19, 20].

REFERENCES

[1] Hirai M, Nakamura K, Imai T, Sato T. cDNAs encoding for storage proteins in the tubers of taro (*Colocasia esculenta* Schott). *Jpn J Genet* 68(3), 229-236, 1993.

[2] Brown AC, Valiere A. The medicinal uses of poi. *Nutr Clin Care* 7(2), 69-74, 2004.

[3] Brown AC, Shovic A, Ibrahim SA, Holck P, Huang A. A non-dairy probiotic's (poi) influence on changing the gastrointestinal tract's microflora environment. *Altern Ther Health Med* 11(1), 58-64, 2005.

[4] Splittstoesser WE. Protein quality and quantity of tropical roots and tubers. *Hortscience* 12, 294-298, 1977.

[5] Pollock NJ. Taro. In: Kiple KF, Ornelas KC (eds). T*he Cambridge World History of Food* (2-Volume Set). Cambridge: Cambridge University Press, 218-230, 2000.

[6] Shewry PR. Tuber storage proteins. *Ann Bot* 91(7), 755-769, 2003.

[7] Singh S, Singh DR, Salim KM, Srivastava A, Singh LB, Srivastava RC. Estimation of proximate composition, micronutrients and phytochemical compounds in traditional vegetables from Andaman and Nicobar Islands. *Int J Food Sci Nutr* 62(7), 765-773, 2011.

[8] Khare CP. *Indian Medicinal Plants: An Illustrated Dictionary*. p.167, New Delhi, Springer, 2008.

[9] Rakesh Prajapati, Manisha Kalariya, Rahul Umbarkar, Sachin Parmar, Navin Sheth. Colocasia esculenta: A potent indigenous plant. *Internatl J Nutrition, Pharmacology, Neurological Diseases.* 1(2), 90-96, 2011.

[10] Bezerra IC, Castro LA, Neshich G, de Almeida ER, de Sá MF, Mello LV, Monte-Neshich DC. A corm-specific gene encodes tarin, a major globulin of taro (*Colocasia esculenta* L. Schott). *Plant Mol Biol* 28(1), 137-144, 1995.

[11] http://medical-dictionary.thefreedictionary.com/plasma+protein

[12] Iwashina T, Konishi T, Takayama A, Fukada M, Ootani S. Isolation and identification of the flavonoids in the leaves of taro. *Ann Tsukuba Botanical Garden* 18, 71-74, 1999.

[13] Terasawa N, Saotome A, Tachimura Y, Mochizuki A, Ono H, Takenaka M, Murata M. Identification and some properties of anthocyanin isolated from Zuiki, stalk of *Colocasia esculenta. J Agric Food Chem* 55(10), 4154-4159, 2007.

[14] Chan HT, Kao-Jao THC, Nakayama TOM. Anthocyanin composition of taro. *J Food Sci* 42, 19-21, 1997.

[15] Eiro MJ, Heinonen M. Anthocyanin color behavior and stability during storage: effect of intermolecular copigmentation. *J Agric Food Chem* 50(25), 7461-7466, 2002.

[16] Cambie RC, Ferguson LR. Potential functional foods in the traditional Maori diet. *Mutat Res* 523-524, 109-117, 2003.

[17] Keppler K, Humpf HU. Metabolism of anthocyanins and their phenolic degradation products by the intestinal microflora. *Bioorg Med Chem* 13(17), 5195-5205, 2005.

[18] Aura AM, Martin-Lopez P, O'Leary KA, Williamson G, Oksman-Caldentey KM, Poutanen K, Santos-Buelga C. *In vitro* metabolism of anthocyanins by human gut microflora. *Eur J Nutr* 44(3), 133-142, 2005.

[19] Fleschhut J, Kratzer F, Rechkemmer G, Kulling SE. Stability and biotransformation of various dietary anthocyanins *in vitro*. *Eur J Nutr* 45(1), 7-18, 2006.

[20] Galvano F, Fauci LL, Vitaglione P, Fogliano V, Vanella L, Felgines C. Bioavailability, antioxidant and biological properties of the natural free-radical scavengers cyanidin and related glycosides. *Ann Ist Super Sanità* 43(4), 382-393, 2007.

INDEX

B

C

D

E

F

G

H

I

J

K

L

M

N

O

P

Q

R

S

T

U

V

W

Y